D1242148

Radiation Detectors

Radiation Detectors

Physical Principles and Applications

C.F.G. DELANEY

Fellow Emeritus, Trinity College, University of Dublin

and

E.C. FINCH

Senior Lecturer in Physics, Trinity College, University of Dublin

CLARENDON PRESS · OXFORD
1992

Oxford University Press, Walton Street, Oxford OX2 6DP

Oxford   New York   Toronto
Delhi   Bombay   Calcutta   Madras   Karachi
Petaling Jaya   Singapore   Hong Kong   Tokyo
Nairobi   Dar es Salaam   Cape Town
Melbourne   Auckland

and associated companies in
Berlin   Ibadan

Oxford is a trade mark of Oxford University Press

Published in the United States
by Oxford University Press, New York

A catalogue record for this book is available from the British Library

Library of Congress Cataloging in Publication Data
(Data available)

ISBN 0 19 853923 1
Typeset by Colset Pte Ltd, Singapore
Printed and bound in Great Britain
by Bookcraft (Bath) Ltd
Midsomer Norton, Avon

Preface

This book had its origins in lectures given to final year undergraduate students of an experimental physics course. It is thus intended to be neither a comprehensive manual on all aspects of nuclear radiation detection nor a mere technical account of the mode of action of the detectors used. Instead the emphasis is placed on simple but thorough explanations of the underlying physics of each detector, and the applications to which detectors can be put.

Dealt with are not only all the widely used conventional detectors but also, as far as is possible in such a rapidly advancing field, many of the exciting new devices now emerging. In order to keep the book to manageable proportions, the treatment is largely restricted to the detection of radiations below 100 MeV. The important topic of electronic noise and its influence on the ultimate performance of detectors is given special attention in a separate chapter.

Although reference is made to other textbooks and scientific journals for those seeking further details (without attempting any complete survey of the literature), nevertheless the book is complete in itself in that both an introduction to the types of radiation involved and their interaction with matter are included, as well as a chapter on the electronics associated with nuclear radiation detectors. Furthermore, the essential prerequisite for the book is merely an understanding of a.c. circuit theory. Consequently, it is hoped that the text may be of interest not only to physics students in their final undergraduate year and upwards, but also to workers in disciplines other than physics, who are involved with the use and detection of ionizing radiations.

We wish to acknowledge the help and advice of our Trinity colleagues, particularly Ian McAulay, Sara McMurry, Iggy McGovern, and John M. Kelly, during the preparation of this book. We are grateful to the many authors and journals who have kindly allowed us to use figures from their publications, especially to those who also provided art work. Individual acknowledgements accompany each figure. Professor Glenn Knoll of the University of Michigan kindly helped us to locate one of our references. Jennifer White and Heather Browne gave excellent assistance with the preparation of the typescript, and John Kelly with the photography.

The expertise and ready advice of our publishers at all stages were much appreciated. But most of all we thank our wives, Mary and Jean, and our children, for their tolerance of our long working hours while this book was being written.

Dublin   C.F.G.D.
November 1991   E.C.F.

Contents

1
Introduction

1.1 Aims and objectives

Radiation from atomic nuclei, apart from one or two special cases, cannot be directly seen, heard, smelt, tasted, felt, or in any other way detected by the human senses. In this, nuclear radiation differs from light, heat, sound, vibrations, and many other phenomena which can be directly perceived by us. The purpose of this book is to describe and explain the physics of the different ways in which nuclear and other similar radiation can be indirectly sensed by us, that is, detected and measured.

It is fair to say that the development and use of the methods of nuclear radiation detection have led to some of the most fundamental advances ever made in our understanding of physics. These advances range from the discoveries of the atomic nucleus itself and of elementary particles, through to many stringent tests of the predictions of quantum mechanics, relativity, and other branches of theoretical physics. Radiation detectors have in addition helped to lead to many new developments in industry, materials, medicine, archaeology, geophysics, and the space sciences. They are of course an essential part of nuclear energy programmes and also of the monitoring of low-level radiations in the environment. In this book we shall give examples of some of the applications to which radiation detectors have been put — applications which in many cases have influenced or even dictated the way in which the twentieth century has evolved.

Historically much of the impetus for the study of the physics of nuclear radiation detectors has come from the needs of nuclear physicists for devices that are faster, more efficient, and of higher resolution. This continues to be true, but the driving forces are, as it were, not all one way. The interaction of radiation with detectors provides a valuable method for investigating and developing the properties of the solid, liquid, or gas from which a detector is made. We shall see in later chapters examples of how this is so.

1.2 Early history of detectors

In this section we shall outline when and how some of the main principles and techniques of radiation detection were first applied. Most of these techniques continue, sometimes in modified form, to this day, and they will be described in detail in the succeeding chapters of this book. We shall also in this section be referring to several different types of radiation; these will be discussed more fully later in this chapter.

One of the most important methods has been the collection of the ionization created by most radiations. By ionization we mean the liberation by radiation of free electric charges in many gases, liquids, and solids. These charges can be collected and detected by the application of an electric field to the medium concerned.

The earliest radiation detector was effectively the gold-leaf electroscope, which dates from the eighteenth century. It can be used to measure the amount of ionization produced by radiation, although this was not realized, or the instrument used specifically in this way, until after radioactivity was discovered in 1896. Two thin gold leaves in the electroscope, joined at the top, repel each other and stand apart at an angle when like charges are placed on them, but when radiation ionizes the air between the leaves, the charge leaks away and the leaves return towards the parallel uncharged position by an amount related to the quantity of radiation received.

The first pulse ionization chamber was developed in 1908 by Rutherford and Geiger at Manchester. This produced a pulse of charge for each particle of radiation incident on the gas inside, instead of, as with electroscopes, measuring the integrated total charge collected. (Strictly speaking, they made a proportional counter, which is an ionization chamber with internal charge amplification, as described in Chapter 3.) Work on ionization detectors made from solids proceeded much more slowly, and the first really practical counting devices, which were made from materials like silver chloride, cadmium sulphide, and diamond, came only in the 1940s. These were followed in 1948 by the first ionization detectors made from liquids.

The very earliest methods by which ionizing radiation was investigated did not, however, use those techniques. Röntgen's experiments in 1895 with X-rays produced by cathode rays (that is, electrons ejected by thermionic emission from the heated cathode in a discharge tube) showed that materials can emit visible light when struck by these X-rays. Materials such as barium platinocynanide, when brought towards the tube, gave an intense effect.

Obviously this initial work did not reveal individual pulses of radiation. This had to wait until 1903, when Crookes and also Elster and Geitel observed zinc sulphide under low magnification, and found that when it was exposed to alpha-particle radiation a number of short-duration scintillations of light could be seen scattered across the surface, rather than a steady, uniform distribution of light.

Rutherford and his students made many of their discoveries in nuclear physics by observing scintillations in this way. For example, in 1932 Cockcroft and Walton showed that atoms (or rather their nuclei) were being split by observing through a microscope the scintillations caused by the recoil products. As Walton later recounted, it was Rutherford who, by looking at the character of the scintillations, immediately suspected (as was soon confirmed) that these products were alpha particles.

The actual discovery by Becquerel in 1896 of radioactivity from the atomic nucleus was made with photographic plates. Röntgen had already found that photographic emulsions responded to X-rays. In his investigations following up this work Becquerel found that blackening also happened when uranium salt crystals were left for a few days next to plates which were otherwise unexposed. Much later, in the 1930s, emulsions with high spatial resolution were developed which permitted the observation of tracks of individual particles.

The first device for track visualization, however, came much earlier, and was developed in the years up to 1912. This was C. T. R. Wilson's cloud chamber, in which air saturated with water vapour was contained inside a chamber fitted with a movable piston. When the piston descended the air expanded suddenly. This caused water droplets to condense preferentially at the nucleation sites (namely, ions) produced along the track of any ionizing particle passing through the chamber at that time.

No historical survey of detectors, however brief, would be complete without a reference to the dramatic effect of the introduction of electronic amplifiers and counting circuits to go with radiation detectors such as the proportional counter. Their introduction and development by Greinacher, Wynn-Williams and others, which dated from the late 1920s onwards, removed the restrictions and tedium imposed by some of the older methods, like the counting of scintillations through a microscope. In due course scintillations themselves could be counted electrically when photomultipliers were introduced. The application of modern pulse amplification and digital techniques and of computers has continued to help the development of nuclear instrumentation to levels of accuracy and complexity which could never have been foreseen earlier.

1.3 Atoms and nuclei

One of our main concerns in this book will be with radiation originating from the central nuclei of atoms. Sometimes radiation from atoms themselves will also be of interest. In this section we shall briefly survey the important features that we need to know about atoms and nuclei. Books cited at the end of this chapter give more comprehensive information.

An atom consists of a number of negatively charged electrons electrostatically bound to and (in a classical picture) orbiting around a positively charged small central nucleus. The number of electrons in an electrically neutral atom is called its atomic number, Z. This number determines the chemical element concerned: 1 for hydrogen, 2 for helium, 3 for lithium, and so on. Each electron carries a charge of $-e = -1.6 \times 10^{-19}$C. The nucleus carries a charge of $+Ze$, and so the whole atom is indeed electrically neutral. The mass of an electron is 9.1×10^{-31} kg, which as we shall see shortly is very much smaller than that of a nucleus. Provided it has not been 'excited' in the manner to be described in Section 1.4.1, the size of any atom is always of order 10^{-10} m or (if Z is large) just a little more.

As a consequence of being bound within a small size, the electrons are constrained to possess only certain discrete or 'quantized' values of energy. A quantum-mechanical treatment of the atom shows that in fact there exist 'shells' of electrons. Within each shell all the electrons have nearly (but not quite) the same energies, that is, they occupy 'levels', the energies of which are sharply defined and which lie close together. The innermost, most tightly bound shell is called the K shell, and can be occupied by no more than two electrons. If this is full the next eight electrons can occupy the next shell out; this is the less tightly bound L shell. If this in turn is full the M and N shells can be occupied by 18 and 32 electrons, respectively, and so on to the O, P, and Q shells.

The nucleus at the centre of an atom consists of a number of positively charged protons and a similar, or somewhat larger number of electrically neutral neutrons. These nucleons (the collective name for protons and neutrons) are bound to each other by strong forces which overcome the electrostatic repulsion between the protons. These strong binding forces extend over only the size of the nucleus, which is a few femtometres (fm), where 1 fm is equal to 10^{-15} m. The charge on each proton is $e = 1.6 \times 10^{-19}$C, and there are Z protons in the nucleus, giving a total nuclear charge of Ze, as we saw earlier. Because they are bound in the nucleus, the protons and neutrons occupy discrete nuclear energy levels analogous to those in atoms. The masses of the proton and neutron are 1.673×10^{-27} kg and 1.675×10^{-27} kg, respectively, which are each nearly 2000 times greater than that of the electron.

Since the proton and neutron masses are so nearly equal we can devise a mass scale based on a unit approximately equal to the mass of either of them. The modern and most useful definition of the unit, from which other atomic and nuclear masses can be most accurately determined, is one-twelfth of the mass of the neutral carbon atom which has six protons and six neutrons in its central nucleus. This gives the 'unified atomic mass unit', or u for short. It can then be shown from this definition that in all cases the total number of protons and neutrons in a nucleus gives to the nearest whole number the mass in u both of the nucleus concerned, and also (because the electron mass is so small) of the atom containing this nucleus. We therefore call this number the mass number of the nucleus. Its value, A, is given by $A = Z + N$, where N is the number of neutrons present.

The fullest nomenclature for a particular nuclear species is $^A_ZX_N$. Here X is the chemical symbol of the element concerned. Thus, for example, $^{12}_6C_6$ is the carbon nucleus referred to above, and $^{23}_{11}Na_{12}$ is the sodium nucleus with 11 protons and 12 neutrons. Clearly there is redundant information contained in this symbolism, since the element X defines the atomic number Z, and furthermore the neutron number N can be deduced if the mass number A and either Z or X are known. We can thus equally well refer to a nucleus as ^A_ZX (for example $^{12}_6C$ or $^{23}_{11}Na$) or simply ^AX(^{12}C or ^{23}Na) if we do not wish to emphasize the actual values of Z or N. The simplifications have the additional advantage that, by omitting the suffix denoting the neutron number, we ensure that there is no chance of confusion with the more usual meaning of this suffix, namely, the number of atoms in a molecule (as in H_2O or CO_2).

The nuclei of a particular chemical element need not, however, be completely identical to one another. They must, as we have seen, all have the same atomic number Z, but they can have different neutron numbers N, and thus different mass numbers A, from one another. These varieties of nuclei are called isotopes of the element concerned. A naturally occurring sample containing carbon has in it mostly $^{12}_6C$, but there is also always a little $^{13}_6C$ present and sometimes traces of $^{14}_6C$. An extreme case is tin, natural samples of which contain no fewer than ten isotopes. On the other hand a natural sample of sodium contains only the one isotope $^{23}_{11}Na$.

In addition to the naturally occurring isotopes of the 90 or so elements in the environment, additional isotopes can also be prepared artificially by various means. For example at least seven isotopes of sodium have been made, ranging from $^{20}_{11}Na$ to $^{27}_{11}Na$, in addition to the one naturally occurring species. The extra isotopes which do not occur in nature are radioactive, that is, they decay to other nuclear species by emitting radiation. It is the detection of this radiation which forms the major part of the subject of this book. Radioactive isotopes ('radioisotopes' is an alternative term) are also found in nature. $^{14}_6C$, mentioned above, is one example since it is being continuously produced in the upper atmosphere by cosmic rays (these are described in Section 1.6.2). Some elements which occur naturally, such as uranium, have no stable isotopes. In this case two of its radioisotopes, $^{235}_{92}U$ and $^{238}_{92}U$, decay away so slowly that they can still be found in nature, even though they are not being produced in the terrestrial environment. Finally, we should note that there are some elements, for example, all those with $Z > 92$, which do not occur naturally, and which can be prepared only in the form of artificial radioisotopes.

The occurrence of stable and radioactive isotopes is shown on the Segrè chart in Figure 1.1. We can see that stable isotopes up to $^{40}_{20}Ca$ have $N \approx Z$.

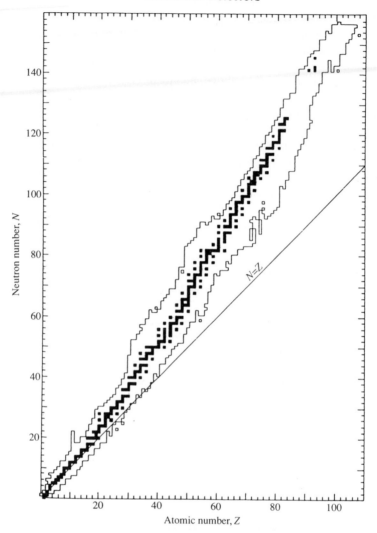

Figure 1.1 The Segrè chart of isotope neutron number N as a function of atomic number Z. Filled squares denote stable and long-lived naturally occurring nuclei. Neighbouring nuclei are unstable: those for which data on masses and decay times are known fill the area bounded by the irregular lines. (Adapted from Cottingham, W. N. and Greenwood, D. A. (1986), *An introduction to nuclear physics*, Cambridge University Press, Figure 4.6, which uses data taken from *Chart of the nuclides* (1977), General Electric Company, Schenectady.)

This can be understood from the way in which protons and neutrons occupy various shells of different energies, analogous to the shells of electrons in an atom. Each shell, and indeed each level within a shell, can be occupied by only a certain number of protons and the same number of neutrons. If $N \neq Z$ then the excess protons or excess neutrons must, for a given mass number A, fill higher energy levels than would be the case were $N = Z$. It thus turns out that an imbalance in N and Z in a nucleus leads to its possession of extra energy. The Segrè chart therefore shows (for A up to 40) stable isotopes around the line $N = Z$, surrounded by radioisotopes with $N \neq Z$ which lose their surplus energy by radioactive decays to product nuclei for which N and Z are more nearly equal.

Above $A = 40$ the Segrè chart shows that N becomes appreciably greater than Z. This can be understood from the nature of the forces experienced by the nucleons inside a nucleus. The strong binding forces within a nucleus, referred to earlier, act essentially between neighbouring pairs of nucleons (regardless of whether they are neutrons or protons) or at least those which are close to each other. On the other hand, the Coulomb forces of repulsion between pairs of protons, although considerably weaker than the strong binding forces, act whether or not the protons are close together in the nucleus. Thus, as we consider heavier (i.e. larger) nuclei, the overall effect of Coulomb repulsion as a proportion of the binding effect of the strong forces becomes larger and larger. As a result a nucleus with $N = Z$ now has surplus energy. It would have less energy if for a given value of A it had more neutrons and fewer protons. This is because, since neutrons are uncharged, Coulomb repulsion effects would be decreased. Hence for stability, as A increases, N becomes larger than Z such that the ratio N/Z itself steadily increases. This is in contrast to light, small nuclei with $A \approx 40$ or less, in which all the nucleons are neighbours (or nearly so) to one another. The effects of Coulomb repulsion as a proportion of those of the strong binding forces thus remain small until about $A \approx 40$, and so there is no tendency for the equality of N and Z for stability to be upset.

One way for a heavy nucleus to lose surplus energy is for it to fragment into smaller parts, for which the Coulomb repulsion is of proportionately less effect. We shall see that this occurs mostly for values of A above about 150. Eventually it must set an upper limit to the values of A which can be achieved; the largest values so far reached are over 260.

In general we can say that an unstable nucleus, that is, one with surplus energy, attains stability through one or more radioactive decays. The nature and energies of the resulting radiations emitted from radioisotopes and the lifetimes for their decay form the subjects of the next two sections.

1.4 General features of radioactivity

1.4.1 *Energies of radiation*

Our first main task in this section is to estimate from the size of a nucleus the order of magnitude that the energies of radiation from it might be expected to have. To do this it is helpful as a preliminary exercise to make the corresponding estimate for energies of electromagnetic radiation (of which light is an example) from an atom.

Such radiation arises essentially as a result of an electron rearranging its position in the atom following some externally imposed disturbance. Let us initially consider for simplicity a hydrogen atom. On a classical picture the electric field strength due to the central nuclear charge e at a distance 10^{-10} m away, which is a characteristic atomic dimension, is about 10^{11} V m^{-1}. A uniform field of this strength would lead over this distance to changes of potential of order $V = 10$ V. Although the field, in fact, varies inversely as the square of the distance from the nucleus, we may nevertheless therefore say that an electron will experience characteristic changes in energy, following any rearrangement on this distance scale, equal to the size of its charge, e, multiplied by this potential V, or an energy change of about 1.6×10^{-18} J. This very crude argument therefore gives the order of magnitude to be expected for the energy taken away from the atom by the resulting electromagnetic radiation. Such an energy can be written of course as 10 eV, where 1 eV (one electron-volt) is defined as the energy acquired by a particle of charge 1.6×10^{-19} C when it is accelerated from rest through a potential difference of 1 V, that is 1.6×10^{-19} J.

Electromagnetic radiation, as implied above, is emitted in discrete bundles or 'quanta'. Their energies are set by the energy lost when an atom moves from a more excited 'state' (to which it has been raised by some previous process) to a less excited state. Using our earlier atomic picture (Section 1.3) we would think, for example, of an electron transferring from a level in one shell to a level in another, inner, shell. Since the energies of the states are quantized, or sharply defined, the energies of the electromagnetic quanta are sharply defined also. The energies of these quanta are in fact given essentially by the energy difference between the two states. Thus, combining the ideas from our classical and quantum pictures, we may expect electromagnetic radiation to be emitted from our hydrogen atom in quanta of order 10 eV or less. This is close to, for example, the energies of 2 to 3 eV which are observed for photons (another name for quanta of electromagnetic radiation) emitted in such rearrangements in the hydrogen atom.

For a heavier atom the same calculation can be applied to one of its outer electrons. This is because such an electron experiences the electric field of the nucleus of positive charge Ze shielded by the negative charge $-(Z-1)e$

of the inner electrons. Since characteristic atomic dimensions do not change substantially as we go to heavier atoms, photons of similar energies are still emitted. However, much larger photon energies are involved for rearrangements of inner electrons, since they experience the much larger electric field of the unshielded or only partly shielded nucleus. Such rearrangements give rise to X-rays (Section 1.7), for which an energy scale in keV, where 1 keV is 10^3 eV, is more appropriate.

In a nucleus, rearrangements of protons and neutrons leading to nuclear radiation take place on a characteristic nuclear distance scale of perhaps 10^{-15} m, or some 10^5 times smaller than in an atom. The fields experienced by the protons and neutrons in a nucleus are more complicated than for the constituents of an atom. As we saw in Section 1.3, there have to be attractive forces which are strong and of short range and which act to bind the nucleus together so as to overcome the electrostatic repulsion between protons. If, nevertheless, we confine ourselves to considering only electric fields, then at a distance of 10^{-15} m from a proton of charge 1.6×10^{-19} C the field (being inversely proportional to the distance squared) is 10^{10} times greater than for a characteristic atomic distance of 10^{-10} m, and potential changes (being inversely proportional to the distance) are 10^5 times greater. When photons are emitted as a result of a nucleus transferring from a more excited to a less excited state, sharply defined photon energies of order 10^6 eV may thus be expected to be observed. This value is in fact typical of the energies of gamma rays, as we call the photons emitted from a nucleus (Section 1.5.3). For charged particle emission (Sections 1.5.1, 1.5.2, and 1.5.5) this picture is complicated by several factors, but energies of this order are often (although not always) still typical. A unit of 1 MeV is equal to 10^6 eV, and is representative of the energies that nuclear radiation detectors need to measure, rather than the few eV calculated earlier for atomic optical radiation.

Each photon of energy E has associated with it an electromagnetic wave of frequency ν and wavelength λ. These are connected by the standard relationships

$$E = h\nu = \frac{hc}{\lambda}$$

where h is Planck's constant and c is the velocity *in vacuo* of light or other electromagnetic radiation. The detectors which we shall be considering in this book (in addition to those which register merely the presence of radiation) essentially can measure at all times the energies of incoming photons or other radiation whatever (within broad limits) the value of these energies. This is in distinction to instruments where the paths of particles or photons are spread out in different directions according to their energies. The resulting spread must be scanned across the range of energies by an appropriate detector. For example, crystals can be used to diffract, and thus spread out, X-rays according to their wavelength. In an analogous way particles like electrons or neutrons can also be diffracted. Here the instruments concerned depend on the 'de Broglie' wavelength λ_b which any moving particle possesses, and which is given by

$$\lambda_b = \frac{h}{p}$$

where p is the particle momentum. Such instruments in general do not come within the scope of this book.

Interestingly we shall see that energies of order 10^4 eV(10 keV) can be shown by both very low-energy nuclear radiation (such as gamma rays) and by high-energy atomic radiation (such as X-rays). Nuclear radiation detectors can thus be used in both cases. In fact the lowest energy deposited in these radiation detectors that can be directly measured is less than 10^3 eV.

The upper limit to the energy deposited that we can measure is, for some detector types, very high. However, for many of the situations we shall discuss in this book the energy deposited is less than 10^8 eV. This corresponds approximately to the highest energy for radiation from radioactive decay. Of course much higher energies can still be used with many of the detectors we shall consider. In these cases the radiation may deposit only a fraction of its energy in a detector. This is a feature which, depending on the detector, can occur for low as well as high energies, as we shall see in succeeding chapters.

1.4.2 *Activity and half-life*

Any radioactive source, regardless of the nature of the radioactivity, is characterized by an 'activity' and a 'half-life'.

The activity of a source is defined as the rate at which the nuclei of the radioisotope species concerned decay with time. The becquerel (Bq), defined as one disintegration per second, has been the official unit of activity since 1975. The older unit of activity, the curie (Ci), is still frequently encountered. 1 Ci is equivalent to 3.7×10^{10} Bq, or (as originally defined) the activity of 1 g of pure radium.

The probability of decay of a radioactive nucleus in a given time interval is independent of its previous history. For an element of time dt this probability (given by the fraction of the number N_t of radioactive nuclei which decay in time dt) is therefore simply $\lambda_c dt$, where the constant of proportionality λ_c is called the decay constant. The number dN_t of N_t radioactive nuclei present at time t that decay in a time interval dt is thus

$$dN_t = -\lambda_c N_t dt$$

the minus sign indicating that N_t decreases with time. Hence the activity, $-dN_t/dt$, is given by

$$-\frac{dN_t}{dt} = \lambda_c N_t. \tag{1.1}$$

From this it follows by integration that

$$N_t = N_0 \exp(-\lambda_c t) \tag{1.2}$$

where N_0 is the number of nuclei at the time $t = 0$. Thus we conclude that both the number of nuclei of a particular type and (from equation (1.1)) the source activity decay exponentially with the constant λ_c.

The mean lifetime τ of a nucleus, in other words, the average time for decay, is given by

$$\tau = \frac{\int t \, dN_t}{\int dN_t}$$

since dN_t is the number with lifetime t. The limits for integration are given by $N_t = N_0$ (when $t = 0$), and $N_t = 0$ (when $t = \infty$). From this expression and equation (1.2), τ can be shown to be equal to $1/\lambda_c$. A time which is more commonly used is the half-life, $t_{1/2}$. This is defined as the time for half the number of the nuclei in a radioisotope to decay. Setting $N_t = N_0/2$ in equation (1.2) therefore gives

$$t_{1/2} = \frac{\ln 2}{\lambda_c} = \tau \ln 2.$$

From equation (1.1) we can see that the half-life is also the time for the activity to halve.

It should be remembered that in certain cases the activity of a source is not the same as the total rate of emission of radiation from it. For example, the daughter product of a radioactive decay may itself be radioactive. In this case the activity is defined (unless otherwise specified) as the decay rate of the parent, whereas with passage of time, as the parent activity decays, the combined emission rate from both isotopes becomes greater than the current parent activity. A full specification of such a source would give separately the parent and daughter activities and also the parent and daughter half-lives.

We shall not give the full theoretical analysis of this situation, but merely quote one important result which follows from it. In many cases the half-life of the daughter is much shorter than that of the parent. Under these circumstances, with no daughter initially present, the parent and daughter activities become approximately equal to each other after a time long compared with the half-life of the daughter. The parent and daughter are now

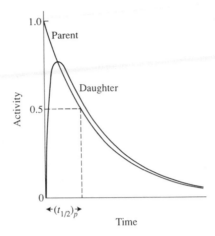

Figure 1.2 The variation with time of the activities (in arbitrary units) of a parent radioisotope, initially pure, and its daughter product when the half-life of the parent is ten times that of the daughter. The parent half-life is indicated by $(t_{1/2})_p$.

said to be in radioactive 'equilibrium', and in such a source the activity of the daughter appears to decay with the half-life of the parent. Figure 1.2 shows the exponential decay of the activity of a parent radioisotope together with how the activity of a daughter first grows and then decays. Note that the actual number of daughter nuclei produced is given from equation (1.1) by the daughter activity divided by the decay constant; since this decay constant is relatively large (the half-life being short), the number of daughter nuclei remains comparatively small.

1.4.3 *Statistics of decays*

We described a moment ago the *probability* of decay of a nucleus. The occurrence of radioactive decay is a statistical process, and this is a consequence of the quantum-mechanical nature of such decays. Indeed for 30 years following the discovery of radioactivity this aspect formed one of radioactivity's early puzzles, one which was quite insoluble prior to the development of quantum mechanics. It therefore follows, for example, that the number of decays in a sequence of equal time intervals will not be the same even if the half-life of the source concerned is much longer than these time intervals. Instead the number of decays, and thus the resulting count in a detector, fluctuates in a random, statistical manner around some mean value, sometimes a little above it and sometimes just below. The mean number of decays per unit time, from what we said earlier, is of course the activity of the source.

This statistical behaviour is not confined to experiments which use radioactive sources. The counts observed from a detector used with accelerators and from cosmic rays can often be governed by similar principles. However, we shall confine our discussion to the counts in a detector viewing the decays in a radioactive source. Only a simplified treatment will be given; for fuller accounts see for example references 1 and 2.

Let us suppose that a very large number of observations of the count in a sequence of equal time intervals leads to an average value, $\bar{s}$, say. Then the fraction, $P(s)$, of this total number of observations actually giving a result s for the number of counts, that is the probability of the result being s, can be shown to be given by the Poisson distribution:

$$P(s) = \frac{\bar{s}^s e^{-\bar{s}}}{s!}. \tag{1.3}$$

Clearly, from its definition, $P(s)$ should satisfy

$$\sum_{s=0}^{\infty} P(s) = 1 \tag{1.4}$$

and this in fact follows from equation (1.3).

Equation (1.3) can be derived on the assumption that the probability that a particular nucleus in the source will decay is very small, but the total number of nuclei is very large. This restriction is normally very easily met, while still maintaining any reasonable number of counts. To verify that the mean of the given Poisson distribution is indeed $\bar{s}$ we can evaluate the usual expression for the mean μ of a distribution,

$$\mu = \frac{\sum_{s=0}^{\infty} s P(s)}{\sum_{s=0}^{\infty} P(s)}$$

From equations (1.3) and (1.4) the result is, as anticipated, $\mu = \bar{s}$. Some examples of the Poisson distribution for various values of $\bar{s}$ are plotted as bar charts in Figure 1.3. The distributions can be seen to be asymmetrical, especially for low $\bar{s}$. The Poisson distribution is an example of a 'discrete' distribution, in which the independent variable (here s) takes only discrete values, with nothing (such as $s = 1.5$ in this case) lying between any two consecutive values.

The widths of the distributions in Figure 1.3 give us the extent to which successive observations of the count fluctuate above or below $\bar{s}$. This is quantified by evaluating the standard deviation σ of the distribution. The standard deviation is defined as the root of the average squared fluctuation from the mean of a distribution. Thus

Radiation Detectors

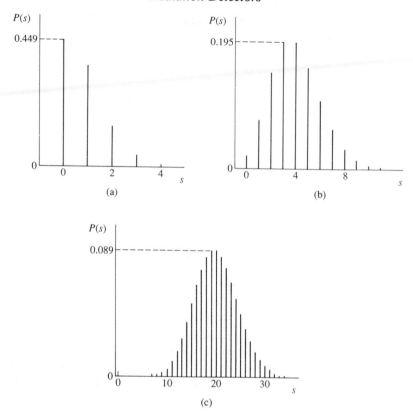

Figure 1.3 The Poisson distribution for values of the mean $\bar{s}$ of (a) 0.8, (b) 4.0, (c) 20.

$$\sigma^2 = \frac{\sum\limits_{s=0}^{\infty} (s - \bar{s})^2 P(s)}{\sum\limits_{s=0}^{\infty} P(s)} \tag{1.5}$$

where σ^2, the standard deviation squared, is known as the variance, V, of a distribution. For a Poisson distribution we can show that the variance is equal to the mean, that is, from equations (1.3), (1.4), and (1.5),

$$V = \sigma^2 = \bar{s}. \tag{1.6}$$

Thus the standard deviation is $\bar{s}^{1/2}$. This result is widely used in counting experiments and will appear frequently in this book. The standard deviation is a measure of how far a single observation of the count is likely to deviate from $\bar{s}$, the mean: it is often called the statistical 'error' in the count.

The fractional error in the measurement of the count is $\sigma/\bar{s}$, and hence

$$\frac{\sigma}{\bar{s}} = \frac{\bar{s}^{1/2}}{\bar{s}} = \frac{1}{\bar{s}^{1/2}}.$$

The relative variance, $\mathcal{V}$, is defined as $(\sigma/\bar{s})^2$, and is thus given by $1/\bar{s}$. The fractional error and the relative variance become smaller, that is, improve, if larger values of $\bar{s}$ can be achieved, hence the importance of obtaining a large count where possible.

We note from Figure 1.3 that the Poisson distribution becomes increasingly symmetrical as $\bar{s}$ increases. Equation (1.3) for a Poisson distribution can be then shown to lead to the Gaussian distribution, such that

$$P(s) = \frac{1}{\sigma(2\pi)^{1/2}} \exp\left(\frac{-(s-\bar{s})^2}{2\sigma^2}\right). \tag{1.7}$$

Here $P(s)$, as before, is the probability of obtaining a value s for a count, $\bar{s}$ and σ (both much greater than 1) give the mean and standard deviation, respectively, of the distribution, and $\sigma^2 = \bar{s}$ since our Gaussian originates from a Poisson distribution. The Poisson distribution shown in Figure 1.3(c), where $\bar{s} = 20$, is already a close approximation to the form of equation (1.7), and the approximation becomes increasingly exact as $\bar{s}$ increases to very large numbers. At the same time the tops of the bars show more and more clearly the form of the continuous smooth curve on which they lie. In fact in other areas outside counting statistics the variable s can take continuous values, such that it is possible to derive a Gaussian distribution which is continuous rather than discrete, and where $\sigma^2 = \bar{s}$ need not necessarily be true (references 1 and 2).

It can be shown that for a Gaussian distribution the probability of obtaining a count between $\bar{s} - \sigma$ and $\bar{s} + \sigma$, that is, within one standard deviation either side of the mean, is 0.685. The probability of the count lying within two standard deviations comes to 0.955, and within three, 0.997. This shows quantitatively what is meant by the common statement that the error in a single measurement s of the mean value can be taken as σ. In particular it shows that the probability of a measurement lying outside such error limits, that is, outside one standard deviation, is far from negligible, contrary to some simple definitions of the word 'error'. A similar cautionary remark applies in the case of all Poisson distributions of counts, of course.

Another quantity of importance is the full width at half maximum, or FWHM. This, as its name implies, is the overall width of the distribution at a level half way down from the peak. For a Gaussian distribution this can be shown to be equal to 2.355σ. Thus the standard deviation is in this case slightly less than the half width at half height. In fact the standard deviation equals the half width at about 60 per cent of the full height.

The Gaussian distribution is much used in other applications to describe distributions of data in which the mean is not related to the standard deviation and the variance in the ordinary Poissonian way. For example, one important case where equation (1.7) can often be taken to hold is for a detector struck by an incident particle or quantum of radiation, where s represents the number of charge carriers collected, which is proportional to the height of the electrical pulse observed in an external circuit. When we come to discuss, for example, X- and gamma-ray spectra in Chapter 5 we shall find that the peaks in the spectra approximate to a Gaussian form.

Finally, it should be pointed out that in any counting system background counts may exist; these are the residual counts remaining when the source being viewed is removed. By subtracting the background count rate from the observed rate the true rate can still be found; however, the effect of background adds to the statistical error obtained. It is for this reason that we make every effort to reduce background (by shielding or other means) when counting low-activity sources. In our discussions we have assumed that the background is negligible.

This completes our brief discussion of some of the general features of radioactive decay. In the next section we proceed to discuss specifically the different decay processes which can occur.

1.5 Types of nuclear decay

1.5.1 *Alpha decay*

We might imagine that, if a heavy nucleus decayed by breaking up into smaller fragments for the reasons outlined in Section 1.3, a whole range of fragment masses would result, with no one mass being especially strongly favoured. For the very heaviest nuclei such a process can indeed occur, as we shall see in Section 1.5.5. However, in practice by far the commonest fragment found in such a break-up is an alpha particle, which is the nucleus $^4_2$He of the helium atom. This is because the configuration of two protons and two neutrons turns out to be particularly strongly bound together. As discussed by most nuclear physics texts, alpha particles can thus be spontaneously emitted from a large number of heavy nuclei with mass numbers above about $A = 150$. Alpha particles were one of the first two types of radiation to be identified; this was achieved by Rutherford in 1899 in his work with the 'Becquerel rays' from uranium salts.

Alpha decay can be written in a general form as

$$^A_Z X \rightarrow \, ^{A-4}_{Z-2}Y + \alpha$$

where it can be seen that a new element Y is formed. One example is

$$^{238}_{92}U \rightarrow \, ^{234}_{90}Th + \alpha.$$

The total energy E_T released is determined by the mass excess, Δm, of X over $(Y + \alpha)$ by, from Einstein's mass–energy relationship, $E_T = c^2 \Delta m$, where c is the velocity of light. For $^{238}_{92}$U this energy E_T comes to 4.27 MeV. There are only two products from an alpha decay, the alpha particle itself and the new nucleus. The energy released is apportioned such that each product has equal and opposite momentum, and in the simplest case this energy appears entirely as the total kinetic energy. From this it can be easily shown that the kinetic energy of each product is inversely proportional to its mass. The alpha particle from $^{238}_{92}$U would hence be of energy $E = 4.20$ MeV (this energy is indeed observed) and that of the recoiling $^{234}_{90}$Th atom 0.071 MeV. Only the former is usually recorded in a detector.

Often in alpha decay there are smaller numbers of alpha particles of slightly different energies from the main one. For example, for a $^{238}_{92}$U source only 77 per cent of the alphas are of energy 4.20 MeV, whereas 23 per cent have the somewhat lower energy of 4.15 MeV. (Section 1.5.3 discusses where the remainder of the energy goes.) As we shall see later in the book, it requires careful measurement to recognize as separate these two closely spaced groups of alpha particles.

Nearly all alpha-particle energies lie in the range of 3 to 9 MeV, whereas the observed half-lives vary between over 10^{15} years and less than 10^{-6} s. It is well known that there is an approximate correlation, first observed by Geiger and Nuttall in 1911, between the alpha energy and half-life. In power law form this can be expressed as $t_{1/2} \propto E^{-x}$, where x is about 80. A practical consequence of this enormously strong variation is that most commercially available alpha sources can give energies of only between about 5 and 6 MeV. Lower-energy sources decay too slowly and have too low an activity; higher-energy sources decay too fast and are soon useless.

We shall see in Chapter 2 that alpha particles slow down continuously in matter and travel only very short distances before they are stopped. These 'ranges' in solids are typically a few tens of micrometres and in air a few tens of millimetres. This means, for example, that alpha sources have to be very thin if the spectrum of sharply defined energies is not to be degraded by self-absorption in the source. Normally in this case the alpha-emitting material is deposited on a suitable substrate to give mechanical strength. If sufficiently precise energy definition can be achieved, an alpha-emitting radioisotope in a sample can be identified from the values of E observed.

1.5.2 *Beta decay*

Beta particles were also first identified by Rutherford in 1899 as a more penetrating component of the 'Becquerel rays'. In fact they must have caused the exposure on Becquerel's photographic plates, since the black paper wrappers around the plates were thick enough to have stopped the alphas. In contrast, as we shall see in Chapter 2, betas slow down less rapidly in matter; their range is typically a few millimetres in solids and of the order of one metre in air.

These beta particles are fast electrons (e^-) and are emitted in the decay of many radioisotopes. In general the process may be written as

$$\,_Z^A X \rightarrow \,_{Z+1}^A Y + e^- + \bar{\nu}.$$

Here $\bar{\nu}$ denotes another particle called an antineutrino, which we believe has little (if any) mass, no charge, and hardly any interaction with matter. (The bar above ν means 'anti', a terminology which we shall discuss in a little more detail shortly). Any ordinary detector will be able to see only the electron. Two examples of this decay are

$$\,_{90}^{234} Th \rightarrow \,_{91}^{234} Pa + e^- + \bar{\nu}$$

and

$$\,_6^{14} C \rightarrow \,_7^{14} N + e^- + \bar{\nu}.$$

The first process is the decay of the daughter product from $\,_{92}^{238} U$ described in Section 1.5.1. It must have been a source of the beta particles detected by Becquerel, as no naturally occurring uranium radioisotope emits betas directly.

The total decay energy released is determined by the mass excess Δm of X over ($Y + e^-$), that is, $E_T = c^2 \Delta m$, similar to the case with alpha decay. In the simplest cases this energy appears entirely as kinetic energy of the products. Unlike alpha decay, however, beta decay leads to three products rather than two. As a result, conservation of momentum this time permits the released decay energy to be shared in any way between the electron and antineutrino. The product nucleus in principle also recoils as in alpha decay, but since the electron mass is only about 1/2000 that of a proton or neutron, and the antineutrino is effectively massless, the nucleus recoil energy is very small. Thus the energies of the electrons, which are the only readily detectable products, stretch continuously from an 'end-point' maximum value, $E_{max} (\approx E_T)$, given by the total energy released (when the antineutrino takes zero energy) all the way down to zero (when the antineutrino takes all the energy).

How are we to describe the spectrum of electron energies, E, which occurs between these limits? Our task is clearly a little more complicated than for the sharply defined energies shown in alpha decay. Suppose we subdivide the energy interval between 0 and E_{max} into many small 'channels' each of the same width ΔE. We count the number, ΔN, of electrons whose energies lie in the channel between E and $E + \Delta E$, doing this exercise for all the channels lying between $E = 0$ to $E = E_{max}$. We can then construct a histogram in which each column, plotted as a function of E, has a width ΔE and an area equal to ΔN, the number of particles with energies in this interval. Clearly the height of the histogram column must therefore be $\Delta N/\Delta E$, and this is what we plot as ordinate. (This is related to what is actually done in experiments, as we shall see in Chapter 6.) The final result for a typical electron energy spectrum from beta decay will look something like Figure 1.4. Note that each value of ΔN, and thus of column height, may suffer from statistical fluctuations just like those described in Section 1.4.3. We assume, however, that the values are large enough for the relative variances in these fluctuations to be negligibly small, as is taken to be the case in Figure 1.4.

If we make ΔE smaller and smaller, the total number of channels becomes, of course, correspondingly very large, and the tops of the histogram columns approximate more and more closely to a smooth curve. (Again we assume that enough counts are taken to make statistical fluctuations negligible.) This curve of $\Delta N/\Delta E$, now dN/dE, against E then shows the 'continuous' electron energy spectrum. Because of the form of the ordinate of the graph the full name 'differential spectrum' is sometimes used. Figure 1.5 shows a typical form of such a spectrum for electrons from beta decay.

Note that the area under such a curve is given by

$$\int y\,dx = \int \frac{dN}{dE}\,dE = \int dN = N$$

and thus represents the number of counts recorded. (Similarly for a spectrum such as we will meet later containing a number of peaks, the number of counts in a particular peak is given by its area, not its height.)

In continuous spectra such as Figure 1.5 the ordinate dN/dE is sometimes

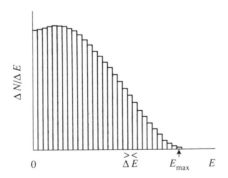

Figure 1.4 A histogram of $\Delta N/\Delta E$ against E as might be obtained in beta decay with value E_{max} for the maximum electron energy.

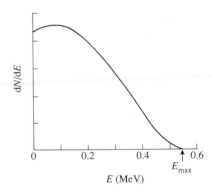

Figure 1.5 As for Figure 1.4, but shown as a continuous energy spectrum.
(Adapted from Evans, R.D. (1955), *The atomic nucleus*, McGraw-Hill, New York,
Chapter 17, Figure 1.4. Reproduced by permission of McGraw–Hill, Inc.)

referred to as the 'distribution function' $\mathcal{N}(E)$. This usage has the disadvantage that $\mathcal{N}(E)$ may be erroneously taken to have the dimensions of a count
number, and we will not make use of it.

Evidence for the existence in beta decay of a neutral particle like an
antineutrino was first suggested by Pauli in 1930 in order to explain (among
other features) the continuous nature of the beta energy spectrum. We note
in passing that monoenergetic electrons, that is, ones with sharply defined
energies, can be obtained from radioactive sources only through another
nuclear decay mode, which we consider in Section 1.5.4.

A continuous energy spectrum of the type just described is observed
with, for example, $^{14}_{6}C$, for which the total energy released is always
156 keV. Many beta emitters have more than one possible value of total
energy release; $^{234}_{90}Th$ releases about 180 keV in 72 per cent of its beta
decays, but only 91 keV in 17 per cent, 90 keV in 7 per cent, and various
totals in the remainder. (In Section 1.5.3 we discuss where the rest of the
energy goes.) Any observed energy spectrum would then be a superposition
of several continuous spectra, with only the highest end-point energy directly
observable.

Total observed energy releases in beta decays range from 18.6 keV
in $^{3}_{1}H$ (tritium), and even lower in one or two cases, up to a few MeV.
Observed half-lives range from microseconds to 10^{16} years or more. High
energy releases tend to go with low half-lives, but in general the relationship
between half-life and energy is not as clear-cut as for alpha decay. Tritium,
for example, has a half-life of only 12 years despite its small energy release.

All the beta decays so far described follow because a neutron inside the radioactive nucleus has been converted into a proton by the exothermic (energy-releasing) process

$$n \rightarrow p + e^- + \bar{\nu}.$$

(Indeed, a free neutron, as produced, for example, in a nuclear reactor, can itself beta decay by this electron emission process, the half-life being about 10 minutes.)

Electron emission in beta decay, while leaving the mass number thus unchanged, reduces the neutron/proton ratio. In line with our discussions in Section 1.3, we now see that it occurs for nuclei that have too many neutrons for stability. Only at most two or three nuclei of a given mass number are stable.

On the other hand, nuclei that have too many protons for stability can increase their neutron/proton ratio by another mode of beta decay. This process may be written in general as

$$_{Z}^{A}X \rightarrow _{Z-1}^{A}Y + e^+ + \nu$$

where e^+ is a positron, the so-called antiparticle of the electron, with identical properties to it (for example, its mass), except that its electric charge is of opposite sign. The symbol ν denotes a neutrino; the antineutrino is its antiparticle, but the two have for our purposes identical properties. Unlike electron emission, positron emission does not occur in nature, and it was in fact the decay process involved when artificial radioactivity was discovered by the Joliot-Curies in 1934. For example, one of the reactions they found was

$$_{15}^{30}P \rightarrow _{14}^{30}Si + e^+ + \nu$$

with a half-life of 150 s.

Most of the remarks about electron emission apply to positron emission, except that the continuous energy spectrum is pushed towards the maximum end-point energy, as shown in Figure 1.6. This is because the positively charged nucleus pushes the positron away, instead of, as with the negative electron, tending to retard its release. A second major difference is that positrons react with ordinary electrons after coming to rest in matter to produce another form of radiation. We consider this in Section 1.5.3.

In positron emission a proton inside the nucleus changes into a neutron through the following endothermic (energy-absorbing) reaction:

$$p \rightarrow n + e^+ + \nu.$$

The energy required for this reaction comes from rearrangements in the structure of the nucleus; it does not, of course, occur for a free proton. The reaction is endothermic because the proton mass is less than the combined masses of the neutron and positron (indeed it is less than that of the

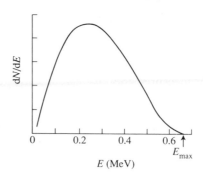

E (MeV)

Figure 1.6 A typical continuous spectrum of positron energies in beta decay with value E_{max} for the maximum positron energy. (Adapted from Evans, R.D. (1955), *The atomic nucleus*, McGraw-Hill, New York, Chapter 17, Figure 1.6. Reproduced by permission of McGraw–Hill, Inc.)

neutron itself): on the other hand the neutron decay given earlier is exothermic because the neutron mass is greater than the combined masses of the proton and electron, as can be verified from the values given in Section 1.3.

As shown in texts on nuclear physics, it is possible for some beta-active radioisotopes to emit both positrons and electrons. $^{64}_{29}$Cu is an example of such a dual beta emitter; in this case about twice as many electrons are emitted as positrons. The energy spectra in Figures 1.5 and 1.6 are in fact those of the electrons and positrons, respectively, which are emitted from this radioisotope. We add here a small point of notation: we shall often refer to decays giving electrons and positrons as β^- and β^+, respectively.

There is a second way in beta decay in which a proton inside the nucleus can change into a neutron, which is

$$p + e^- \rightarrow n + \nu.$$

This is the process of electron capture, in which a proton in a nucleus captures one of the surrounding atomic electrons. Usually an inner shell electron, such as a K electron, is captured. An example is

$$^{55}_{26}Fe + e^- \rightarrow ^{55}_{25}Mn + \nu$$

which has a half-life of 2.6 years. It would therefore appear that an ordinary nuclear radiation detector should give no response with, as in this case, a radioisotope which decays in only this way. This is not so, and we shall find out why in Section 1.7.2.

In many cases a nucleus has a chance of decaying by either electron capture or by positron emission. Electron capture is the preferred and easier route in that the energy released is greater, as can be seen by comparing the masses lost in the two reactions given above for changing a proton into a neutron.

Unlike positron emission, electron capture can therefore sometimes occur by itself.

Because of the continuous distribution of particle energies in beta decay, it is not as simple to identify beta-emitting radioisotopes present in a sample from energy measurements as was the case with alpha emitters. In addition, although beta particles penetrate matter more easily than alpha particles, as our earlier range data indicated, energy losses of betas and their self-absorption in thick sources are nevertheless appreciable. This would also hinder isotope identification in such sources from energy measurements.

1.5.3 *Gamma decay*

Gamma rays were discovered by Villard in 1900 as a penetrating component in the radiation from radium. They are high-energy photons which can be emitted when an excited nucleus formed in a previous nuclear process decays to a less excited state, which may or may not be its ground state:

$$^{A}_{Z}X^* \rightarrow {}^{A}_{Z}X + \gamma.$$

No change in nuclear species therefore occurs. The process is analogous to the decay of an atom in which a low-energy photon is emitted. The energies of gamma rays are, as we discussed in Section 1.4.1, sharply defined, and normally (although not always) much larger than those of photons from atoms. Typically they lie between 10 keV and 3 or 4 MeV. The actual value in a particular decay is given (neglecting the very small effect due to nuclear recoil) by the difference in energies between the two states concerned.

We should remark here that, by extension, the term 'gamma ray' has often come to include any photon of energy above, say, 200 keV, even if it is not produced in the way we have just described. An example of this will be given shortly.

The lifetime for gamma decay, that is, the mean time for an excited nucleus to decay, is usually very short, in the femtosecond (10^{-15} s) to microsecond range. In exceptional circumstances, however, much longer lifetimes, up to greater than 100 years, can occur, and these 'metastable' states or 'isomers' are of importance in various applications.

The excited state which is required for gamma emission from a nucleus is often produced in a previous alpha or beta decay. It is relatively uncommon for these decays to proceed exclusively to the ground, that is, unexcited, state of the daughter nucleus. For example, $^{238}_{92}U$ decays by alpha emission to the ground state of $^{234}_{90}Th$ in 77 per cent of the cases, but 23 per cent of the decays proceed to a $^{234}_{90}Th$ state excited by 49 keV, which then rapidly decays to the ground state by emitting a 49 keV gamma ray. Figure 1.7 shows this information diagrammatically in the form of an energy level diagram.

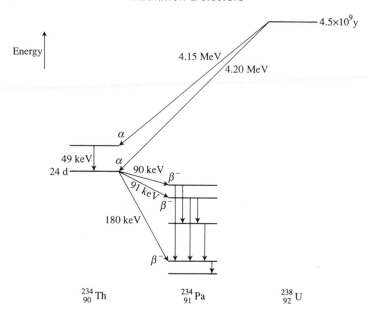

Figure 1.7 An energy level diagram of the decay of $^{238}_{92}$U to $^{234}_{90}$Th and $^{234}_{91}$Pa. The vertical heights of the levels denote the energies of the states relative to one another (not to scale), and the horizontal positions indicate the atomic numbers Z. Some half-lives are also shown.

We can now understand why more than one energy can occur in alpha decays of a given nuclear species. In our example the 49 keV gamma energy corresponds (apart from a small effect caused by the $^{234}_{90}$Th recoil energies) to the difference between the two alpha-particle energies, 4.15 and 4.20 MeV, which were previously given in Section 1.5.1.

$^{234}_{90}$Th later decays by beta emission to one or other excited states of $^{234}_{91}$Pa, again with subsequent gamma emission, as Figure 1.7 also shows. This illustrates why not all beta emitters have a unique total energy release, as was mentioned in Section 1.5.2.

Two other examples will be briefly given: the well-known 1.17 and 1.33 MeV gamma rays from a $^{60}_{27}$Co source come, following beta decay, from decays between different states of the $^{60}_{28}$Ni daughter, and the 662 keV gamma from $^{137}_{55}$Cs (of importance in nuclear fission product measurements) comes from the decay of an excited state of its daughter, $^{137}_{56}$Ba.

In all these four cases the gamma decays, as is usual, have much shorter half-lives than the preceding alpha or beta decays. In each case, therefore, from what we said in Section 1.4.2, the preceding decay controls the rate of gamma emission, which thus appears to decrease with the half-life of the parent concerned.

An example of a pure beta decay, in which no gammas are ever emitted, is the decay of $^{14}_{6}C$; this proceeds directly to the ground state of $^{14}_{7}N$. This is why we were able to state in Section 1.5.2 that beta decays from this radioisotope show a unique value for the total energy released.

There is another common process which produces gamma rays, although in this case they are not directly nuclear in origin. This is the annihilation of positrons in matter. As a positron comes to rest, it reacts with one of the many free electrons in its surroundings, producing two annihilation gamma rays. These, as required by momentum conservation, are of equal energy and recoil in opposite directions. Each gamma ray is of energy 0.511 MeV, such that the two energies add up to the energy equivalent $2\,m_0 c^2$ of the disappearing electron and positron. Here m_0 is the rest mass of either particle and c, as before, is the velocity of light. The existence of annihilation radiation in the spectrum of products from beta decay by positron emission is the second feature referred to in Section 1.5.2 which distinguishes it from electron emission.

Other sources of gamma rays are the nuclear fission process, which is considered in Section 1.5.5, and nuclear reactions induced by charged particles, which are discussed in Section 1.8.2.

Gamma rays, as we shall discuss in Chapter 2, can be much more penetrating than alpha or beta particles. Depending on the energy and the absorber, an individual gamma ray can sometimes travel through several tens of millimetres or more in solids without its energy being affected in any way. Gamma rays do not possess a definite range like alphas, or even an upper limit to their range like betas. Instead each gamma is removed suddenly at some point along its path, such that only a mean length of travel can be defined. Hence gamma sources need not be thin, and they can be obtained sealed in plastic or other light material; most of the gammas escape unaffected whereas any alphas or betas from the decays producing the nuclear excited states can be stopped by a sufficient thickness of material. Furthermore (like alphas, but unlike betas) gamma rays have sharply defined energies (as we noted earlier), which are characteristic of the particular radioisotope involved. For these reasons we shall find that gamma-ray detection is a particularly valuable method for identifying and quantitatively analysing radioisotopes in samples of material. Examples of gamma-ray spectra will be shown in the sections of succeeding chapters where gamma-ray detectors are considered.

1.5.4 *Internal conversion*

There is another mode for an excited nucleus to lose its energy; it is closely related to gamma decay but does not actually lead to gamma-ray emission.

This is the process of internal conversion, in which the excitation energy of a nucleus is transferred directly (not via a gamma ray, as was originally thought) to an inner-shell atomic electron. This electron is then ejected with kinetic energy $E_\gamma - E_B$ where E_γ is the energy of the gamma ray that would have otherwise appeared, and E_B is the binding energy of the ejected electron. By 'binding energy' we mean the minimum amount of energy required to free a bound electron from its parent atom. These conversion electrons are the monoenergetic electrons which were referred to in Section 1.5.2.

In practice, even if E_γ is uniquely defined, we see groups of such electrons in an energy spectrum. Each group corresponds to electrons which have been ejected from a different inner shell—K,L,M, and so on, as we discussed in Section 1.3. Thus E_B has values E_K, E_L, E_M, ..., corresponding to the same K,L, and M shell binding energies that control the energies seen in X-ray spectra (Section 1.7).

Internal conversion is clearly an alternative to gamma emission, and both processes may occur in the same source. Internal conversion can be shown to be most likely for low-energy transitions in heavy nuclei, and in the 100 keV region it is sometimes, depending on the circumstances, more probable than gamma emission. In the 1 MeV region, however, internal conversion is usually very improbable.

1.5.5 *Spontaneous fission*

We have by now seen that when a radioisotope decays a variety of products often appears. It is quite possible for alphas, betas, and gammas to be emitted from a single source, that is, from the parent radioisotope itself and from its daughters. The nuclear fission process, in which a heavy nucleus splits into two roughly equal fragments, leads to an even greater profusion of possible decay products. The most familiar example of the process is the reaction of $^{235}_{92}$U with a neutron, in which the $^{236}_{92}$U nucleus formed can immediately undergo fission; this is the process which occurs in nuclear reactors, and which, as we shall see in a moment, gives such complexity to the decay products and radiations associated with used nuclear fuel.

Fission can, however, also occur spontaneously as a mode of radioactive decay for some isotopes of mass number greater than about 230. One or two of these are particularly useful as convenient small-scale sources of fission fragments. These fragments (which are similar to those produced in neutron-induced fission) are heavy ions of high energy, having a range of masses between about 75 and 160 u and energies between 50 and 115 MeV. Few fragments are emitted in the region around 120 u; most fission events are asymmetric, leading to a light fragment in the region of 100 u and 100 MeV and a heavy fragment around 140 u and 80 MeV, as Figures 1.8 and 1.9 show. There is thus a whole spectrum of masses and energies, and in fact a spon-

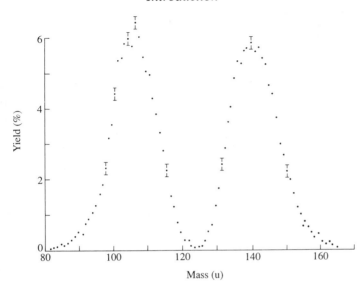

Figure 1.8 A spectrum of fragment masses from the spontaneous fission of $^{252}_{98}$Cf. The ordinate gives the percentage yield per mass unit, where *each* peak totals 100 per cent. (Adapted from Schmitt, H.W., Kiker, W.E. and Williams, C.W. (1965), *Physical Review*, **137**, B837–47, Figure 8.)

taneously fissile radioisotope is the only radioactive source of such heavy and energetic ions. By far the commonest laboratory source is $^{252}_{98}$Cf, which has a 2.6 year half-life. In this case fission occurs for 3 per cent of the decays; in the other 97 per cent alpha particles of energy 6.1 MeV are emitted.

In order to conserve momentum, the two fragments from a fission event are emitted in roughly opposite directions. The range of fragments in solids is only of the order of 10 μm, and it therefore follows that the source must be very thin. Even then half of them are lost in the source backing unless this also is very thin. A very fragile source now results, but is, however, needed if, as in some experiments, it is required to observe simultaneously a pair of fragments, both fragments originating from the same fission event.

Spontaneous fission also provides the only radioactive decay mode with a reasonably long half-life in which neutrons are emitted. Each fission event yields on average about three or four fast neutrons, the energies of which stretch up to several MeV. They are sometimes referred to as 'prompt' neutrons in that they are emitted almost simultaneously (within 10^{-16}s) with the fission event itself. If only the neutrons are required the source need not be thin and it can be sealed, since (as we shall see in Chapter 2) the neutrons are not absorbed by the source itself or by the can. In this sense such neutron

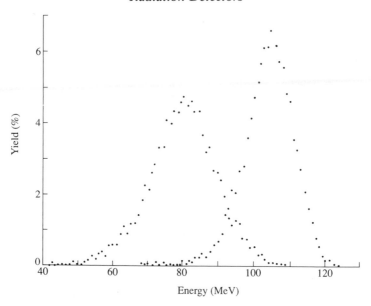

Figure 1.9 A spectrum of kinetic energies for light fragments (right-hand peak) and heavy fragments (left-hand peak) from the spontaneous fission of $^{252}_{98}$Cf. The ordinate gives the percentage yield per MeV. (The separation into light and heavy fragments was made by time-of-flight measurements like those referred to in Chapter 6.) (Adapted from Schmitt, H. W. *et al.* (1965), *Physical Review*, **137**, B837–47, Figure 5.)

sources resemble gamma sources. $^{252}_{98}$Cf is the fission neutron source which is normally used. (Note that the fragment mass and energy spectra in Figures 1.8 and 1.9 show the situation after the emission of these prompt neutrons.)

(Another source of neutrons, nuclear reactions, is described in Section 1.8. Neutron emission can also be observed as a result of a few instances of beta decay in which the product nucleus is excited and happens to be unstable against neutron emission. This is the 'delayed neutron emission' process, which, when it occurs in neutron-induced fission, is of central importance to the stability of the running of nuclear reactors. However, the half-life of the previous beta decay leading to one of these excited states is always short — a minute at most — and the excited state itself decays very quickly.)

The fission process, whether spontaneous or induced by neutrons, produces fragments which, if prompt neutrons were not emitted, would have roughly the same ratio N/Z of neutrons to protons às that of the original fissioning heavy nucleus. We can see from Section 1.3, however, that the value of N/Z for stability against decay falls as N and Z fall. Neutron emission thus occurs because it reduces the N/Z ratio for the fragments, and thus directs them towards stability. Even after this, however, the fragments are always neutron-rich, and so each decays to stability through a chain of perhaps four or five beta decays with electron emission (accompanied very occasionally by delayed neutron emission, as previously described). Half-lives for these beta decays vary from small fractions of a second to tens of years. Several gamma rays are also emitted, both from each fission and from each decaying fragment.

The radiations emitted from fission are therefore in total very complex, especially when we appreciate that there are over 80 mass numbers that have been observed for fragments. However, relatively few of the hundreds of possible decays are long lived. One of the long-lived products, $^{137}_{55}$Cs, which has a 30 year half-life, has already been referred to in Section 1.5.3. In Chapter 5 we describe how, thanks largely to detector developments, even very low activities of this and other fission products can be detected and assayed in environmental samples.

1.5.6 *Decay series of the heavy elements*

Most if not all of the isotopes with atomic numbers greater than lead (82) are radioactive. One of the earliest achievements in the study of radioactivity was to piece together chains or sequences of successive alpha and beta decays for these isotopes, slightly analogous to the shorter chains of beta decays observed for fission fragments. Since alpha decay reduces the mass number of a radioisotope by four, and beta decay leaves the mass number unchanged, it follows that there are four decay chains:

1. The thorium series, for which the mass number $A = 4n$, where n is some integer. The longest-lived member is $^{232}_{90}$Th (half-life $t_{1/2} = 1.41 \times 10^{10}$ years) and the isotopes decay finally to the stable nucleus $^{208}_{82}$Pb.

2. The neptunium series: $A = 4n + 1$: longest-lived member, $^{237}_{93}$Np ($t_{1/2} = 2.14 \times 10^6$ years): final nucleus (certainly extremely long lived and possibly stable), $^{209}_{83}$Bi.

3. The uranium series: $A = 4n + 2$: longest-lived member, $^{238}_{92}$U ($t_{1/2} = 4.47 \times 10^9$ years): final stable nucleus, $^{206}_{82}$Pb.

4. The actinium series: $A = 4n + 3$: longest lived member, $^{235}_{92}$U ($t_{1/2} = 7.04 \times 10^8$ years): final stable nucleus, $^{207}_{82}$Pb.

In the case of the uranium series ($A = 4n + 2$) the first few decays have already been shown in Figure 1.7. A Segrè-type plot of the whole series is given in Figure 1.10.

Thorium mineral ores contain in them all the daughters from the thorium series, and uranium ores contain all the daughters from the uranium and

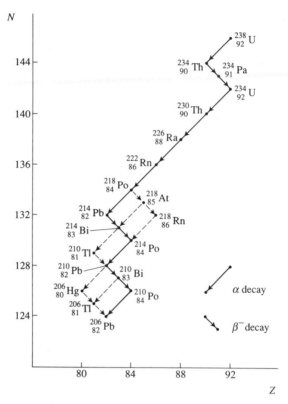

Figure 1.10 A Segrè plot of neutron number N against atomic number Z for the uranium series of heavy radioisotopes. The main sequence of decays is shown by full lines: broken lines indicate branches which are followed by only a small percentage of the decays.

actinium series. Within a series it is found that in equilibrium all the radioactive daughters each contribute in such ores equal activities (disintegration rates) to that of the parent. This result holds even though the daughters are of short half-life and are present in only very small quantities. As can be seen from texts in nuclear physics, this is an extension of the principles we outlined in Section 1.4.2. The radiations detected from these ores are therefore a complex mixture of alphas, betas, and gammas.

Sometimes one of the radioisotopes in a series has two alternative decay modes, one through alpha emission and the other by beta decay. Examples occur for $^{218}_{84}$Po, $^{218}_{85}$At, $^{214}_{83}$Bi, $^{210}_{82}$Pb, and $^{210}_{83}$Bi in Figure 1.10. The equality of activities referred to a moment ago obviously does not hold for the two daughters produced in this way from the same parent radioisotope. Note that even with this 'branching' the final stable nucleus in a given series will always be the same whichever resulting chain of decays is taken.

The neptunium series does not contain a radioisotope of long enough life for it to occur in nature. It was therefore discovered later than the others, the entire series having to be produced artificially.

New short-lived members of each series continue to be discovered. They add side chains to the main sequences, and transuranic ($Z > 92$) radioiso-topes at the tops. These achievements are due in no small measure to the development of new or better detectors over the years.

1.6 Accelerated charged particles

1.6.1 *Laboratory accelerator sources*

The beam of fast charged particles produced by an accelerator is one of the most important forms of radiation available. In this way energies can be reached and particle types produced which cannot be obtained from radio-isotope sources. Many accelerator types exist, but all require a source of charged particles and an electric field to accelerate them. Electrons are pro-duced from a hot filament, whereas ions are obtained from an electrical discharge, intense radiofrequency field or some other means. Acceleration is achieved by passing the electrons or ions one or more times through a potential difference. One single acceleration through something like 10^5–10^7 V is used in Cockcroft–Walton and simple van de Graaff accele-rators; multiple accelerations through a smaller difference are achieved in circular accelerators such as cyclotrons and in linear accelerators or 'linacs'.

The detailed principles of the various types of accelerators are beyond the scope of this book. However, the books by Burge and by Krane (listed as Further Reading at the end of this chapter) each have a complete chapter on the topic, as does that by England (which is listed in the Gene-ral Bibliography). The choice of accelerator type depends on the particles required and their energy and intensity. It was with one of the first single-stage accelerators that Cockcroft and Walton in 1932 'split the atom' (or rather its nucleus) by directing a proton beam of only a few hundred keV at a target of lithium. In contrast, energies of up to 10^{11} or 10^{12} eV are now available with multiple accelerations, although at these energies accelera-tors are so large that only a few exist in the world. Some are kilometres across and one (the SPS accelerator at CERN, Geneva) straddles a national frontier.

Particle beam intensities, depending on the accelerator type, can be up to $10^{12} s^{-1}$, and sometimes very much more, which is far too intense for direct measurements by simple detectors designed for single-particle work.

Normally of course, as in Cockcroft and Walton's work, what our detector will monitor will be the products of a nuclear or particle reaction produced by the beam from a target, and these will be very much fewer in number per second. If for some reason we wish to experiment with the particles in the beam itself this can be achieved if a small fraction is first scattered out of the beam towards the detector by a thin gold foil or other similar arrangement, as the large reduction in beam intensity otherwise required is usually difficult.

Nowadays in high-energy accelerators the target is sometimes a large detector designed to make visible the tracks of the particles incident on it and produced in it. (A simple precursor was the Wilson cloud chamber referred to in Section 1.2.) Most subatomic particles discovered since about 1950 have been found in this way. Complex high-energy detectors such as these are largely outside the scope of the book, but one or two of the simpler ones will be referred to in Chapter 8.

Sometimes the reaction products themselves form a usable source of radiation for another experiment. Neutrons and gamma rays, for example, can be produced in this way, as discussed in Section 1.8. Section 1.7 describes how accelerated particles can also be used to produce electromagnetic radiation such as X-rays.

1.6.2 *Cosmic-ray sources*

Cosmic rays are high-energy radiation of extraterrestrial origin, and consist principally of accelerated charged particles from galactic and possibly extragalactic sources, and also from the sun. Elster and Geitel and also Wilson in 1900 observed their ionizing effects, but it was not until 1911 that Hess realized that they could not originate from the earth.

Nuclei of all atomic numbers have been observed, but most of them are protons, and to a smaller extent, helium nuclei. The energy range observed is extremely large, from a few MeV up to very occasionally 10^{19} eV or more. Electrons, positrons, and neutrinos are also present. In addition to these primary cosmic rays travelling through space there is also secondary radiation produced when an incoming primary interacts with the atmosphere of the earth. These secondaries, which are often observed arriving together in showers, as it were, consist of electrons and photons together with other particles such as muons and neutrons which are unstable.

This flux of cosmic rays is far lower than that of particles from an accelerator. Despite the enormous energies often involved, some of the detectors to be described in this book have been used with cosmic rays in earth-based, balloon, rocket, and satellite experiments. Before the advent of high-energy accelerators the detection of cosmic rays and their secondaries provided one of the most important means of discovery of elementary particles. The study of the primaries continues to be of great importance to astronomy and astrophysics.

The occurrence of X- and gamma rays in cosmic radiation is described in Section 1.7.

1.7 X-rays

1.7.1 *Nature of X-rays*

X-rays can be described as fairly high-energy photons which are non-nuclear in origin. Their energies, around 10^2–10^5 eV, tend to be lower than those of gamma rays. The term 'X-rays' can embrace radiation from several types of sources, but they usually (though not always) involve an electron changing its energy or momentum. In practice the term can be defined in more than one way. Sometimes we think in terms of their origin (for example, as attempted in the initial description above), at other times in terms of an appropriate energy range for the photons. It is probably best not to attempt to formulate one single, rigorous definition.

1.7.2 *X-rays from atoms*

One of the best known methods of X-ray production occurs when an electron fills a vacancy in an inner shell of an atom, the vacancy having been previously created by the disruption caused by some excitation process. In Röntgen's experiments (Section 1.2) this excitation came from the accelerated electrons in cathode rays, but the excitation can come from other external radiation (alphas, betas, gammas, and so on—even other X-rays), or from the radioactive decay of the central nucleus of the atom concerned.

The energies observed for these atomic X-rays depend on the binding energy for an electron in the inner shell concerned and on where the electron filling the vacancy comes from. An electron coming from the L shell and filling a vacancy in the innermost K shell would give a so-called K_α X-ray of energy equal to the difference in binding energies for the K and L shells; an electron coming from the M shell into the K shell would give a K_β X-ray of slightly higher energy, and so on. Vacancies in the outer L, M (and so on) shells lead to lower energy, L_α, L_β, M_α, etc., X-rays.

All the X-ray energies depend on and are characteristic of the element concerned. For the lightest elements even K binding energies are very low (13.6, 25, and 55 eV for hydrogen, helium, and lithium, for example) but for most elements the K energies are high enough to be measurable by nuclear radiation detectors. For sodium (of atomic number $Z = 11$) this energy is about 1.0 keV, for silicon ($Z = 14$), 1.8 keV, for germanium ($Z = 32$),

11 keV, and for uranium ($Z = 92$), 116 keV. L X-ray energies exceed 1 keV for nickel ($Z = 28$) and above.

A convenient X-ray source excited by radioactive decay of the central nucleus of the atom concerned entails the electron capture process (Section 1.5.2). This leaves an inner (usually K) shell vacancy in the atom, which, when filled, leads to X-ray emission characteristic of the atom produced, but with a half-life set by the decaying nucleus. The advantage of this mode of excitation is that essentially no other radiation is emitted, except possibly either positrons, if their emission is also energetically allowed, or gamma rays if the daughter nucleus is formed in an excited state, or both. In fact $^{55}_{26}$Fe, which was the example of electron capture we discussed earlier, emits only X-rays characteristic of $^{55}_{25}$Mn.

Internal conversion (Section 1.5.4) also leads to X-ray production because of the vacancy created in an inner (again usually K) shell by the ejection of an electron. Gamma rays usually accompany the X-rays since gamma decay is almost always an alternative to internal conversion, as we saw earlier.

Self-absorption is a problem with these 'radioactive' types of X-ray sources. This is because the distance travelled by an individual X-ray before it interacts with matter is much shorter than for gamma rays—perhaps only 10^{-3} to 1 mm, depending (as we shall see in Chapter 2) on the absorber and the X-ray energy. It limits the effective strength of such sources, since (assuming a chemically and isotopically pure sample to start with) increasing the strength of a source of given area can be achieved only by increasing its thickness and hence the self-absorption.

X-rays, as we saw earlier, are also produced if a target is struck by external radiation from an alpha, beta, or gamma source. Alternatively an accelerated electron beam (as in Röntgen's original experiments) or other X-rays (creating X-ray fluorescence) may be used. In all cases the incident radiation must have a higher energy than the X-ray required from the target.

1.7.3 *Bremsstrahlung*

Many of the X-ray sources just described show in an energy spectrum from a detector not only the characteristic lines of the material concerned but also a background X-ray continuum. This is an example of 'bremsstrahlung', a word which in general is applied to the electromagnetic radiation arising from the acceleration of a charged particle as it is deflected from its straight-line motion by the electrical action of a nucleus near its path in a stopping material. The word is German in origin, and means 'braking radiation', which refers to the slowing down caused by the emission of the radiation. The particle most likely to show bremsstrahlung is an electron, because its mass is so low that it can be deflected very easily. At MeV energies and below it is in fact the only particle which gives significant bremsstrahlung. Even for electrons, however, the kinetic energy loss caused by bremsstrahlung is usually very much less than that due to other mechanisms for energy loss, unless the energy of the electrons is well above 10 MeV and the stopping material has a high mass number (like, for example, lead).

X-ray tubes which use fast electrons for excitation often give particularly prominent bremsstrahlung stretching in energy up to that of the electrons themselves. Even electron capture and internal conversion X-ray sources show a little bremsstrahlung along with the primary monoenergetic photons. Sources excited by external alpha particles show no bremsstrahlung, and X-ray fluorescence sources show only a little bremsstrahlung scattered from the original source.

1.7.4 *Synchrotron radiation*

Synchrotron radiation is emitted by a beam of fast electrons when they are deflected by magnetic fields, and is so called after the type of accelerator in which it was first observed. This previously unwanted electromagnetic radiation is now produced in accelerators specially constructed for the purpose. These accelerators give intense, highly collimated beams of radiation both in the conventional X-ray energy region and outside it. Like bremsstrahlung, synchrotron radiation has a continuous distribution of energies. These features make this radiation of great value in many fields of physics and other branches of science.

1.7.5 *X-ray astronomical sources*

X-ray astronomy is the study of X-rays that originate above the atmosphere. As with cosmic rays, they are both solar and galactic in origin. Bremsstrahlung and synchrotron radiation emitted by cosmic-ray electrons are thought to be among the physical processes responsible for cosmic X-rays. Because they penetrate the atmosphere to only a limited extent, it is necessary to detect them from balloons, rockets, or satellites. When energies are observed which are higher than those normally associated with X-rays, these photons are often called gamma rays regardless of their origin.

1.7.6 *X-rays and Auger electrons*

Just as the emission of internal conversion electrons is an alternative to gamma-ray emission from a nucleus, so there is an alternative decay mode to X-ray emission for the atom. In this alternative the excitation energy of the atom arising as a result of the vacancy created in an inner shell is transferred directly (not via a photon) to an outer electron. This electron is then ejected with kinetic energy $E_x - E_B$, where E_x is the energy of the X-ray that would have otherwise appeared and E_B is the binding energy of the ejected outer electron. This is called an Auger electron. Emission of such electrons occurs chiefly for elements of low atomic number. The electrons are therefore typically only a few keV in energy, and so are easily slowed or stopped by the source material itself or by any protective material or other 'dead' layer in front of the active region of a detector.

1.8 Neutrons and other radiations from nuclear reactions

1.8.1 *General principles*

The use of nuclear reactions extends the availability and energies of nuclear radiations beyond what is accessible with radioactive decay. In this approach a target nucleus 'X' is bombarded with radiation 'a', giving a product nucleus 'Y' and the required radiation 'b'. We write

$$X + a \rightarrow Y + b$$

or X(a,b)Y for short. The incoming radiation 'a' can come from, for example, a radioactive decay, a particle accelerator or a nuclear reactor. In a sense this is a formal analogue to the production of X-rays by external excitation of atoms with alphas, betas, or cathode rays.

Most of the reactions we shall consider have a charged particle (for example, an alpha particle) as the incident radiation 'a'. It is important, however, for the incident particle to be able to overcome the Coulomb 'barrier' or repulsion between its own nuclear charge ze and the charge Ze of the target nucleus 'X'. Otherwise 'a' will not be able to approach 'X' closely enough to have a sufficiently high likelihood of inducing the nuclear reaction. Since the force of repulsion at a separation r is proportional to Zze^2/r^2, the barrier effects are least and reaction yields greatest if the atomic numbers Z and z are low. Thus alpha particles (for which $z = 2$) and light targets are often used, as we shall see in Section 1.8.2. In Section 1.8.3 we consider how the extreme case of $Z = z = 1$ can be realized in practice. Particles with higher kinetic energies are, of course, also able to penetrate Coulomb barriers more easily.

Nuclear reactions are useful in obtaining neutrons, since only a few neutron-emitting radioisotopes exist, such as $^{252}_{98}$Cf (Section 1.5.5). Also, being uncharged, neutrons cannot be directly accelerated by a potential difference in the way that fast charged particles can be produced in an accelerator. Neutron production will therefore form the main concern of the present section.

1.8.2 *Reactions using radioisotopes*

The radiation from radioisotopes is often used to produce neutrons from reactions induced in a target by the radiation. For example, if an alpha particle is incident on beryllium, the following (α, n) reaction occurs:

$$^9_4\text{Be} + \alpha \rightarrow {}^{12}_6\text{C} + \text{n}.$$

The energy released by this reaction is 5.7 MeV — we call this energy the Q value of the reaction. When Q is positive, as in this case, the reaction is said to be 'exothermic' (a term we first introduced in Section 1.5.2 in connection with beta decay). When, as sometimes occurs in this case, the products are formed in their ground (unexcited) states, the Q value also gives the difference between the kinetic energies of the products and the initial reacting nuclei. Most of the Q value release here goes into the kinetic energy of the neutron, as it is much lighter.

Normally the alpha emitter required to initiate the above reaction is chosen to be a metal which can be alloyed with beryllium; the source is then sealed in a metal container. Alpha emitters used include $^{239}_{94}\text{Pu}$, which is probably the commonest, and also $^{241}_{95}\text{Am}$, $^{238}_{94}\text{Pu}$, $^{227}_{89}\text{Ac}$, $^{226}_{88}\text{Ra}$, and $^{210}_{84}\text{Po}$. The choice depends on source strength, half-life, cost considerations, and the effect of any other radiations present. Strengths of 10^7 or more neutrons per second are possible.

The alphas are slowed down by different amounts depending on the distance between the emitting nucleus and a beryllium target nucleus. However, many of those which actually succeed in initiating an (α, n) reaction must have previously suffered only fairly small energy losses in the source, since lower-energy alphas cannot overcome Coulomb barrier effects as easily. The alphas therefore add their kinetic energies, which fall within a somewhat limited range up to about 5 MeV, to the Q value of the reaction. If the product ^{12}C nucleus is formed in its ground state, a neutron with an energy somewhere around 10 MeV is therefore obtained. Lower-energy neutrons are, however, also produced, since sometimes a ^{12}C nucleus is formed in one or other of its excited states, which leaves less energy available for an emitted neutron. The overall result is that a fairly continuous distribution of neutron energies up to 10 MeV appears.

It should be noted that $^{226}_{88}\text{Ra}$ and $^{227}_{89}\text{Ac}$, although they give particularly high neutron yields for a given alpha activity, also give much stronger gamma-ray backgrounds than those from the other isotopes. As well as possibly interfering with an experiment, the gammas increase the radiation hazards associated with the handling of such strong sources.

Neutrons of lower energies can be obtained from sources which use (α, n) reactions on certain light targets other than beryllium. ^7_3Li and $^{19}_9\text{F}$, for example, give Q values for the (α,n) reaction of as low as −2.8 and

−1.9 MeV, respectively. The negative sign means that the alphas must have a certain threshold value of energy to cause the reaction to proceed. Thus, for example, a 2.8 MeV alpha incident on ^{7}Li would produce a neutron of zero energy; at higher alpha energies E_α the kinetic energy of the products would be $(E_\alpha - 2.8)$ MeV. (This is an example of an endothermic reaction.) The neutron source strengths obtainable with ^{7}Li and ^{19}F are, however, less than with ^{9}Be.

Not all the reaction energy released necessarily goes to the neutron. The product nucleus, apart from receiving a small recoil kinetic energy from the reaction (additional to that absorbed from the original alpha kinetic energy), is, as we saw a moment ago, sometimes formed in an excited state. This then gamma decays to the ground state. In the two cases of $^9_4$Be and $^{13}_6$C the gamma rays emitted after an (α, n) reaction are, at 4.4 and 6.1 MeV, respectively, much higher than can be obtained from radioisotopes following beta decay (Section 1.5.3). They are thus useful for producing high-energy gamma rays as well as being neutron sources.

Neutrons can also be obtained from the (γ, n) reaction. Photoneutron sources, as they are called, are available using either deuterium or beryllium as a target:

$$^2_1\text{H} + \gamma \rightarrow ^1_1\text{H} + \text{n}$$

$$^9_4\text{Be} + \gamma \rightarrow ^8_4\text{Be} + \text{n}$$

(the $^8_4$Be very quickly splits into two alphas). Provided that the gamma energies more than offset the Q values of −2.2 and −1.7 MeV, respectively, neutrons are emitted which are almost monoenergetic if the gammas are too. This advantage is offset by the high activities of the gamma radioisotopes which are required by such a source, and which supply an unwanted intense gamma background to the neutrons. Gamma emitters commonly used include $^{24}_{11}$Na, which gives with $^9_4$Be neutron energies of 0.8 MeV, but with a half-life of only 15 hours. $^{124}_{51}$Sb has a longer half-life of 60 days and leads to 24 keV neutrons with $^9_4$Be. Neutron source strengths of at least 10^6 s^{-1} are possible.

1.8.3 Reactions with accelerated ions

When accelerated charged particles are incident on a target, neutrons are very often emitted from nuclear reactions. As we saw in Section 1.8.1, Coulomb barrier effects are least and neutron yields likely to be highest for target and particle atomic numbers Z and z both equal to one. Two particularly useful reactions are therefore

$$^2_1\text{H} + \text{d} \rightarrow ^3_2\text{He} + \text{n} \quad (Q = 3.3\,\text{MeV})$$

$$^3_1\text{H} + \text{d} \rightarrow ^4_2\text{He} + \text{n} \quad (Q = 17.6\,\text{MeV})$$

where d indicates a deuteron, that is, an accelerated deuterium nucleus, and ^2_1H and ^3_1H indicates the target deuterium and tritium, respectively. A beam of about 10^{15} deuterons s^{-1} of only a few hundred keV energy will produce about 10^8 and 10^{10} neutrons s^{-1}, respectively, from these two reactions. With such low deuteron energies the neutrons produced are nearly monoenergetic. Their energies are about 2.5 MeV and 14 MeV, respectively, as given by momentum conservation.

If targets with larger atomic numbers are used then particle energies have to be higher in order to overcome Coulomb barrier effects. For example, the Rutherford–Appleton Laboratory, UK, uses accelerated protons of energies approaching 10^9 eV to induce 'spallation' (i.e. fragmentation) reactions in a uranium target. A single reaction produces about 25 neutrons each with energy in the 1 MeV region. These energies can be reduced or 'moderated' using the type of techniques described in Section 1.8.4. Although expensive, these procedures enable research using neutrons to be carried out in several scientific areas.

1.8.4 *Nuclear reactors*

Another complex and expensive, but excellent, source of neutrons comes from the fission reactions induced by neutrons in a nuclear reactor. The reaction with $^{235}_{92}U$ (Section 1.5.5) is effectively

$$^{235}_{92}U + n \rightarrow F + F' + xn$$

where F and F' denote fission fragments and x is a number which is on average about 2.5. For example, in order to produce radioisotopes the neutrons surplus to those required to sustain the chain reaction can be used to irradiate a sample placed in the reactor core. Alternatively the neutrons can be allowed to emerge from the core as a beam. The latter method has the advantage that the accompanying gamma radiation can be reduced or even eliminated.

The neutrons have energies up to several MeV if (as in fast reactors) they come more or less straight from the reaction. These energies are similar to those obtained from spontaneous fission. If low-energy 'thermal' neutrons are required, the neutrons can be moderated down to thermal energies in elastic collisions with low-atomic-number nuclei in a suitable material. In this way the neutron energies are brought down to the region of 0.025 eV (at room temperature) over a path length of not less than several tens of millimetres. The materials concerned are called 'moderators', and include light water, H_2O, heavy water, D_2O (D stands for deuterium, ^2_1H), and graphite, $^{12}_6C$. In any case neutron moderation is an integral part of the operation of many reactors because low-energy neutrons have a much larger probability of inducing the further fissions required to keep the reactor running.

Moderators can, of course, also be used with any other neutron source.

Reactors designed for electricity generation are not normally used as neutron sources, since the volume density of neutrons produced is relatively low so as to keep the overall power density down. Special reactors, such as that at the Institut Laue-Langevin, Grenoble, France, are instead designed to have very high power and neutron densities in a relatively small volume. 10^{14}–10^{15} neutrons can cross $1\,\mathrm{cm}^2$ in the core of this type of reactor in 1 s. Other special types of reactors include, for example, assemblies which operate in pulsed modes, producing bursts of neutrons. Some of the most significant recent advances in pure and applied physics, as well as in other disciplines, have been made with neutrons derived from reactors. These studies range from previously unobserved quantum-mechanical effects as revealed by neutrons through to neutron diffraction studies of the structure of complex biological molecules.

References

1. Barlow, R. J. (1989). *Statistics*, Chapter 3. Wiley, Chichester.
2. Lyons, L. (1986). *Statistics for nuclear and particle physicists*, Chapter 3. Cambridge University Press.

Further reading

We list here some recently published books on nuclear physics; the bibliographies of these books should be consulted for details of older works.

Blin-Stoyle, R. J. (1991). *Nuclear and particle physics*. Chapman and Hall, London. An introduction, written from a theoretical viewpoint.

Burge, E. J. (1988). *Atomic nuclei and their particles*. (2nd edn). Clarendon Press, Oxford. An introductory text.

Jelley, N. A. (1990). *Fundamentals of nuclear physics*. Cambridge University Press. Takes the subject up to graduate level.

Jones, G. A. (1987). *The properties of nuclei*, (2nd edn). Clarendon Press, Oxford. Designed to take over where the book by Burge (see above) leaves off.

Krane, K. S. (1988). *Introductory nuclear physics*. Wiley, New York. A more substantial and comprehensive text than its title might suggest. There are helpful comments accompanying the references at the end of each chapter.

Pais, A. (1986). *Inward bound*. Clarendon Press, Oxford. The history from 1895 of atomic, nuclear, and particle physics.

Williams, W. S. C. (1991). *Nuclear and particle physics*. Clarendon Press, Oxford. An introductory account.

The bibliography at the end of this book gives details of more general texts, each covering the topics treated in this and other chapters.

2

Interactions of radiations with matter

2.1 Introduction

Before we can begin to understand the actual operation of radiation detectors we must first discuss the ways in which various radiations interact with matter. We shall deal first with the directly ionizing particles like protons, alpha particles, and electrons, then with gamma and X-rays, and finally with neutrons, which as we saw in Chapter 1 differ, because of their neutrality, from other particles. The treatment which follows will be largely descriptive: for a fuller discussion reference 1 at the end of the chapter should be consulted.

2.2 Interaction of directly ionizing particles

2.2.1 *The Bethe equation*

Ionization, as we mentioned in Section 1.2, is the liberation by radiation of free electric charges — electrons and positive ions in the case of a gas — in the medium through which the radiation passes. (Incidentally many of the electrons produced will be energetic enough to themselves produce further free charges, and the process will continue until the electron energies drop below the level required for ionization.) The loss of energy through ionization represents in general the principal mechanism by which a charged particle is slowed down in the medium. The collection of this ionization by an applied electric field provides, as we saw, one of the most important methods for particle detection. We thus start by discussing the loss of energy by ionization for the case of such a particle.

The required equation, first derived by Bethe, gives us an expression for $-dE/dx$, the rate of energy loss with distance (also known as the 'specific energy loss' or 'stopping power', S). It is particularly appropriate for ions such as protons or alpha particles of a few MeV energy, which are totally stripped of their atomic electrons over nearly all of their 'range', or path length before they are stopped. The ionic charge of these particles (the 'missiles') is therefore simply ze, where z is the atomic number of the particle and the symbol e represents the electronic charge. (Later we shall discuss two more cases, namely those in which the particles concerned are, respectively,

electrons and partially stripped ions.) Bethe's equation for the stopping
power can be written as follows:

$$S = -\frac{dE}{dx} = \frac{z^2 NZ}{v^2} \phi(v) \qquad (2.1)$$

where

$$\phi(v) = K[\ln(2m_0 v^2/I) - \ln(1 - \beta^2) - \beta^2] \qquad (2.2)$$

and

$$K = \frac{e^4}{4\pi\epsilon_0^2 m_0}. \qquad (2.3)$$

The quantities z and e have already been defined: the particle velocity is
represented by v, the electron rest mass is indicated by m_0, and N and Z
are, respectively, the number of atoms present per unit volume and the
atomic number for the target material through which the particle is passing.
As usual ϵ_0 represents the permittivity of free space. In equation (2.2) the
ratio of the velocity v to that of light is indicated by the usual symbol β, and
the quantity I takes care of the fact that the rate of the loss of energy must
depend to some extent on how easily the electron in the target material
can be excited and ionized. We call I the 'mean excitation energy' and as it
name implies it is in principle a sort of average over all possible excitation
and ionization processes in the target atom. It is, however, usually viewed
as a parameter which can be experimentally adjusted for best fit, and for
example is about 150 eV for aluminium, 280 eV for copper, and 600 eV for
gold. We can see that it is roughly proportional to the atomic number, as we
might expect. We note finally that, although the mass of the electrons with
which the missile interacts appears in the equations, the mass of the missile
itself, M say, does not, although we shall introduce it in later calculations.

We have written the Bethe equation in the form shown in equation (2.1)
to allow us to focus on its main features, that is its dependence on z, Z, N,
and v, while separating out the effect due to $\phi(v)$. This function $\phi(v)$ is less
important, firstly because the term $\ln(2m_0 v^2/I)$, being logarithmic, varies
only slowly with v, while the last two items in $\phi(v)$ are relativistic terms
which, for heavy particles, are usually only important at high energies,
100 MeV for protons for example.

Before going on to discuss equation (2.1) in more detail, we must make
two further points. Firstly one need not expect the Bethe equation to hold
down to very low missile velocities, not merely because of the difficulties that
will eventually arise with the v^2 in the denominator of the main part of the
equation, but also because the term $\ln(2m_0 v^2/I)$ in $\phi(v)$ becomes negative
for $m_0 v^2 < I/2$. In fact it turns out that we need to keep the velocity of

the particle large compared with $(I/m_0)^{1/2}$—which effectively means large compared with the orbital velocity of the electrons in the simple Bohr picture. Secondly we must consider what modifications are needed to the Bethe equation in order to deal with electrons as missiles. (We consider partially stripped ions in Section 2.2.2.) The first of these is merely a recognition that, as the mass of the electron is so much smaller than say a proton, relativistic terms in the equation can not be neglected even for electrons of moderate energy. The second is more fundamental, and concerns the principle of 'indistinguishability' between two electrons. To take an extreme example, suppose an incoming electron of energy E interacts with an electron effectively at rest, and the result is an electron emerging with the same energy E. As electrons can not be labelled, it is not possible to decide whether this is the electron initially at rest (i.e. an energy E has been transferred) or whether it is the other electron (i.e. no energy has been transferred). Putting this another way we have to assume that no incoming electron can lose more than half its energy in a collision. We expect then that equation (2.1) for example should now be written

$$S = -\frac{dE}{dx} = \frac{z^2 NZ}{v^2} \psi(v) \tag{2.4}$$

with $z^2 = 1$ for an electron of course, but also with $\psi(v)$ a different and more complex function than the $\phi(v)$ for heavy particles. Indeed we might expect that $\psi(v)$ would be a very different function from $\phi(v)$, and while this is in general true, it is worth noting that for an electron of energy about 10 keV, where relativistic effects have not set in, yet where the condition $v \gg (I/m)^{1/2}$ is satisfied, then $\phi(v)$ and $\psi(v)$ become very similar, as we shall show later by a numerical example. The indistinguishability condition thus does not always result in the drastic change we might have expected—as indeed we know can happen also, for example, in statistical physics.

2.2.2 *Consequences of the Bethe equation*

We now look at equation (2.1) to examine the physical consequences arising from the various terms. Let us start with the term in z^2 which governs the charge on the incoming particle. Obviously the greater the value of z the larger is the rate of loss of energy, so we would correctly expect that an alpha particle ($z = 2$) would have a shorter and more densely ionizing track than a proton of similar energy. Fission fragments, in spite of their very high energies, have very short tracks because of their large value of ionic charge. We shall return to a quantitative discussion of this problem of ranges later in this section.

Now let us look at the effect of the velocity terms on the rate of energy loss. Clearly if we consider missiles of higher and higher energies then the rate of energy loss goes down as $1/v^2$, that is as the inverse of the energy. However, at very high energies (i.e. where the velocity starts to be comparable with the velocity of light and the quantity β becomes non-negligible) then the relativistic corrections in $\phi(v)$ — or $\psi(v)$ — in fact cause the rate of energy loss to slowly rise again. This is shown in Figure 2.1 for a variety of particles in air: note, as we mentioned previously, how much earlier the $1/v^2$ law breaks down for electrons than it does for, say, protons or alpha particles.

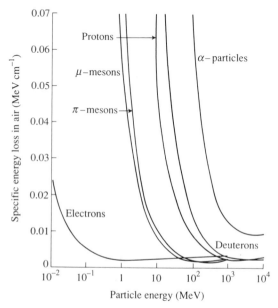

Figure 2.1 Specific energy loss of various particle in air as a function of energy. (From Beiser, A. (1952), *Reviews of Modern Physics*, **24**, 273–311, Figure 2-1.)

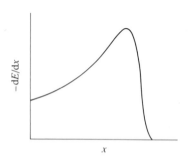

Figure 2.2 Bragg curve for an alpha particle.

Another way of displaying the same result is by considering the famous 'Bragg curve' for an alpha particle. This gives (see Figure 2.2) the rate of loss of energy by an alpha particle (typically of initial energy, say, 5 MeV) against the distance travelled. As we go to the right along the curve of increasing distance, the energy drops, and since at 5 MeV initial energy the $1/v^2$ law is well obeyed, the rate of energy loss should rise — as indeed the graph shows. Eventually, of course, the curve must turn over and drop back to the baseline as the alpha particle is neutralized electrically as it picks up electrons in coming to rest.

We can take the next two terms N and Z together in our discussion of their effect on the stopping power. Now Z for moderate-mass elements (in fact even up to $Z \approx 30$) is fairly well proportional to A, the mass number, so $NZ \propto NA$, and remembering that N is the number of target atoms per unit volume and that A gives the atomic mass (which is proportional to the actual mass of the atom) we see that NZ is proportional to ρ, the density of the material through which the missile is passing. We can thus write equation (2.1) as:

$$-\frac{dE}{dx} \propto \frac{\rho z^2 \phi(v)}{v^2}$$

or

$$-\frac{dE}{d(\rho x)} \propto \frac{z^2 \phi(v)}{v^2}$$

where there is now nothing on the right-hand side about the absorbing material. This is an extremely helpful result. What it means in practical terms is that if instead of expressing the distance of material traversed by the missile, x, in centimetres (or metres) we express it in terms of ρx which has the dimensions of $g\,cm^{-2}$ or $kg\,m^{-2}$, (the former still the most common) then the rate of loss of energy of the particle in question is (at least approximately) independent of the absorber material, and consequently so is the total particle range.

For example the measured range of a 5 MeV alpha particle in air is 35 mm, while in silicon it is very different at 23 μm. However, if we express the range in air in the $g\,cm^{-2}$ form it becomes $\rho x = (1.29 \times 10^{-3}) \times 3.5$ $g\,cm^{-2}$, while in silicon it would be $2.33 \times (23 \times 10^{-4})\,g\,cm^{-2}$ using the densities of air and silicon as 1.29×10^{-3} and $2.33\,g\,cm^{-3}$, respectively. The ranges in air and silicon in the new units work out at $4.5 \times 10^{-3}\,g\,cm^{-2}$ of air and $5.4 \times 10^{-3}\,g\,cm^{-2}$ of silicon, which are reasonably close. Similar remarks apply to the ranges of electrons in materials, so much so that it is rare for an electron range to be quoted in anything but mass per unit area units. (The use of these mass/unit area units has also an interesting practical bonus. If one requires the thickness of a thin absorbing sheet for an experiment, the simplest way to determine it is to use a balance to find its mass.

A measurement of its linear dimensions then immediately gives its thickness in $g\,cm^{-2}$ or $kg\,m^{-2}$.)

The last topic we deal with here is whether we should be able to obtain particle ranges directly from the Bethe equation, by integrating the expression for dE/dx from $x = 0$ to $x = R$, where R is the range. Unfortunately the answer is no, because of the complexity of the function $\phi(v)$. We can, however, do two things. The first is of limited application. If we consider only small changes in v (that is a particle passing through a thin absorber rather than being fully stopped in a thick slice) then since $\phi(v)$ only varies slowly with v we can consider it approximately constant. Then we can write equation (2.1), for a particular z, Z, and N:

$$-\frac{dE}{dx} \propto \frac{1}{v^2} \propto \frac{1}{E}$$

as E, the energy of the missile, is $Mv^2/2$ (where M is its mass): that is

$$-\frac{dE}{dx} = \frac{constant}{E} = \frac{a}{E}$$

say. Integrating we obtain

$$E_1^2 - E_2^2 = 2at$$

where E_1 is the energy before, and E_2 the energy after, passing through an absorber of thickness t. The relation $E_1^2 - E_2^2 \propto t$ is known as 'Whiddington's law'. Unfortunately it still does not help us to predict the full range of the particle, as it only holds for thin absorbers and small energy changes.

The second thing we can do, and in fact the best we can achieve, is to predict something about the *relative* ranges of various particles, which we do as follows. Starting again with our equation (2.1) we rewrite it as

$$-\frac{v^2}{\phi(v)}\frac{dE}{dx} = z^2NZ$$

or

$$-\int_{E_0}^{0}\frac{v^2}{\phi(v)}\,dE = z^2NZ\int_{0}^{R}dx$$

where the left-hand integral runs from the initial particle energy E_0 to its final value of zero, with the distance x going correspondingly from zero to its final value, the range R. Taking for simplicity the non-relativistic case where $E = Mv^2/2$ (where as before M is the mass of the missile) we have $dE = Mv\,dv$ and our previous equation reads

$$M\int_{0}^{v_0}\frac{v^3}{\phi(v)}\,dv = z^2NZ\int_{0}^{R}dx$$

where v_0 is now the initial particle velocity and we have reversed both the limits of integration and the sign on the left-hand side. The integral on the right is simply R: the integral on the left can not of course be evaluated, but at least we know it will be some function of v_0, which we will write $F(v_0)$. So

$$MF(v_0) = z^2 NZR$$

or

$$R = \frac{M}{z^2} \frac{F(v_0)}{NZ}.$$

Consider now two different types of particles (say a proton and an alpha particle) which we will distinguish by the suffices A and B, traversing the same absorbing material, and having started with the same initial velocity v_0. Then

$$R_A = \frac{M_A}{z_A^2} \frac{F(v_0)}{NZ}$$

and

$$R_B = \frac{M_B}{z_B^2} \frac{F(v_0)}{NZ}$$

(where of course NZ, which refers to the target material, is the same in both cases, and v_0 we have arranged to be the same). Hence

$$\frac{R_A}{R_B} = \frac{M_A}{M_B} \left(\frac{z_B}{z_A} \right)^2.$$

Thus although we have not been able to absolutely predict ranges from the Bethe equation, we are now able to make important comparisons between the ranges for two different types of particle, provided they originally start with the same velocity (and we stress it is velocity and not energy).

Let us illustrate our previous calculation with some numerical examples, and we start with an actual comparison between alpha particle and proton ranges. We remember that we must compare particles with equal initial *velocities*, and since the mass of the alpha particle is four times that of the proton we must therefore take the alpha energy to be four times greater ($E = Mv^2/2$). Let us compare then a 4 MeV alpha and a 1 MeV proton. From our previous formula, and remembering that the alpha-particle charge is twice the proton charge, we have

$$\frac{R_p}{R_\alpha} = \frac{M_p}{M_\alpha} \left(\frac{z_\alpha}{z_p} \right)^2 = \frac{1}{4} \left(\frac{2}{1} \right)^2 = 1.$$

So the ranges are equal; that is, a 4 MeV alpha particle should have the same range as a 1 MeV proton, and we can check from experimental measurements that this is true. For example in silicon both have a range of approximately 16 μm. This calculation also underlines our earlier qualitative statement that particles with a larger z have smaller ranges: quite clearly a 4 MeV alpha particle would have a much shorter range than a 4 MeV proton.

As a contrast let us compare the range of a fission fragment with that of the appropriate alpha particle. For the fission fragment one could typically take the mass M as 100 u, the energy E as 100 MeV, and the effective z as 20. We talk about an effective z because some, but not all, of the electrons normally belonging to the atom will have been stripped off in the fission process. The effective z is thus the number of protons in the fragment nucleus less the number of remaining electrons still attached, and is, as indicated, typically in the region of 20 for a mass of 100 u and energy 100 MeV. This compares with the value of about 40 we might have expected for a similar but fully stripped fragment. Using the previous formula for the ratio of the ranges one finds that this 100 MeV fission fragment should have one quarter of the range of a 4 MeV alpha particle; in fact the two ranges are not far from equality. This is because the fission fragment energy loss process is rather different from that for an alpha particle, as we shall now briefly indicate. Firstly because of its high initial charge state, the fragment tends to pick up electrons all along its track (rather than just at the end like an alpha particle). This means that the effective z drops steadily, and from this the Bethe equation implies a decreasing rate of loss of energy along its path, which overshadows the normal increase due to decreasing velocity. Furthermore the effect of non-ionizing collisions with, and energy loss to, atoms, particularly near the end of its track is much more significant than for alphas. Indeed a separate theory on energy loss for such particles has been formulated, but is outside the scope of this book: it will be alluded to briefly in Chapters 5 and 8.

Finally let us compare the range of an electron and the appropriate alpha particle, to see if we can deal with two such disparate entities. For the reason we mentioned previously when discussing the Bethe formula for electrons, we will choose an electron of energy 10 keV. An alpha particle with the same velocity must have an energy of $(4 \times 1840) \times 10$ keV = 74 MeV (remembering the figure for the proton/electron mass ratio of about 1840). As before

$$\frac{R_e}{R_\alpha} = \frac{m_e}{M_\alpha}\left(\frac{z_\alpha}{z_e}\right)^2 = \frac{1}{4 \times 1840} \times \left(\frac{2}{1}\right)^2 = \frac{1}{1840}.$$

Now the range of a 74 MeV alpha particle, (accelerated of course), from experimental data, is about 2000 μm in silicon, so we would expect that of a 10 keV electron to be 1/1840 of this, or just over 1 μm. Experimental data on electron ranges in silicon give a result quite close to this, which is gratifying but a bit fortuitous, because of the difficulty (which we shall talk about shortly) of defining just what we mean in practice by the range of an electron. In any case our result again underlines the fact that for the case of equal *energies* an electron will obviously have a much greater range than an alpha particle.

2.2.3 Bremsstrahlung

There is one further mode of energy loss for charged particles which is worth alluding to briefly — 'bremsstrahlung' or 'braking radiation'. When a charged particle is accelerated by being deflected in the field of a nucleus, then classical electromagnetic theory predicts, as we saw in Section 1.7.3, that radiation (that is photons) should be produced. All we need to say about this method of energy loss is that for it

$$-\frac{dE}{dx} \propto \frac{ENZ^2}{M^2} \qquad (2.5)$$

where E is the energy of the incident particle, M its mass, and N and Z refer as usual to the absorbing material. We note at once that because of the factor of M^2 underneath the expression this type of energy loss will be far more important for electrons than for more massive particles. Furthermore because of the dependence of the energy loss on E, and more particularly on Z, the losses, even for electrons, will only be important for high-energy particles impinging on targets of large atomic number. For electrons in the MeV range passing through materials of moderate Z the bremsstrahlung losses are always small compared with those from ionization, and will not be considered further here.

2.2.4 Ranges

Although we have already talked quite a bit about the ranges of particles, we have done so in a rather general way. In this section we examine more closely what we mean by the range of a particle, and even whether the concept of a range in the ordinary sense may always exist. If by means of some of the methods discussed in Chapter 8 we render visible the tracks of alpha particles from a speck of radioactive material we shall see that they have a well-defined range, that is, every particle travels, to a good approximation, in a straight line and for a definite distance before coming to rest. Alternatively if we plot a graph showing the fraction of alpha particles which emerges from various thicknesses of absorber, we obtain a curve like that in Figure 2.3. Although it is clear that the alphas do not all travel exactly the same distance, yet the 'straggle' is small compared with the mean value of the

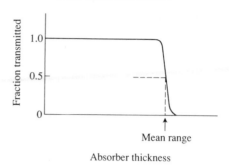

Figure 2.3 Transmission curve for alpha particles.

range. This behaviour follows from the fact that an alpha particle (or any other heavy particle) loses its energy in a large number of interactions with electrons, each of which involves, because of the much greater alpha-particle mass, the transfer of only a small fraction of its energy. (Very occasionally, particularly in high-Z materials, a heavy particle will suffer a large deflection in its path, and related energy loss, as a result of a close encounter with an atomic nucleus, but these are relatively rare events.)

Electrons are of course very much more penetrating than heavy particles, but in addition the nature of their paths is very different. While very high-energy electrons do travel in straight lines through an absorbing material, even an electron with an energy of 1 MeV may suffer many deflections as it travels along, and these deflections become more frequent as the energy is further reduced. This arises from two causes: first of all deflections by nuclei are more important than for the case of heavy particles, and secondly an electron interacting with an orbital electron can, because of their equality of mass, transfer considerable energy to it and suffer a correspondingly large deflection. As a consequence the paths of electrons will in general be tortuous (even re-emerging sometimes from the absorbing material on the side where they entered), and we thus can not speak of a range in the same sense we did for alpha particles. If then we plot a graph of the fraction of electrons, originally monoenergetic, which emerge through various absorber thicknesses we obtain a curve (Figure 2.4) which is very different in character from that of Figure 2.3 for alpha particles. The point where the curve reaches the x axis represents the maximum range R_{max}, and corresponds to those electrons which have suffered little or no large deflections. Other electrons with more tortuous tracks can penetrate only smaller thicknesses of absorber. Now it is not easy, because of the shape of the curve, to pinpoint exactly the precise value of R_{max}, so instead we often characterize the penetrating

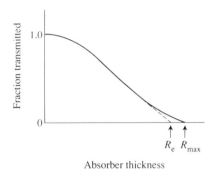

Figure 2.4 Transmission curve for monoenergetic electrons.

power of the electrons by an 'effective' (or 'extrapolated') range R_e, which is formed by extrapolating the straight part of the curve as shown. The effective range thus corresponds to the absorber thickness which will stop the bulk of electrons of a given initial energy.

For beta particles coming from a radioactive source there is an additional complication, as we know that these are not monoenergetic but have a continuous distribution in energy up to a limiting value E_{max} characteristic of the emitting isotope. The shape of the transmission curve is now as shown in Figure 2.5 and the precise estimation from the curve of the maximum range, R_{max}, is even more difficult than before. However, it turns out that we can find in most cases a simple empirical parameter to characterize the absorption, for the reason that quite fortuitously the early part of the absorption curve is approximately exponential. This implies that if n_0 beta particles are incident per second on a sheet of material of thickness x then the number n emerging per second on the far side is given approximately by the formula $n = n_0 \exp(-\mu x)$ where μ, the absorption coefficient, is characteristic of the particular group of beta particles (and incidentally almost independent of the absorber material if x is measured in units of $g\,cm^{-2}$ and μ therefore in units of $cm^2\,g^{-1}$). Alternatively one can use the parameter $d_{1/2}$, the half-thickness, that is the thickness of material required to reduce n_0 to $n_0/2$. Clearly $d_{1/2} = (\ln 2)/\mu$. We stress again that this exponential law is only a useful approximation for the early part of the curve, and can not hold for large x, as the true curve of n against x must ultimately reach the x axis where $x = R_{max}$. For a group of betas with $E_{max} = 1\,MeV$ the half thickness is about $40\,mg\,cm^{-2}$ (or 0.17 mm in silicon), the exponential law would be quite well obeyed over about five half-thicknesses, and the maximum range would be about ten times the half-thickness—corresponding to a few metres in air.

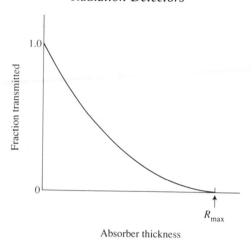

Figure 2.5 Transmission curve for beta particles.

When we come in the next section to discuss the absorption of gamma rays, we shall find that the concept of a range, that is the thickness of material which is sufficient to stop all of a group of particles, has entirely disappeared. Instead we find a true exponential law followed for absorption, this being characterized by an absorption coefficient μ (or a half-thickness $d_{1/2}$) just as occurred, approximately, for beta particles. No parameter corresponding to a range, however, exists. The description of the interaction of neutrons with matter has much formally in common with that for gamma rays, their absorption being characterized in the same sort of way without any definite range parameter.

2.3 Gamma-ray interactions with matter

2.3.1 *General considerations*

When a beam of gamma rays traverses matter the individual quanta do not fritter away their energies in a long series of interactions as do charged particles, but are either absorbed totally in a single interaction, or else lost to the beam by being scattered out of it (by a different mechanism) also in a single interaction. That this implies an exponential type of absorption can be seen as follows. Figure 2.6 shows a thin slice of the absorbing material of thickness dx and of area A. It is part of a larger slab of material (shown dashed) whose absorbing properties we wish to investigate. There are as

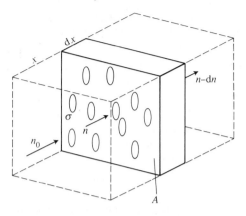

Figure 2.6 Diagram for gamma-ray absorption calculations.

usual N absorber atoms per unit volume, each of effective cross section σ. The term 'cross section' refers to the area through which a gamma quantum must pass in order to interact with the atom. A number n of gammas are shown entering normal to the slice per second, of which dn are removed by interactions in this particular slice.

Clearly the fraction dn/n of the quanta removed from the beam by the slice is given by the total cross sectional area of all the atoms divided by the cross sectional area of the slice. The total number of absorber atoms is $N(A\,dx)$ and therefore

$$-\frac{dn}{n} = \frac{N(A\,dx)\sigma}{A}$$

or

$$-\frac{dn}{n} = N\sigma dx. \tag{2.6}$$

Hence

$$n = n_0 \exp[(-N\sigma)x]. \tag{2.7}$$

Here n_0 is the number of quanta incident per second on the front face of the absorber at $x = 0$, and n is the number per second emerging through a finite thickness x. (Note incidentally that some textbooks use I and I_0 for our n and n_0.)

Alternatively one can write equation (2.7) as

$$n = n_0 \exp(-\mu x) \tag{2.8}$$

where μ ($= N\sigma$) is the same sort of quantity as the absorption coefficient we met earlier for beta particles. However, in this case it is more appropriate to refer to it as an 'attenuation' coefficient, since, as we noted earlier, gammas are lost to the beam not only by absorption but also by scattering from it. However, we may use either term if the sense is made clear by the context. Physically μ represents, as can be seen from equation (2.6), the probability of interaction per unit length for a gamma. It can be expressed in cm^{-1} or m^{-1} (linear absorption coefficient) or in $cm^2 g^{-1}$ or $m^2 kg^{-1}$ (mass absorption coefficient) — although the latter form will not in general for photons ensure its independence of the nature of the attenuating material, as we make clear in Section 2.3.3. The half-thickness $d_{1/2} = (\ln 2)/\mu$ can as before also be used to characterize the penetrating power of the gammas, and may of course be quoted in centimetres (or metres) or as $g \, cm^{-2}$ (or $kg \, m^{-2}$). In numerical terms μ for a 1 MeV gamma ray would be about $0.15 \, cm^{-1}$ for silicon, with a corresponding value for $d_{1/2}$ of 4.6 cm. For lead the figures would be $0.8 \, cm^{-1}$ and 0.87 cm. The much greater penetrating power of gammas as compared with betas is obvious.

We must now consider in more detail the precise ways in which gamma rays may interact with matter. There are in fact three main mechanisms whereby gamma quanta may be absorbed or scattered by materials — photoelectric absorption, Compton scattering, and pair production. In terms of the previous discussion this means that the total cross section σ can be broken up into three components, i.e. $\sigma = \sigma_{PE} + \sigma_C + \sigma_P$ where the subscripts should be self-explanatory. Similarly we can write $\mu = \mu_{PE} + \mu_C + \mu_P$.

We now proceed to give a brief account of each of these processes.

2.3.2 *Photoelectric absorption*

This is analogous to the ordinary photoelectric effect with light, where a photon interacts with an atom, ejecting an outer electron, only in this case the more energetic gamma ejects one of the more tightly bound electrons near the nucleus, especially in the inner K and L shells. The energy of the electron dislodged is given as usual by $E_e = E_\gamma - E_B$ where E_γ is the incoming gamma energy $h\nu$, and E_B the binding energy of the electron to the atom. E_B for a K electron ranges from about 2 keV for silicon through 11 keV for germanium to over 100 keV for uranium. For a gamma ray of say 100 keV energy in germanium the ejected electron thus carries away the bulk of the incident energy. However, because the penetrating power of such an electron is, as we saw, much smaller than that of the parent gamma ray it will be stopped relatively quickly in the absorber. As a result of the interaction and the consequent vacancy in the inner shell, the atom is left in an excited state. This excitation energy may be got rid of, as the atom rearranges its shells, by the emission of one or more characteristic X-ray photons. An external electron must also be captured to replace the one originally lost, and return the atom to neutrality. As an alternative to X-ray emission, the atom may de-excite by the emission of an outer-shell electron: such electrons are known as Auger electrons. As a low-energy X-ray will normally be stopped quickly in the absorber, as will an Auger electron, the whole energy of the incoming gamma is usually deposited there. It is nevertheless possible to observe in detectors a few events where the X-rays escape, as we shall see in Chapters 3 and 5.

The exact expression for μ_{PE}, the absorption coefficient which gives the probability of interaction by the photoelectric effect is complex, but can be approximated for our purposes by the formula:

$$\mu_{PE} = N\sigma_{PE} \propto N(Z^5/E_\gamma^3)$$

where N and Z refer to the usual absorber parameters. We note two striking things about this formula. Firstly, because of the high power to which Z is raised, the photoelectric interaction for a given gamma-ray energy increases very sharply for heavier elements, and secondly because of the relatively high power to which the energy E_γ in the denominator is raised, the photoelectric interaction will, for a given absorber, be much more important at low gamma-ray energies than at higher energies.

2.3.3 *The Compton interaction*

For gamma energies comparable with or even considerably above the K shell binding energy of an atom the photoelectric effect remains the dominant mode of interaction. However, for higher gamma energies the binding energy of the electrons to the atom becomes increasingly irrelevant, and the Compton interaction, which is directly between the gamma quantum and an essentially free electron, becomes important. Note that in the photoelectric case we were able to absorb all the gamma energy because we had a third body present — the atom — which obligingly absorbed surplus momentum, while taking away a negligible amount of energy. Here with only two bodies, the gamma quantum and the electron present, we can not expect to do so.

We will now outline as much of the relevant theory of the Compton effect as will be needed for our treatment of radiation detectors. Figure 2.7 shows an incoming gamma, energy $h\nu$, interacting with an electron essentially at rest initially, and as a result a scattered gamma (of reduced energy $h\nu'$) and the recoiling electron moving out at angles θ and ϕ, respectively. The appropriate equations governing the collision are

$$h\nu + m_0c^2 = h\nu' + mc^2 \tag{2.9}$$

$$h\nu/c = (h\nu'/c)\cos\theta + mv\cos\phi \tag{2.10}$$

and

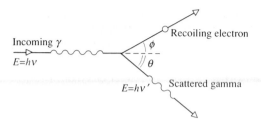

Figure 2.7 The Compton interaction.

$$0 = - (h\nu'/c)\sin\theta + mv\sin\phi. \qquad (2.11)$$

In the first of these equations m_0 and m represent the rest mass and the mass $(= m_0/(1 - v^2/c^2)^{1/2})$ of the electron, respectively, and the equation merely records the necessity for conservation of energy. Similarly equations (2.10) and (2.11) record the conservation of momentum in the direction of the incoming quantum and perpendicular to it, respectively, and we remember that the momentum of a photon of energy $h\nu$ is $h\nu/c$. Eliminating in turn the terms in ϕ and v we obtain

$$\frac{1}{\nu'} - \frac{1}{\nu} = \frac{h}{m_0c^2}(1 - \cos\theta) \qquad (2.12)$$

or

$$\lambda' - \lambda = \frac{h}{m_0c}(1 - \cos\theta) \qquad (2.13)$$

which gives the increase in wavelength of the gamma ray after scattering, and is the form in which the result is often quoted. For our purposes we will find it more instructive to solve for ν' in terms of ν and write it as

$$h\nu' = \frac{h\nu}{1 + (h\nu/m_0c^2)(1 - \cos\theta)}. \qquad (2.14)$$

We note first that, depending on the nature of the collision and therefore the angle θ at which the electron recoils, various amounts of energy will be carried away by the outgoing photon. We will not pursue here the difficult problem of how often various values of θ occur, but we will, however, enquire about the minimum energy that the scattered gamma photon can carry off. This will be important when we discuss detectors because quite often the scattered gamma escapes from a detector, leaving there only the energy of the recoil electron. Naturally we would like the deposited energy to be as large as possible—that is the energy of the scattered gamma ray should be a minimum. Now from equation (2.14) the fraction of the energy lost is given by

$$\frac{h\nu'}{h\nu} = \left[1 + \left(\frac{h\nu}{m_0 c^2}\right)(1 - \cos\theta)\right]^{-1} \tag{2.15}$$

and the most favourable condition for making this as small as possible is when $\cos\theta = -1$, that is the case of a 'head-on' collision, with the scattered gamma returning along its original path. In this case, (which we will concentrate on)

$$\frac{h\nu'}{h\nu} = \left[1 + \frac{2h\nu}{m_0 c^2}\right]^{-1}. \tag{2.16}$$

Note that at low gamma energies ($h\nu \ll m_0 c^2$ where $m_0 c^2 = 0.51\,\text{MeV}$) this fraction is close to unity, while at higher energies it is approximately equal to $m_0 c^2 / 2h\nu$ and goes towards zero at very high energies. In other words the Compton interaction can be viewed as producing largely scattering of the incoming gamma-ray energy if this is small, and as largely an absorption of it if it is big. In no circumstance, however, as we envisaged earlier, can all the energy of the gamma ray be deposited in the absorber by a Compton interaction — a significant difference from the photoelectric case. The best we can do is at very high energies where the fraction lost is, as we have seen, $m_0 c^2 / 2h\nu$, and therefore the energy lost is $(m_0 c^2 / 2h\nu) \times h\nu = m_0 c^2 / 2 \approx 0.25\,\text{MeV}$. (For a 0.5 MeV gamma ray the corresponding fraction lost is, from equation (2.16), approximately one third.)

From what we have said previously about the Compton effect the cross-section per atom for interaction must be proportional to the number of electrons present, that is to Z. Thus $\mu_C(= N\sigma_C)$ must be proportional to NZ. Our experience with a similar factor in the Bethe equation suggests that as an alternative to μ_C (in units of cm^{-1}) we use μ_C/ρ (in units of $\text{cm}^2\,\text{g}^{-1}$). We will then find that, just as in the Bethe case, the attenuation coefficient in these units will be approximately independent of the absorber material for moderate values of Z. This is the only one of the three attenuation coefficients of the required form for this to be true. The dependence of μ_C on the energy of the gamma ray is often quoted as roughly proportional to $1/E_\gamma$ giving $\mu_C \propto NZ/E_\gamma$ overall. However, the actual energy dependence is a good deal more complex and will not be discussed here. We will merely note that for low values of gamma energy μ_C is proportional to $1 - (2E_\gamma/m_0 c^2)$ where $m_0 c^2$ is as usual the electron rest energy, while for larger values of E it is proportional to $(1/E_\gamma)[\ln(2E_\gamma/m_0 c^2) + 1/2]$. However, the main thing to remember is that compared with the photoelectric case the Compton attenuation coefficient lacks the rapid changes with both Z and E that obtained there: we shall see shortly that it is dominant at energies higher than those where photoelectric absorption is important, but lower than those where pair production, which we now discuss, takes over.

2.3.4 *Pair production*

Pair production, the third major way in which gamma rays can lose energy, refers to the total disappearance of a gamma photon and its replacement by a positive and negative electron (see Figure 2.8). Note firstly that charge is conserved in this interaction, being zero both before and afterwards. Secondly we recognize that the energy of the photon reappears not only as the rest mass of the electron and the positron, but also in the kinetic energy which they carry away. We can write

$$hv = m_0c^2 + m_0c^2 + E_{kin}^+ + E_{kin}^-$$

in an obvious notation. Clearly, unlike the two other modes of gamma interaction, a threshold exists for pair formation. Unless the gamma-ray energy reaches $2m_0c^2$, to provide at least the rest mass of the particles, no reaction can take place. So below an energy of $2m_0c^2$ or 1.02 MeV the absorption coefficient μ_P is zero. A subsidiary consideration in connection with pair production becomes very obvious for the case of a gamma photon of exactly $2m_0c^2$, which of course would produce an electron and a positron at rest. Although energy is conserved, it appears that the incoming momentum of the photon, hv/c, has no counterpart after the interaction, and thus the momentum conservation principle would seem to be breached in our simple picture. The necessary modification does not in fact disturb our simple ideas very much. It involves merely assuming that the pair formation takes place in the neighbourhood of an atom. The atom, being so much more massive than the electron or positron, is again obligingly able to take up

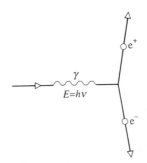

Figure 2.8 Pair production.

the necessary momentum while at the same time stealing only a very small amount of the available energy – rather like what occurred in the photoelectric case.

We must now follow the fate of the electron and positron involved in pair production. As they will both be rapidly brought to rest in the absorber, one might assume that pair production shares with the photoelectric effect the property of permitting all the energy of the incoming gamma to be completely absorbed. This is not necessarily so, and in fact pair production has in this respect more in common with the Compton effect, for the following reason. A positron is an unstable particle, and after coming to rest will quickly combine with an electron (also at rest) to produce gamma radiation once again. To ensure momentum conservation on this occasion it is sufficient for two gammas to be produced moving in opposite directions (see Figure 2.9). Charge is of course once more conserved, and as a mass of $2m_0$ has disappeared each gamma ray will have energy equivalent to m_0c^2 or 0.51 MeV. Depending on the size, geometry, and nature of the absorber one or both of these gamma rays may escape, so, as in the Compton case, not all the initial energy of the gamma may be retained. As we mentioned earlier this fact has important implications for gamma-ray measurements, and will be returned to when we discuss detectors.

Finally, as for the other modes of interaction, we enquire how the attenuation coefficient varies with absorber properties and gamma-ray energy. Again we cannot quote a formula which is applicable for all energies; however, near threshold $\mu_P \approx NZ^2(E_\gamma - 2m_0c^2)$. We note that the attenuation coefficient rises with both increasing absorber atomic number and gamma

Figure 2.9 Positron annihilation.

energy (and quite strongly with the former), and that it has the required property of vanishing when E_γ drops to $2m_0c^2$. For higher energy ranges μ_P
increases logarithmically with E_γ, that is, more slowly than before. None
the less as the only attenuation mechanism of the three which increases with
E_γ it must ultimately dominate at higher energies, as we shall illustrate in
the next sub-section.

2.3.5 *Overall view of gamma-ray attenuation coefficients*

In this section we will look graphically at how the overall attenuation
coefficient μ (made up of $\mu_{PE} + \mu_C + \mu_P$) varies with the gamma energy E_γ
and the absorber atomic number Z. Figure 2.10(a) shows how μ (expressed
in cm^{-1}) and its components μ_{PE}, μ_C, and μ_P vary with gamma energy for
a sodium iodide absorber. (Although the computation for sodium iodide is
more complex than for a simple material like lead or aluminium as it involves
taking account of the different cross-sections for the sodium and iodine
atoms, we choose it because, as we shall find later, it is an important detector
of gamma rays.) In the diagram we can see many of the features we have
already commented on. At low energies the photoelectric absorption is dominant, but drops off sharply with energy because of the factor E_γ^3 in the
denominator of the equation for μ_{PE}. At very high energies pair production
accounts for nearly all the attenuation, because as we saw, μ_P is the only
one of the three constituent coefficients which increases at the highest
energies. In the middle energy range Compton attenuation is the main contributor. Note incidentally that both the horizontal and vertical scales in this
figure are logarithmic, and that the latter prevents us seeing the exact point
at energy $2m_0c^2$ ($= 1.02\,\text{MeV}$) where μ_P goes to zero. The sharp rise in the
photoelectric absorption at slightly over 0.03 MeV is due to the gamma ray
at this point just having enough energy to eject electrons from the tightly
bound K shell in the iodine atoms. Below this energy only the L (and higher)
shells can interact. The same discontinuity for the sodium atoms will,
because of their lower atomic number, occur at a much smaller energy value,
and be off the scale to the left. In fact we can be sure because of the Z^5 factor occurring in the formula for the photoelectric absorption coefficient,
that the bulk of this type of interaction will be due to the iodine atoms in the
sodium iodide.

If we plot graphs for other materials, say silicon and lead, (Figures 2.10(b)
and 2.10(c)) these will be of the same general shape as Figure 2.10(a), but the
contributions from the various mechanisms at particular energies will differ.
For silicon with its low Z the photoelectric effect will be important only for
quite low energies, while for lead its predominance will extend to much higher
energies. Finally if we wish to encapsulate our whole discussion in a simple,
but necessarily crude, statement one could say that for moderate-atomic-

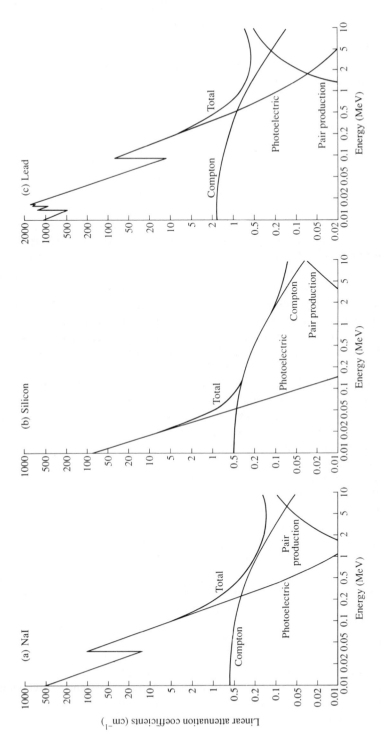

Figure 2.10 Variation of the total gamma-ray attenuation coefficient and its components with energy for: (a) NaI, (b) silicon, and (c) lead. (Graphs from the catalogue of Harshaw Chemie B.V., by permission of Quartz and Silice Holland B.V.)

number materials, gamma-ray interaction is due largely to the photoelectric effect below about 0.1 MeV, to the Compton interaction around 1 MeV, and to pair production above 10 MeV.

2.4 Neutron interactions

Neutrons, unlike alpha particles or electrons, carry no electric charge, and unlike gamma rays, are obviously not a form of electromagnetic radiation. Consequently none of the interactions with atoms that we have been discussing in the preceding sections apply to them. Neutrons, we know, occur in the nucleus of the atom and are bound there to the protons and other neutrons by nuclear, non-electromagnetic forces. We might expect then that the interaction of a neutron beam passing through a material would be with the nuclei of the atoms there, and this in fact turns out to be the case. Neutrons can be scattered from the nucleus in an elastic 'billiard ball' type of collision, or else captured by the nucleus to produce a nuclear reaction, with the subsequent emission of a gamma photon or an ionizing particle. (It is hardly necessary to say that the capture by the nucleus of low-energy neutrons is only possible at all because of the lack of charge on the neutron. A positively charged particle like a proton entering a nucleus must first, as we noted in Chapter 1, have sufficient energy to surmount the 'Coulomb barrier' arising from the positive nuclear charge.) Both the energy of the neutron and the nature of material through which it is passing determine which of the two types of interaction will be predominant in any particular case.

Scattering occurs at all energies, and is for example the mechanism by which the fast neutrons produced in a nuclear reactor (with energies in the MeV range) are slowed down or 'moderated' by a succession of collisions, until their energy becomes comparable with the thermal energy of the atoms of the moderating material. These 'thermal neutrons' are those with energies around kT (where k is Boltzmann's constant and T the absolute temperature corresponding to 20°C) or about 1/40 eV. Obviously the slowing down process is more efficient if the mass of the moderating nucleus is comparable with that of the neutron, as elementary mechanics indicates that more energy per collision can then be transferred. Most energy transfer takes place for equal masses, that is if the moderator used is hydrogen — or a material containing hydrogen atoms. We shall be particularly interested later in these 'knock-on protons' produced by a collision between a neutron and a hydrogen nucleus, as a means of detecting fast neutrons. Depending on the nature of the collision, protons can carry off any fraction of a neutron's energy right up to its full value (for a head-on collision), and therefore most will be easily detectable in any device depending on ionization processes — in which of course the parent neutrons would produce no response.

Scattering of slower neutrons (say in the keV energy range) will also produce recoil nuclei, but these, because of their lower energy, will be detectable with difficulty, if at all. It is fortunate therefore that the second neutron interaction we mentioned — neutron capture followed by a nuclear reaction — while possible at all neutron energies, is most important at the lower ranges. For many materials important for neutron detection the cross-section for such a capture reaction (and we use the term 'cross-section' in exactly the same sense as we did for gamma rays) varies *inversely* with the neutron velocity, and this relationship holds quite well from around 100 keV right down to thermal energies. The consequent large absorption at lower energies can be thought of as following from the strongly wave-like nature of the neutron in this region. (Indeed these wave-like properties show up in an entirely different context in the diffraction of neutrons by the regular arrays of atoms in a crystalline material. It is again the nuclei which are responsible for this effect rather than the electrons as in the diffraction of X-rays.)

Incidentally the attenuation coefficient μ which we met for gamma rays, which was equal to $N\sigma$, is for neutrons called the 'macroscopic cross-section' and denoted by the symbol Σ. Just as μ could broken up into its components μ_{PE}, μ_C, and μ_P so Σ can be split into $\Sigma_{scattering}$ and $\Sigma_{capture}$. The mean free path $l = 1/\Sigma$ is also used as an alternative to Σ to characterize the absorption of a beam of neutrons, and is the same sort of parameter as $d_{1/2}$. (Note, however, that the simple picture we are implying here of a beam of neutrons being absorbed in, or scattered out of an absorbing sheet of material is quite a special situation. The case of neutrons being moderated by making collisions many times in an extended body of material and then diffusing around as thermal neutrons until they are ultimately captured is, for example, a much more complex situation, and needs additional parameters to describe it.)

We mentioned earlier that the production of a nuclear reaction by the absorption of a neutron is followed very quickly (usually in picoseconds) by the emission of a gamma photon or a heavy particle. The former is by far the more common (and indeed is what usually happens when radioisotopes are produced in a reactor) but the latter is of more importance for our interest here — the detection of slow neutrons. This is so because the detection of a heavy particle such as an alpha particle or a proton is usually more efficient and gives a larger signal than that of a gamma photon. We list here for future reference three reactions producing particles which will be important in our discussion of detectors. We can write these concisely as $^{10}B(n,\alpha)^7Li$, $^6Li(n,\alpha)^3H$ and $^3He(n,p)^3H$. The notation is as in Section 1.8.1 (where the *production* of neutrons was being discussed). The first symbol gives the target nucleus, i.e. for the first reaction, ^{10}B, one of the stable isotopes of boron, for the second, ^6Li, a moderately abundant isotope of lithium, and for the third, ^3He, a rare isotope of ordinary helium. The first symbol inside the brackets defines the incoming particle (neutrons, naturally, in all our examples here) while the second indicates the nature of the projectile emitted from the reaction — alpha particles in the first two cases, a proton in the third. Finally the last symbol indicates the nature of the residual nucleus, ^7Li, the most abundant lithium isotope in the first two cases, and ^3H, the heaviest hydrogen isotope, tritium, in the third. In all cases the reaction is so energetic that even for a thermal neutron the outgoing projectiles have total energies of the order of 1 MeV. Remember that the residual nuclei, being also quite light, will recoil with energies which will not be negligible with respect to the outgoing projectile, and for the second reaction will be actually comparable with it. We will return to these matters in our discussion on neutron detectors.

The final neutron interaction to be mentioned is a nuclear reaction (cf. Section 1.8.4) which in some heavy nuclei results in the fission of the nucleus into two fragments of roughly equal mass (and with the production, incidentally, of some additional neutrons). The energy of each of these 'fission fragments' is of the order of 100 MeV, thus obviously providing a basis for a simple detection system for neutrons. The cross-section for fission depends not only on the neutron energy but also on the particular target nucleus involved. At MeV energies the cross-sections for typical fissionable nuclei such as ^{235}U and ^{238}U are quite small and roughly constant with energy. ^{235}U, however, has a cross-section which increases rapidly at lower energies to values comparable with those for the other three reactions that we have just been discussing, and thus can be the basis for a very efficient detector of thermal neutrons. ^{238}U, on the other hand, has a cross-section which drops to negligible values below 1 MeV, and as such can be the basis of an excellent 'threshold detector' for fast neutrons. ^{239}Pu behaves in the same general way as ^{235}U, while ^{237}Np has similar characteristics to ^{238}U. We shall return to all these matters later in our discussion of neutron detection by various methods.

2.5 Implications for detection methods

The discussion so far in this chapter has, we hope, laid the foundation for our investigation of various types of detectors in Chapters 3, 4, and 5. In Chapter 3 we talk about gas-filled detectors. These are often cheap to purchase and simple to operate. However, it is clear from what we have said about ranges and absorption coefficients that these devices will be suitable for detection of alpha and possibly beta particles, but hardly very efficient for gammas (or X-rays) unless these are of quite low energy. For more penetrating gammas of moderate and higher energies it is clear that the detector itself should be a solid, in order for a reasonable fraction of the gamma rays passing through it to interact there. Such devices can be either scintillation detectors, discussed in Chapter 4, or semiconductor detectors, discussed in Chapter 5. What we have said in this chapter about the various modes of gamma interaction with matter will be of prime importance in evaluating the performance of these detectors. It must not be deduced, however, from what we have said above that 'solid' detectors are useful only for gamma rays. In fact scintillation and semiconductor detectors (of a type a little different from those used in gamma-ray detection) are widely used for alpha particles and other ions, and electrons, while a form of scintillation counter using a liquid rather than a solid as the active element is the optimum device for the detection of low-energy beta particles. Detectors for neutrons using both gas-filled and solid devices are available. In Chapter 8 we briefly survey a number of other types of detection devices whose operation in general depends on principles rather different from those discussed in this chapter. Their mode of action will be investigated case by case there.

Reference

1. Evans R. D. (1955). *The atomic nucleus*, pp. 567–745. McGraw-Hill, New York. Still one of the most thorough textbook treatments available on the interaction of radiation with matter.

3
Gas-filled detectors

3.1 Introduction

In the previous chapter we have been discussing what happens to various types of radiation as they are stopped in or pass through absorbing material. For the absorbing material itself, whether it be solid, liquid, or gas, the result is the production in it of ionization — that is the appearance of free positive and negative charge carriers — from part of the incoming particle or photon energy, with further energy going to produce excitation of various kinds in the material. The ionization is the basis for the great majority of detectors with which we shall be dealing in this text. We remember that alpha and beta particles produce ionization in a very direct fashion, while gammas and neutrons do so indirectly, the former usually through the production of an energetic secondary electron, and the latter by, for example, the production in a suitable material of a knock-on proton.

In this chapter we shall deal with detectors which depend on ionization in gases. These were historically the earliest of the ionization detectors, and are usually also the simplest to understand. We shall meet solid detectors in later chapters, as well as a mention of liquid devices. The three gas-filled detectors we shall deal with here are the ionization chamber, the proportional counter, and the Geiger counter. All three, naturally, depend on an initial ionization of the gas, but differ in the subsequent history of the ions produced. We start with the ionization chamber.

3.2 Ionization chambers

3.2.1 *The d.c. chamber*

One method of indicating the presence and magnitude of radiation through the ionization it creates has already been mentioned in Chapter 1. This consists in allowing the conductivity produced in the air by the ions to discharge a previously charged electroscope — and indeed a modified form of this device can be used as a pocket dosemeter to give an indication of the radiation dose received by the wearer. However, for more accurate and informative measurements it is normal to drive the charged ions to electrodes by means of a continuously maintained external voltage, and make subsequent current or charge measurements. The simplest such arrangement one could picture might resemble that shown in Figure 3.1 This consists of a pair of plates of linear dimensions say about 100 mm (to exceed the range of the radiation being detected) spaced apart say 10 mm in air, with radiation from a source entering from the side as shown. The potential applied to the plates might typically be 100 V. The radiation will produce free negative electrons and positive ions in the air of the chamber, and these will be swept by the electric field between the plates to their appropriate electrode (and before any appreciable number of them have had time to recombine if the field is sufficiently strong). A steady flux of radiation through the chamber thus produces a steady current round the circuit (hence the name 'd.c. ionization chamber') and this can be measured by the meter. (The positive ions, of course, are neutralized at the negative electrode by conduction electrons arriving there along the wire.) Let us see what sort of currents we might expect, using in the first instance radiation in the form of alpha particles, which, as we know, produce large amounts of ionization per unit path length.

First we must know the average energy needed to produce an ion pair. For air this is 35 eV (and for other gases lies in the 20–40 eV range). The reader must not confuse this figure of 35 eV with the actual energy to remove an outer electron from an air molecule, which is considerably less. The value of 35 eV takes account of the fact that not all of the radiation energy goes into producing ion pairs, but some (from the point of view of ionization) is

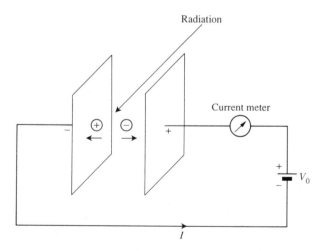

Figure 3.1 Basic d.c. ionization chamber (schematic).

'wasted' in producing other effects like excitation of the molecules. Let us take then the energy of the incoming alpha particle as 3.5 MeV for convenience which, making the reasonable assumption that the total range of the alpha is accommodated in the chamber, will provide us with 10^5 ion pairs per alpha. A modest laboratory source might emit 10^4 particles per second, giving us 10^9 ion pairs per second overall, and producing a current of $1.6 \times 10^{-19} \times 10^9 = 1.6 \times 10^{-10}$ A in our meter. This current, while small, is still measurable on a not too sophisticated laboratory multimeter.

In practice the main use today for d.c. ionization chambers is in gamma-ray monitoring, but here we have a different story. This is because of the low probability of absorption of gammas in the gas of the chamber. Thus both a large chamber and more sophisticated amplification to readable values of the resulting d.c. ion current are needed to measure even moderate levels of gamma radiation. Let us illustrate this by an example, and in doing so we will not consider a collimated beam of radiation as we did for the case of alpha particles: instead we will specify the energy deposited per unit volume of the air by gammas coming from all directions, as we might have for example if we were monitoring a radioactive environment with our chamber. The SI unit used in such circumstances is the 'gray' (Gy) which corresponds to the deposition of $1 \, \text{J kg}^{-1}$ in the absorbing material. The background gamma flux in everyday surroundings corresponds to about 10^{-3} Gy year^{-1}. Using the fact that the energy to produce an ion pair in air is 35 eV and that the density of air is $1.29 \, \text{kg m}^{-3}$ we can expect a current of only about 10^{-15} A in a chamber of volume one litre. To measure accurately currents near this value means, as we have said, not only relatively large chambers and high-gain stable d.c. amplifiers, but also careful design to minimize any leakage currents across the electrode insulation which could mask the genuine signal.

Before we conclude this section we must discuss briefly what might at first seem a rather unexpected topic — the role of the walls of the ionization chamber during gamma-ray detection. (For a more detailed account see reference 1.) Real ionization chambers, unlike our crude symbolic structure of Figure 3.1, will normally be entirely enclosed, if only because we may want to use another gas inside rather than air (for reasons that will be clearer in later sections). If the chamber is designed for alphas (or betas) then of course we will need to provide a thin entrance window: for gammas this will be unnecessary. Ionization chambers can take on a whole variety of shapes, including cylindrical and spherical; however, we will continue to use the simple configuration of Figure 3.1, and imagine that the other four sides required to close our 'box' are made of some plastic insulating material.

Our interest in the walls arises because of the long range in air (or other gas) of the 'secondary' electrons which are produced by the interaction of gammas with the gas in the chamber. These electrons, we know, can have energies up to almost the full energy of the incoming photons, and therefore ranges even for moderate-energy gammas of up to a metre in air. Consequently in any normal-sized chamber electrons produced inside the chamber will spend only a small fraction of their range and energy producing ions in the gas, with the rest absorbed in the chamber walls. (We assume here that these are at least as thick as the range in them of the most energetic electrons we are dealing with.) The other side of the coin is that there are many gammas absorbed in the walls of the chamber which produce electrons able to emerge into the chamber gas and contribute to the ionization there. Indeed we can considerably increase the efficiency of the chamber for gammas by making use of high-atomic-number materials (with their greater absorption for gammas) in the walls — a stratagem we will return to later for other detectors. However, this is often not the requirement in environmental gamma measurements. Instead we wish the ionization in the chamber to be a typical sample of that in similar volumes of air close by — rather as if the chamber had no walls at all and was losing electrons to, and receiving electrons from, the surrounding air rather than the walls. It is not hard to see that an equivalent situation to this can be achieved by having walls of atomic number close to that of air, and as we specified previously, of thickness (in $g \, cm^{-2}$) equal to or greater than the maximum range of secondary electrons (although thin enough to produce little attenuation of the incoming gammas). Such a chamber is said to be 'air equivalent' and the rather simple-minded calculation which we made earlier in this sub-section, ignoring wall effects, now applies directly to it. Materials like aluminium and plastics are sufficiently close in atomic number to air to make reasonably air-equivalent chambers.

3.2.2 *Pulse ionization chambers*

The d.c. ionization chambers we dealt with in the previous sub-section are used, as we have said, for monitoring a continuous radiation background (usually gammas) by means, ultimately, of a current meter. The question arises as to whether it would be possible to detect the arrival in a chamber of a single particle or photon by means of the pulse of charge produced on the electrodes by the movement and collection of the trail of ions formed. Let us make a calculation on this using again an alpha with its intense ionization, and our simple chamber of Figure 3.1, modified as shown in Figure 3.2 In this somewhat idealized experiment a voltage V_0 is applied to the chamber by closing the switch S. This is then opened and any change in the chamber voltage when an alpha particle enters it can be monitored between the points A and B — in principle by a voltmeter, but in practice in a rather more sophisticated fashion. When this has been done the chamber can be recharged by briefly reclosing the switch S, and the experiment repeated with

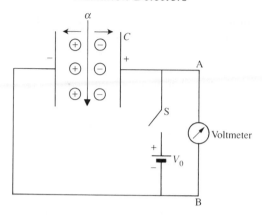

Figure 3.2 Idealized pulse ionization chamber.

a further alpha if required. The quantity C shown in Figure 3.2 represents the electrical capacitance of the chamber.

When an alpha particle arrives, a trail of electrons and positive ions will be produced in the chamber and will drift under the influence of the field between the plates to their respective electrodes. We shall shortly discuss in some detail this movement of the charges: for the moment, however, let us concentrate on what happens when all the carriers have arrived at the electrodes. If N ions of either sign have been formed by the alpha particle then the N electrons arriving at the positive plate will reduce the charge there by an amount Ne (where $e = 1.6 \times 10^{-19}$ C) and a similar reduction will be caused by the positive ions at the other plate. Since the potential on a capacitor is equal to the charge divided by the capacitance, the drop in potential between the plates is given by Ne/C where C is the capacitance of the chamber. For a chamber of dimensions $10 \times 10 \times 1$ cm we have $C = 8.85$ pF, and for an alpha particle of energy 3.5 MeV we have, as before, $N = 10^5$, giving a calculated voltage drop of 1.8×10^{-3} V, from its original level of say, 100 V. (This value of 1.8 mV is somewhat optimistic because of our neglect of 'stray' capacitance which we shall discuss later. However, the order of magnitude is correct.) With appropriate blocking off of V_0 in a manner to be indicated later, amplification of a few thousand will make the drop in voltage a convenient size for measuring on a simple voltmeter, or displaying in some other way, for example on an oscilloscope. So there are no real problems in detecting individual alpha particles, but detection of individual betas or gammas is virtually impossible because of the very small fraction of their energy which fast electrons will leave in any reasonable-sized chamber – as we noted previously.

We move on then directly to discuss the exact shape of the voltage pulse produced in an ionization chamber as the carriers are collected. We shall find that our treatment has implications not only for the ionization chamber itself but also for its derivatives – the proportional and Geiger counter – and indeed for the solid-state chambers we will meet in Chapter 5.

3.2.3 *Pulse shapes in ionization chambers*

We shall use Figure 3.3 in our discussion on pulse shapes. It is basically Figure 3.2 again with some changes and additions. The first of these is the replacement of the switch S by a resistor R. Sophisticated electronic versions of our simple switch are available, and indeed are used in some critical applications of solid-state ionization chambers to be met later. However, for most normal uses a simple resistor is preferred, and for our air-filled ionization chamber we will, for a reason which will be clear later, need one of very high resistance, say $10^{10}\,\Omega$. The second change from Figure 3.2 is that we have shown explicitly in the new diagram the first transistor and some components of the amplifier which we require to bring the small signals from the chamber up to a manageable size. We can deal later with any modification needed due to the amplifier, so for the moment let us imagine the connection at the point X broken. The alpha particle enters as before parallel to the electrodes and its distance from the positive plate will be denoted by L. The total width of the chamber we will call L_0: its capacitance is C, as previously.

When the voltage V_0 is applied to the chamber its capacitance will charge through R towards V_0 with a time constant $T = RC \approx 10^{10} \times (9 \times 10^{-12})$ s $= 100\,\text{ms} = 0.1\,\text{s}$, so after a few seconds we can start operating. Now we shall see in a moment that the drop in potential caused by the carrier movements in which we are interested will take place in times much smaller than

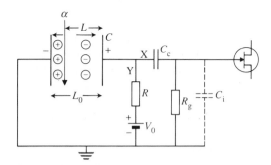

Figure 3.3 More practical ionization chamber circuit, showing field effect transistor (FET) and other components at input of amplifier.

RC. Consequently, since the recharge process is on such a long time-scale, only a negligible amount of it takes place during the collection of the carriers. Thus we can break the process into two parts, firstly the movement and collection of the electrons and positive ions, with the related drop in potential, followed on a longer time-scale by the recharging of the plate capacitance through the resistor *R* ready for the next alpha particle. Putting this another way, we can, after the capacitance has been charged up, ignore for the purposes of our discussion the role of *R*. Indeed we can imagine the connection at the point Y effectively severed — just as if we had a switch there!

How then in these circumstances does the potential between the plates change? Our first instinct might be to discuss this in terms of actual ion currents, and indeed we will return to this briefly further on. However, it is probably simpler, both for the present case and for some of the detectors we will be discussing later in this chapter, to use the more conventional approach in terms of energy. In this treatment we recognize that because the chamber is, as we pointed out, effectively isolated during the movement and collection of the charges, the work done by the carriers can only be supplied from the energy stored in the charged chamber capacitance. An electron (or ion) on its passage across the chamber does work by being accelerated in the applied electric field, delivering energy (either kinetic or excitation) to an atom or molecule of the gas during a collision and repeating the process again and again, with the energy abstracted from the field eventually ending up as heat. (The fields and the accelerations we are talking about here are too small for the carriers themselves to produce ionization, but this is a possibility which will be extremely important in detectors we shall be discussing later.)

The mean velocity u of a carrier across the chamber is proportional to the field $\mathscr{E}$ (as is usual for motion in a resistive medium), and inversely proportional to the pressure p (since this quantity controls the number of atoms or molecules per unit volume). We write then $u = \mu \mathscr{E}/p$ where μ is called the 'mobility' of the carrier and has dimensions usually given as m^2 atm $V^{-1} s^{-1}$. For most gases, including air, μ for positive ions has a value around $10^{-4} m^2$ atm $V^{-1} s^{-1}$. Thus for the sort of fields we are assuming, that is $100 \, V \, cm^{-1}$ or $10^4 V \, m^{-1}$, we expect velocities, with $p = 1$ atmosphere, of the order of $1 \, m \, s^{-1}$. This would correspond to a transit time of 10 ms for a positive ion across our full chamber width of 10 mm. Because of their much smaller mass we expect the electrons to have considerably greater mobilities than the positive ions, and indeed they normally correspond to velocities of the order of a thousand or more times the positive ion velocities (although we have a caveat to make later in this section on this regarding certain gases, including air).

Returning then to the question of the transfer of energy from the charged capacitance of the chamber to the carriers, and remembering that the energy stored in a capacitance is given by half the capacitance multiplied by the potential squared, we can write

$$\frac{1}{2}CV_0^2 - \frac{1}{2}CV^2 = Ne\mathscr{E}x_+ + Ne\mathscr{E}x_-. \tag{3.1}$$

The left-hand side of this equation represents the energy lost by the capacitance in dropping its potential from the initial V_0 to its new value V in a time t. The right-hand side represents the work done by the carriers up to this time t, when the ions have travelled a distance x_+ and the electrons a (greater) distance x_-.

The term $e\mathscr{E}$ gives in each case the force involved and N, as before, represents the number of ion pairs. (It is quite clear that the two terms on the right-hand side should add, as work is done independently by both sets of ions, so we have no trouble with the signs. However, because things may not be quite so clear cut when we are dealing later with similar terms in a cylindrical chamber it is worth pointing out that the second term on the right should in the first instance be written $N(-e)\mathscr{E}(-x_-)$, which of course gives us the same result as before.) Factorizing the left-hand side of Equation (3.1) and remembering that on the right $\mathscr{E} = V_0/L_0$ gives us

$$\frac{1}{2}C(V_0 - V)(V_0 + V) = Ne(V_0/L_0)(x_+ + x_-). \tag{3.2}$$

Now we already know that $V_0 - V$, the drop in potential, is at best very small compared with V_0 (millivolts against about 100 V), so we can write $V_0 + V$ as $2V_0$ quite closely. Equation (3.2) then becomes

$$V_0 - V = (Ne/CL_0)(x_+ + x_-) \tag{3.3}$$

or if we like to express this in terms of velocities and time

$$V_0 - V = (Ne/CL_0)(u_+ + u_-)t. \tag{3.4}$$

Because of the much greater mobility of electrons compared with positive ions, the normal course of events is for the electrons to reach their electrode and be collected first: this will happen at a time $t = L/u_-$. After this time the electrons have no further contribution to make to the drop in voltage on the plates and equation (3.4) can be written (remembering that the term previously u_-t will remain fixed at L)

$$V_0 - V = (Ne/CL_0)(u_+t + L). \tag{3.5}$$

The slower-moving ions will will not be collected until time $t = (L_0 - L)/u_+$ when the voltage drop is given by

$$V_0 - V = (Ne/CL_0)[(L_0 - L) + L] \tag{3.6}$$

$$= Ne/C \tag{3.7}$$

which of course is the voltage drop we expected from our earlier calculation of the total charge collected on each plate.

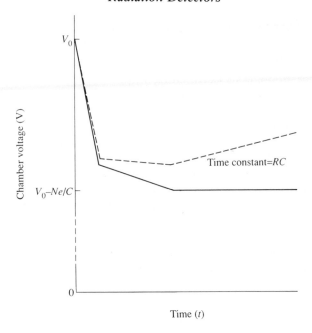

Figure 3.4 Pulse shapes from an ionization chamber. Relative velocities of electrons and positive ions not to scale. Note also the break in the voltage axis.

Figure 3.4 shows us the sort of pulse shape we would expect from an alpha particle passing down not too far from the negative electrode. The full line shows the initial rapid change in voltage due primarily to the large value of u_- which occurs until all the electrons are collected, followed by the much slower change as the positive ions continue to move towards their electrode. (Note that both electrons and positive ions contribute to the first rapid change, but because of the much lower velocity of the ions, their contribution is almost entirely overshadowed by that of the electrons.)

If, as we have been doing, we neglect the role of the resistor, then the chamber will remain at a voltage $V_0 - V = V_0 - Ne/C$ indefinitely, as the full line shows. The role of R as we mentioned previously, is to restore the chamber potential to its original value of V_0 in readiness for the next alpha pulse. This it does with a time constant RC, where C is the chamber capacitance (providing we are still ignoring any effects due to the amplifying transistor and its components). The value of RC should be made — at least in this first approach — considerably larger than the total collection time of the positive ions, and with the values we have assumed this will be the case. The signal under these circumstances is shown by the broken line in Figure 3.4 Note that with such a large but finite value of RC the maximum change in output will not quite reach Ne/C. We will return to this point in more detail later.

This is probably the appropriate place to mention the role of the components C_c, R_g and C_i in Figure 3.3 Assuming the connection is remade at X, then in the absence of the capacitor C_c the full voltage V_0 of the chamber supply would be applied directly to the transistor input — with catastrophic results! C_c is thus a 'coupling' capacitor which blocks off the steady voltage V_0, but hopefully will correctly transmit the changing voltage due to the alpha-particle pulse. To perform this role C_c has to be large enough so that it can not charge appreciably, that is the voltage across it appreciably alter, in a time corresponding to the formation and subsequent decay of the alpha pulse. If this is so then the changing voltage of the pulse appearing on the left-hand plate of C_c must also appear on the right-hand plate — that is, be faithfully transmitted through to the transistor input. Such a capacitor can thus be considered as a direct connection for pulses, while still remaining as a block on steady voltages.

While this 'direct connection' is what we have been anxious to achieve, two things deriving from it must be noted. First of all the resistor R_g (which is needed for the correct operation of the field-effect transistor (FET) we would normally use as the first amplifier element) is now, from the point of view of signals, directly in parallel with R. Secondly the capacitor C_i in Figure 3.3 is effectively in parallel with the chamber capacitance C. This capacitance C_i is shown dashed because it is not an actual capacitor, but represents the 'stray' irremoveable capacitance between the input of the FET and the other electrodes. Together with the stray wiring capacitance to ground it could typically be about 10 pF, while R_g could conveniently be made about the same size as R, say $10^{10}\,\Omega$, so the previous RC value for the decay time of the pulse from our chamber is now replaced by: $(\frac{1}{2} \times 10^{10})(10 + 9) \times 10^{-12}$s or 95 ms. While the decay time has not changed much from our previous calculation, a considerable change has taken place in the size of the output signal. Previously it was approximately Ne/C but now it is approximately $Ne/(C + C_i)$ or somewhat less than half of its previous value. Clearly to maximize the output we must select an FET with a low value of C_i (although a more sophisticated criterion for this choice will be discussed in another context in Chapter 7). The signal at the input of the FET now looks as shown by the full line in Figure 3.5. (There is incidentally a way in which we can avoid the complication of a blocking capacitor — shown in Figure 3.6. Here the point X is at ground potential when the chamber is charged, so no capacitor is required, nor indeed a separate resistor R_g. However, this arrangement has the counterbalancing disadvantage that neither electrode of the chamber is earthed: a grounded electrode, particularly the outer electrode in the case of a cylindrical chamber, is helpful in avoiding pick-up of interference. The circuit of Figure 3.6,

been collected the signal will still be *Ne* divided by the total capacitance. We must stress that this last result depends on the collection of all the carriers, and this is not something that we may always want to do. For example it is not possible with the arrangement we have been discussing to count alpha particles arriving at anything but an extremely slow rate, because of the long collection time for positive ions, and the necessarily large return time which this implies. If we choose to ignore the bulk of the contribution from the positive ions and 'clip' the pulse with a much shorter time constant as shown by the broken line in Figure 3.5 then much faster counting rates can be achieved. However, at time $t = L/u_-$ when the electrons have been collected, equation (3.5) tells us that the output pulse is a function of L, so if the tracks are inclined at an arbitrary angle to the plates, or even if they are parallel to the plates but enter at different places, then the signals, although coming from identical alpha particles, will have different amplitudes. We have thus gained the ability to count faster only at the expense of information on pulse amplitude, that is alpha energy. There is a more sophisticated method using a chamber with an internal grid by which this difficulty can be overcome, but we will leave the interested reader to consult reference 2, for example, for further details.

There is an additional caution to be mentioned in connection with electron collection. In some gases electrons can rapidly attach themselves to neutral atoms or molecules, thus producing negative ions. These ions like their positive counterparts have a low mobility, and the advantage of a fast electron pulse rise-time disappears. Oxygen is a serious offender in this regard, so air is not a particularly good gas for the present purpose. We will be returning to the question of counter gases later in the chapter.

The second important result we wish to spell out concerns the relative contribution to the output from the electrons and positive ions under various circumstances. It is clear from equation (3.3) and also from equation (3.6) that the contribution to the signal in our particular ionization chamber by an electron or a positive ion is in direct proportion to the respective distance each has travelled. For geometries other than the simple planar case we have taken the result can be more generally stated as ' . . . . to the respective potential through which each has fallen'. We will be making use of the more general statement later in the chapter: in the meantime we give an illustration of the simpler version for the case of a parallel-plate chamber.

Let us think again then of alpha particles incident parallel to the plates, and consider what happens if one of them enters very close to the negative electrode. Clearly the positive ions now have essentially zero distance to travel before collection and they will contribute nothing to the output. The signal is thus attributable to electrons only, and its rise-time will be correspondingly rapid. Such an arrangement, if it were otherwise feasible, would allow counting rates which would be fast in the present context, while at the

same time producing pulse amplitudes of the full amount. For an alpha particle entering close to the positive electrode, the signal would be due entirely
to the positive ions and its rise would be correspondingly slower. We develop
these ideas further in the next sub-section.

3.2.4 *Some additional remarks on pulse shapes in ionization chambers*

We conclude our treatment of pulse ionization chambers with two further
topics. The first is a rather more sophisticated discussion on pulse shapes,
and the second deals with the alternative method, which we mentioned
earlier, of obtaining our basic equation.

It is not easy to sketch accurately the pulse shapes under various
conditions in a gas-filled ionization chamber because of the very big dis-
parity (three orders of magnitude, as we saw) between the mobilities. So let
us for illustration imagine that the electrons have only three times the mobility of the positive ions. (In fact this is about the mobility ratio we shall find
in Chapter 5 between the carriers in the solid-state equivalent of our ionization chamber, so what we say here will have direct significance at that later
stage.) Once more we take the simple case of an alpha particle entering from
the side and parallel to the chamber plates: we shall investigate how the pulse
shape depends on the exact point between the electrodes where it enters —
that is the value of L. Figure 3.7 shows an alpha entering at a selection
of such points, and below the resulting pulse shapes which we should
expect — as we now demonstrate.

In these diagrams three slopes should occur in the rising part of the curves,
corresponding to three classes of current flowing. The first, and numerically
the largest, is when both electrons and positive ions are in motion, and we
have labelled it I in the diagrams. The second, of smaller magnitude, occurs
when positive ions alone are in motion, and is labelled III. Both these slopes
have already appeared in our earlier Figure 3.4. We should also in principle
expect, in the appropriate circumstances, to see a slope corresponding to
electrons only in motion: its magnitude would lie between those of slopes I
and III, but with the values we have chosen, nearer to slope I: obviously it
will be labelled with the symbol II. Let us now take a look at the actual
diagrams.

In diagram (a) the alpha particle enters close to the negative electrode and,
as we explained earlier, the positive ions make no contribution to the output
pulse, so the slope of the curve will at all times be that due to the electrons
alone as shown (labelled II). In diagram (e) we have the complementary case
where the pulse is due purely to positive ions and has the correspondingly
slow rise-time (III). Now let us consider case (c) where with the position of
the alpha particle as indicated, and with the assumed ratio of 3:1 for the
mobilities, it is clear that the electrons and positive ions arrive simulta-

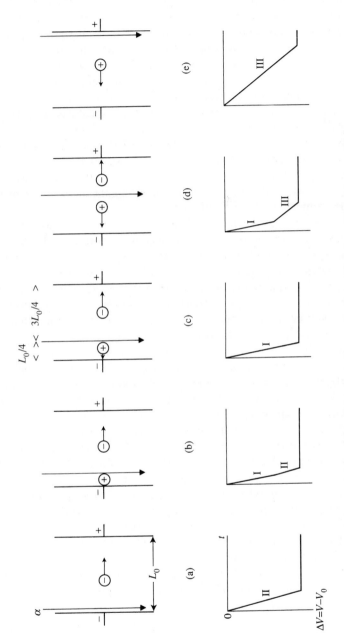

Figure 3.7 Dependence of pulse shape on place of entry of ionizing particle.

neously at their respective electrodes. Consequently the slope of the voltage versus time curve will, throughout its rise, have the maximum value I. In case (d) with the alpha entering closer to the positive plate than in case (c) both the carriers are in motion until the electrons are all collected, giving a slope of I, followed by the smaller slope III as the remaining positive ions go to their electrode. This is, of course the usual scenario which we illustrated earlier in Figure 3.4. However, for a track as in diagram (b), that is lying closer to the negative electrode than in the critical case (c), we will clearly have a first rise I due to both carriers until all the positive ions are collected, followed by a rise of intermediate slope II while the remaining electrons travel to their plate. All these five cases (a) to (e) are quite different for the ratio of mobilities we have assumed here for illustration purposes. In reality for gases, where the mobilities, as we know, are three orders of magnitude apart, then in case (c) the alpha particle must be only about a thousandth of the chamber width away from the negative electrode, with the result that diagrams (a), (b), and (c) effectively coalesce into one. Thus for gas-filled chambers (but not for their solid-state counterparts) we have in practice just three cases to consider. It is worth reiterating before we finish with this particular topic, that while our discussion may have been helpful in trying to understand the principles underlying pulse formation, in practice the diagrams will be smeared out for tracks in any other direction than the special one we have chosen — as indeed we found out earlier.

In the final part of this sub-section we return to a suggestion made earlier, that there might be an alternative method of obtaining the basic equations for the pulse shape by considering the actual currents flowing in the chamber, rather than using as we did an energy argument. To develop this idea fully and in particular to apply it to geometries other than the planar one would take us too far afield and involve recourse to a very general statement known as Ramo's theorem (reference 3). However, we can, in a simple and plausible way, obtain the result for the planar case as follows. We know that in general whenever electric charges are in motion the current density j is given by $j = nqu$, where n is the number of charges per unit volume, q the charge on each, and u their velocity. If in our ionization chamber A is the area of each plate, and L_0, as before, the distance between them, then for a single charge moving across $n = 1/AL_0$ and $j = qu/AL_0$. Now the current i is just jA, so we have immediately $i = qu/L_0$ (or Nqu/L_0 if N charges are involved). In our case we have currents due to both sets of carriers so we have for the magnitude of the total current (putting $q = e$)

$$i = (Ne/L_0)(u_+ + u_-). \tag{3.8}$$

Now the rate at which the charge Q on each electrode is changed (that is reduced) due to this current will be given by $dQ/dt = -i$ so the change in Q will be $\Delta Q = -it$, where t is the time the charges have been in motion. Thus we can write

$$\Delta Q = - (Ne/L_0)(u_+ + u_-)t \qquad (3.9)$$

or for the change in voltage ΔV, given by $\Delta Q/C$

$$\Delta V = - (Ne/L_0C)(u_+ + u_-)t. \qquad (3.10)$$

Now $\Delta V = V - V_0$, so changing signs, finally

$$V_0 - V = (Ne/L_0C)(u_+ + u_-)t. \qquad (3.11)$$

This is the same as equation (3.4) which we obtained by energy considerations, and from which our subsequent discussion on pulse shapes derived.

However, we can go directly from our equation $i = qu/L_0$ to an expression for the pulse shape, and it will be instructive to consider this. Figure 3.8(a) shows our basic set-up again (without bothering with the modifications due to the amplifier), while Figure 3.8(b) shows its 'equivalent' circuit. The equivalent circuit is one which contains only those components which directly contribute to the signal in question. For example the steady voltage V_0 is absent from Figure 3.8(b) since its role is just to provide the field necessary for the movement of the carriers. As a corollary to this we expect to obtain from our equivalent circuit, not V, as in the left-hand diagram, but the change $\Delta V(=v$, say) in this quantity. The alpha particle in Figure 3.8(a) provides a source of current in the ionization chamber, and this is modelled in our equivalent circuit by a current source of size Neu/L_0 in parallel with C (and of course R). Just to simplify the problem a little we will imagine that the alpha enters close to the negative plate so that we can consider the current as coming only from the movement of electrons in the chamber, so u here represents the electron velocity.

We can deduce the output signal from our equivalent circuit as follows

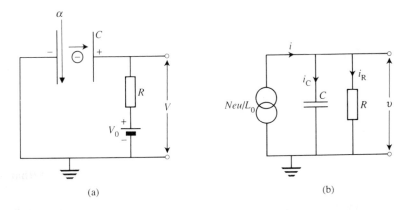

(a)　　　　　　　　　　　(b)

Figure 3.8 (a) Pulse ionization chamber and components. (b) Equivalent circuit.

$$i = i_C + i_R = \frac{dQ}{dt} + \frac{v}{R}$$

$$= C\frac{dv}{dt} + \frac{v}{R}.$$

So

$$\frac{dv}{dt} + \frac{v}{RC} = \frac{i}{C} = -\frac{Neu}{CL_0}$$

(where the minus sign indicates that the current flow is in a sense to reduce the capacitor voltage) giving

$$v = -\left(\frac{Neu}{L_0}\right) R \left[1 - \exp\left(\frac{-t}{RC}\right)\right]$$

remembering that $v = 0$ at $t = 0$. This equation holds up to a time $t = L_0/u$ when the electrons reach the positive plate: at that stage the current source can be removed from Figure 3.8(b) and the voltage already on the capacitor simply decays away with time constant RC. The overall result is shown in Figure 3.9 (full curve). For RC very large compared with the electron transit time (so that $\exp(-t/RC) \sim 1 - t/RC$) the expression for v becomes $v = -(Neu/L_0C)t$ and its value at $t = L_0/u$ is $-Ne/C$ as expected. The broken line shows this approximation. Incidentally we note for future reference in Chapter 5 that the formula $i = qu/L_0$ can be shown to be true for a planar semiconductor detector, even though, as we shall see, the electric field in this type of detector is very different from that in a gas-filled device.

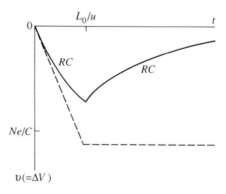

Figure 3.9 Pulse shapes computed directly from current considerations.

3.2.5 *Applications of pulse ionization chambers*

From what we have said earlier, pulse ionization chambers are usable only for the detection of alpha and other heavily ionizing particles such as protons and fission fragments. For such particles, of course, we can, in addition to merely detecting them, also (under appropriate conditions) deduce their energy via the size of the output pulse, from the number N of ion pairs formed. We can thus, for example, identify various alpha-emitting isotopes by the energy of the alpha particles coming from them. It may be worth mentioning here the concept of 'energy resolution'. The energy of an alpha particle, as we saw earlier, is shared between producing ionization and other effects such as excitation of the gas. This sharing is a statistical process and consequently the number of ions produced by a succession of alpha particles will not be exactly the same in each case, but will fluctuate around a mean number $\bar{N}$. The consequent variation in the output pulse is further increased during the process of amplification. What this means is that because of the finite spread in the pulse size there will be a lower limit to the energy difference we can detect between two groups of alphas of different energies, and we refer to this as the 'energy resolution'. We can in practice with a gas ionization chamber distinguish the signals from a group of alphas of say 5 MeV from those of another group only 0.02 MeV different in energy. The whole question of energy resolution in radiation detectors is of such importance that we have devoted all of Chapter 7 to this topic, so we will leave further discussion until then. With the advent of solid-state (i.e. semiconductor) ionization chambers—which are dealt with in Chapter 5—the role of gas-filled pulse ionization chambers has become restricted to a few specialized areas such as, for example, operation in a high flux of heavily ionizing particles where the crystal structure of a semiconductor detector might be degraded, or where special fill gases might be helpful. However, the principles we have been developing here for this type of ionization chamber will prove to be of importance not only for solid-state detectors, but more immediately for our treatment of proportional and Geiger counters.

3.3 Proportional counters

3.3.1 *Introduction*

If it were possible to increase the field in a gas-filled ionization chamber to such an extent that the carriers would gain enough energy to ionize further atoms or molecules, than clearly a cumulative chain of multiplication of charge would take place inside the chamber. This would in turn simplify matters by reducing the amount of external amplification needed for the output signal. We note at once that it is on the electrons rather than the positive ions that we would be depending to produce this effect, as their very much lower mass implies that they will be able first to achieve sufficient energy between collisions for this secondary ionization of the gas.

Proportional counters and Geiger counters both rely for their operation on this gas multiplication: they are distinguished in the first instance by the different overall charge multiplication which occurs in the device. Before we look at the mechanism of these individual devices it will be worth while to take a general view of what happens as we arrange for greater and greater multiplication of the ions originally formed in the counter. This is illustrated in Figure 3.10. In this hypothetical experiment we compare the result of forming 10^2 ion pairs in our counter by some ionizing particle with what would happen if we formed 10^4 ion pairs by a more strongly ionizing particle. (These are both small numbers of carriers compared with those we talked about when considering alpha particles in ionization chambers, but of course now we are going to increase these numbers by a factor equal to the gas multiplication.) The final number of ion pairs collected in various circumstances is displayed logarithmically on the y axis, and the x axis indicates the counter voltage. We have not calibrated this last axis because the multiplication we produce depends, as we saw, on the electric field, which in turn depends not only on the applied voltage, but also on the geometry (planar, cylindrical, or whatever), and on the nature and pressure of the gas filling. However, we can say roughly that in region II we are talking of the order of hundreds of volts, while in the higher regions we might be talking of a thousand volts or more under similar conditions. We shall indicate later values used in practice.

Region II is just the ionization chamber region we discussed at length

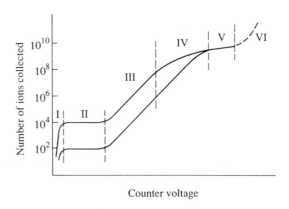

Figure 3.10 Dependence of gas multiplication on counter voltage.

earlier. Here if we form 10^2 ion pairs initially in the chamber we collect 10^2 ion pairs, as there is no multiplication: with 10^4 ion pairs we naturally collect 10^4 ion pairs. Also it is clear that the operating voltage is not critical over quite a wide range — its role is primarily to collect the carriers. However, we do note that there is a region I where the number of ion pairs collected is less than the number produced originally. This is due to the fact that at the low voltage applied here the ions have a finite chance of recombining before they can be collected. At the boundary between region II and region III the curves begin to rise, showing that gas multiplication has commenced, and as we move through region III the overall multiplication increases in value, as naturally does the number of ions collected. This region is known as the region of proportionality — because for any particular voltage, the number of ion pairs originally produced is always increased in the same proportion, independent of the original number. For example if 10^2 ion pairs were originally created and were multiplied by a factor of 1000, then if instead 10^4 ion pairs were originally produced, they too would be multiplied by 1000. (If this perhaps seems a self-evident requirement for a radiation detector, the reader should note that we will almost immediately be mentioning a case where it does not hold!) This region of proportionality represents the regime under which the proportional counter operates: it is characterized in our diagram by the fact that the two curves there run parallel over the region in question (which with the logarithmic scale guarantees the proportionality property). In region IV, on the other hand, the upper curve moves relatively closer to, and finally joins, the lower curve. What this means is that proportionality no longer exists, or in other words when the initial number of ion pairs is small, the multiplication is larger than when this is a big number. Finally when we reach the border of region V the number of ion pairs eventually produced in the counter becomes the same no matter what number of pairs were first present. Region IV is known as the region of 'limited proportionality' and what is happening is that some form of saturation is appearing in our counter, which prevents a sufficiently large initial number of ion pairs from achieving its expected multiplication. This effect becomes more marked as the voltage is increased until, at the start of region V, any initial ionization, no matter how small (even down to a single ion pair), will produce the same final number of ions. This section of the curve is known as the 'Geiger', 'Geiger–Müller' or simply GM region, and characterizes the regime under which this counter operates. If we continue to increase the voltage beyond that of region V, our counter will go into continuous discharge without any trigger from radiation (region VI), and its usefulness as a detector will have ceased.

3.3.2 *Proportional counter mechanisms*

The electric field to begin the multiplication process, that is to arrive at
the border between region II and region III in Figure 3.10, is a few times
10^6 V m^{-1} for most gases at atmospheric pressure. This is 10^4 V cm^{-1}, so if
we stick to a parallel-plate chamber of the dimensions we have talked about
so far we will be dealing with voltages which are usually inconveniently high
(although such devices do exist and will be mentioned briefly later). We can
escape from this difficulty by remembering the good physical principle that
a modest voltage applied between two electrodes can produce quite a high
field near one of them provided this particular electrode is sharply curved.
As a consequence by far the most common form for a proportional counter
is one with a cylindrical cathode, having as anode a thin wire running down
the axis (and of course insulated from the cathode). This is shown schematically in Figure 3.11 with some details that we will need for our calculations.
Before we become involved with the multiplication process in proportional
counters, however, we need to develop some basic equations for pulse formation in cylindrical geometry corresponding to those we deduced earlier for
the planar arrangement.

So, let us imagine that the voltage V_0 applied in Figure 3.11 is below the
value necessary to produce multiplication in the gas, so our counter behaves
simply as a cylindrical ionization chamber. Let us also imagine in this preliminary discussion that an alpha particle enters through a thin window at
the end as shown (although side entry is the more normal mode) leaving a
trail of ions and electrons conveniently at a constant distance r_0 from the

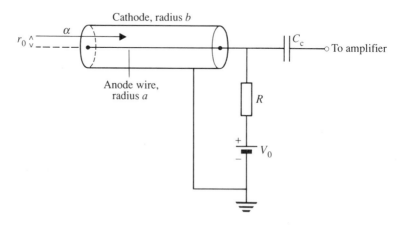

Figure 3.11 Proportional counter (schematic).

axis. Now the electric field in the cylindrical arrangement is, by Gauss' theorem, proportional to $1/r$, or $\mathscr{E} = k/r$ say. Integrating to obtain the potential V, and substituting the boundary conditions of $V = 0$ at $r = b$, and $V = V_0$ at $r = a$, allows us to determine k, and hence find $\mathscr{E} = [V_0/\ln(b/a)](1/r)$. We can now write down a statement analogous to equation (3.1), equating the work done by the ions and electrons to the loss in energy of the chamber capacitance (again assuming the resistor to be large). Here, however, since the field is not constant, the result has to be in integral form. So we have

$$\frac{1}{2}CV_0^2 - \frac{1}{2}CV^2 = \int_{r_0}^{r_+} N\mathscr{E}e\, dr_+ + \int_{r_0}^{r_-} N\mathscr{E}(-e)\, dr_- \tag{3.12}$$

where r_+ represents the position of the positive ions at time t and similarly r_- represents the position of the electrons. (Note that although a minus sign appears before the charge e in the second integral, both integrals do in fact add, as dr_- is basically negative in going from r_0 to a smaller value r_-.) Putting in our previous value for $\mathscr{E}$ in terms of r and remembering that $\frac{1}{2}CV_0^2 - \frac{1}{2}CV^2 \approx CV_0(V - V_0)$ gives us

$$V_0 - V = \frac{Ne}{C\ln(b/a)}\left[\int_{r_0}^{r_+}\frac{dr_+}{r_+} - \int_{r_0}^{r_-}\frac{dr_-}{r_-}\right]$$

or

$$V_0 - V = \frac{Ne}{C\ln(b/a)}\left[\ln\left(\frac{r_+}{r_0}\right) + \ln\left(\frac{r_0}{r_-}\right)\right] \tag{3.13}$$

(and we note as expected that since $r_0 > r_-$ the two contributions add). Equation (3.13) can be written finally as

$$V_0 - V = \left(\frac{Ne}{C\ln(b/a)}\right)\ln\left(\frac{r_+}{r_-}\right) \tag{3.14}$$

and of course when all the carriers are collected we have $r_+ = b$ and $r_- = a$ so the maximum output is given by

$$V_0 - V = \left(\frac{Ne}{C\ln(b/a)}\right)\ln\left(\frac{b}{a}\right) = \frac{Ne}{C}. \tag{3.15}$$

This is the usual result, and is independent of r_0 as expected when we collect all the carriers.

We can now turn to a discussion of the multiplication in proportional counters, and our first task will to make a simple estimate of the value in specified circumstances of the multiplication, or 'gas gain' as it is also known. As in Figure 3.11 we take the cathode and wire radii to be respectively b and a, and the applied voltage V_0. The multiplication process in the gas is usually specified by two parameters K and ΔV. The parameter K defines the critical field at which multiplication commences for the particular gas used, and it turns out that this field is directly proportional to the gas pressure. So we define K from the equation $\mathscr{E}_c = Kp$, where $\mathscr{E}_c$ is the critical field and p the pressure, giving us dimensions for K of V m^{-1} Pa^{-1} or more commonly V m^{-1} atm^{-1}. Values for K for common fill gases are a few times 10^6 V m^{-1} atm^{-1}, giving critical fields of the same numerical value for proportional counters operating at 1 atm. The parameter ΔV represents the mean potential through which an electron falls between successive ionizing collisions, and again of course depends on the gas in use. (Indeed we might expect numerical values for ΔV to be the same sort of sizes as those we quoted for the energy lost by, say, an alpha particle to produce ion pairs in various gases, and this is generally, but not exactly, true.)

Let us imagine then that a single ion pair is created in a proportional counter with the positive ion moving out to the wall and the electron going towards the wire. At a certain radius r_c (which we will find out later will be quite close to the wire), the electron arrives at a place where the field has reached the critical value $\mathscr{E}_c$, and multiplication occurs. Multiplication occurs again and again in its further travels, but we note that as the field is increasing sharply towards the anode wire, the radial distances through which the electron must move to fall through a potential of ΔV continually decrease. This is indicated schematically in Figure 3.12 where we have

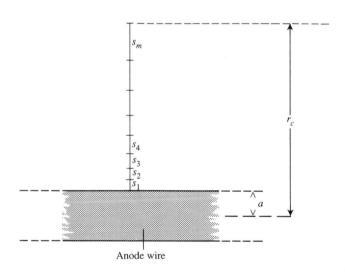

Figure 3.12 Diagram for multiplication calculation in the proportional counter (not to scale).

labelled these various distances s sequentially from the wire. (Of course we are not implying that these segments necessarily represent the electron's actual path — merely that these are the distances measured towards the anode through which it must move to fall through ΔV.)

We can estimate the size of s_1, s_2, s_3, ..., etc. as follows. At a typical distance r the field is given by $\mathscr{E} = [V_0/\ln(b/a)](1/r)$ and as over the small distance s the field variation is small, we can write (apart from sign), $\mathscr{E} = \Delta V/s$ approximately, and thus $s = [(\Delta V/V_0)\ln(b/a)]r = Gr$, say. So for the first segment $s_1 = Ga$, for the second, $s_2 = G(a + s_1) = G(a + Ga) = Ga(1 + G)$, and for the third, $s_3 = G(a + s_1 + s_2)$ which, after inserting the values for s_1 and s_2, turns out to be $Ga(1 + G)^2$. The various segments therefore form an increasing geometric series which we can easily sum to n terms (where n is the number of segments between radius a and radius r_c). Equating this result to $r_c - a$ and remembering that r_c can be obtained from $\mathscr{E}_c = Kp = [V_0/\ln(b/a)](1/r_c)$ allows us to solve for n. The result is

$$n = \frac{\ln[V_0/aKp\ln(b/a)]}{\ln(1 + G)}. \tag{3.16}$$

Now the constant $G = (\Delta V/V_0)\ln(b/a)$ is usually much less than unity ($\Delta V/V_0 \approx 10^{-2}$ while $\ln(b/a) \approx 7$ typically) so we can write $\ln(1 + G) \sim G$ to give us

$$n = \frac{V_0}{\Delta V\ln(b/a)}\ln\left(\frac{V_0}{aKp\ln(b/a)}\right). \tag{3.17}$$

Now the overall multiplication M is given by $M = 2^n$ since a doubling occurs at every stage so

$$\ln M = \frac{V_0\ln 2}{\Delta V\ln(b/a)}\ln\left(\frac{V_0}{Kpa\ln(b/a)}\right). \tag{3.18}$$

This formula was originally derived by Diethorn (reference 4) by a slightly different method, and is found to fit well with experimental results.

We could alternatively write equation (3.18) as

$$M = \exp\left[\frac{V_0\ln 2}{\Delta V\ln(b/a)}\ln\left(\frac{V_0}{Kpa\ln(b/a)}\right)\right]. \tag{3.19}$$

which underlines the very rapid dependence of the gas gain on applied voltage, and incidentally points to the need for a well-stabilized power supply for proportional counters in order to inhibit changes in M due to supply fluctuations.

What does equation (3.19) predict in practice? Let us take $a = 2.5 \times 10^{-5}$m, $b = 2.5 \times 10^{-2}$m (making $b/a = 1000$), a counter gas fill (of which more later) having $K = 5 \times 10^6 \,\text{V m}^{-1}\,\text{atm}^{-1}$, a gas pressure of 1 atm, and a value of ΔV of 25 V. This gives a calculated gas gain of 848 at 2000 V applied voltage, 3860 at 2200 V, and 18 908 at 2400 V, dramatically illustrating the sharp rise in multiplication with voltage. For purposes of further discussion of the mechanism of the proportional counter it is important to have a feel for the size of r_c, the critical radius where multiplication begins, as well as for the value of s_1, the distance between ionizing events close to the wire. For the counter parameters we have just been using $r_c = [V_0/\ln(b/a)]/Kp$ turns out, at $V_0 = 2200\,\text{V}$ to be $6.37 \times 10^{-5}\,\text{m}$, while $s_1 = Ga = [(\Delta V/V_0)\ln(b/a)]a$ is $1.96 \times 10^{-6}\,\text{m}$. (For comparison the value of s just below r_c is $4.63 \times 10^{-6}\,\text{m}$.) Note how small r_c is compared with the cathode radius b—in fact it is only a few times the radius of the wire, a. The volume in which multiplication takes place is thus a small fraction of that of the counter. The volume in which the bulk of the multiplication takes place, that is within a radius of a few times s_1, say, is even smaller.

We are now in a position to make two important observations about the type of pulses we obtain from a proportional counter. The first is a direct result of our remarks just above concerning the very small fraction of the total volume in which multiplication takes place. Because of this it follows that an electron formed practically anywhere in the counter volume undergoes the same multiplication. Perhaps we can see this most clearly if we imagine what would happen if r_c were *not* small compared with b, but equal say to $b/2$. In that case the multiplication undergone by an electron formed between $r = a$ and $r = b/2 = r_c$ would depend precisely how far away from the wire it began its life—that is on how many stages of multiplication occurred on its journey to the anode. For those electrons formed outside $r = b/2$ their history would consist of a slow drift inwards until they reached the critical radius r_c, followed by the full multiplication appropriate to the applied voltage. Returning now to the counter as it actually behaves in practice, we see that because of the tiny value of r_c compared with b, all but a negligible fraction of any electrons formed by ionizing radiation in the counter will be in the outer region beyond r_c where the multiplication is independent of the point of formation. So if, for example, an X-ray enters the counter and produces a photoelectron in the gas, then the direction of the electron track will make no difference to the size of the output pulse (provided of course the electron expends all its energy in the gas).

Our second important point is at first sight rather surprising in view of the prominence we have so far given to the role of the electrons. It is this: the output pulse from a proportional counter is due almost entirely to the movement of the positive ions. However, when we remember our previous result with ionization chambers that the contribution to the output pulse from an ion or electron depends on the distance it travels (or more accurately on the potential it moves through) on its way to the electrode, our statement becomes more meaningful. The great bulk of the electrons are produced in the final stages of multiplication quite close to the wire, and so it is not hard to accept that (in spite of the high field there) their contribution will be much less than that of the positive ions starting at the same place and moving through a very much greater distance (albeit through smaller fields). We can get a rough estimate of the relative contributions by assuming that the bulk of the ions and electrons are all produced at some convenient distance from the wire, say $2s_1$, and making use of equation (3.13). From that equation we can see that the ratio of the electron to ion contribution is in general $[\ln(r_0/r_-)]/[\ln(r_+/r_0)]$ and for our particular case $r_0 = a + 2s_1$, and of course $r_+ = b$ and $r_- = a$ when the carriers reach their respective electrodes. Putting in the numerical values for a, b, and s_1 previously used gives us a ratio of 0.02 or 2 per cent. (Note that this result is only true for electrons formed close to the wire. In a cylindrical *ionization* chamber, since the bulk of the volume occurs at large values of r, most signals will be largely due to electron movement—a fact we will need in Section 5.11.)

The fact that it is positive ions, with their low mobility, rather than electrons that are largely responsible for the output pulse from the proportional counter might seem to carry with it the implication that the output rise would be slow. However, we must remember that, unlike the case of the parallel-plate ionization chamber, the positive ions in a cylindrical proportional counter are moving, at least initially, in a very high field, so the rise-time may be quite reasonable after all. Let us make a calculation on this point. Now we remember from our earlier discussions that the velocity of an ion is given by $u = \mu_+ \mathscr{E}/p$ where μ_+, the positive ion mobility, has a value for most gases around $10^{-4}\,\mathrm{m^2\,atm\,V^{-1}s^{-1}}$. So for an ion moving radially outwards from the anode wire to the cathode we can write

$$u_+ = \frac{dr_+}{dt} = \frac{\mu_+ \mathscr{E}}{p} = \frac{\mu_+}{p} \frac{V_0}{r_+ \ln(b/a)}.$$

Therefore

$$r_+ \frac{dr_+}{dt} = \frac{\mu_+ V_0}{p \ln(b/a)}$$

and integrating from where the ion is formed initially near the wire ($r_+ \approx a$) to a typical value r_+ which the ion has reached by time t gives us

$$\int_a^{r_+} r_+\, dr_+ = \frac{\mu_+ V_0}{p \ln(b/a)} \int_0^t dt$$

or

$$r_+^2 - a^2 = \frac{2\mu_+ V_0 t}{p \ln(b/a)}$$

or

$$t = \frac{p \ln(b/a)}{2\mu_+ V_0} (r_+^2 - a^2) . \tag{3.20}$$

So finally the collection time t_c when $r_+ = b$ is given by

$$t_c = \frac{p \ln(b/a)}{2\mu_+ V_0} (b^2 - a^2) . \tag{3.21}$$

Inserting into equation (3.21) the usual values for a and b, and remembering the value for μ_+ quoted above, gives us for a counter with $p = 1$ atm and $V_0 = 2200$ V a value for the collection time of 9.8 ms. This seems disappointingly long, but it is a reasonable supposition that most of the total collection time is due to the slow final travel of the ion in the weak field near the cathode. So let us instead ask how long it takes the contribution from the positive ions to reach half its final value. From equation (3.13) this happens when $\ln(r_+/r_0)$ reaches half of $\ln(b/r_0)$ that is when $(r_+/r_0)^2 = b/r_0$ or $r_+^2 = br_0$. Now in our case $r_0 \approx a$ so we have $r_+^2 \approx ab$, giving $r_+ \approx 7.9 \times 10^{-4}$m: putting this value into equation (3.20) yields a value for t of 9.8 μs. This is three orders of magnitude less than t_c, as indeed we might have expected because we have in equation (3.21) replaced $b^2 - a^2 \approx b^2$ by ab.

Indeed with parameters more suitably chosen for fast rise-time we might have done even better. Reducing p and/or b in equation (3.21) would certainly improve matters, although we might have to decrease V_0 somewhat to prevent the gain from becoming impossibly high. The time to reach 50 per cent of the signal's final value can thus be reduced to a few microseconds. By 'clipping' the signal with a time constant large compared with this time, say 20 μs (in an analogous fashion to what we did when selecting only the electron contribution in an ionization chamber) we have a device capable of counting at quite fast rates while still retaining the ability to produce fairly large pulses of equal amplitude from the same initial ionization, no matter where deposited in the chamber.

There is one effect that could lead in principle to a degrading of the good rise-times we have just been talking about. This occurs because, if the original electrons are formed at different distances from the wire (as for example if the initiating particle track is at an arbitrary angle to the counter axis) then, although we know that the gas gain will be the same for all these electrons, the time taken by them to drift inwards to the critical radius where multiplication begins will vary. Consequently the rising edge of the output signal could be smeared out by the superposition of contributions arriving at slightly different times. However, for some of the special counter gas mixtures to be dealt with in the next sub-section, the electron drift velocity can be as high as $4 \times 10^4 \mathrm{m\,s^{-1}}$ at fields of $10^4 \mathrm{Vm^{-1}}$ and 1 atm pressure, so the travel time even of an electron starting near the cathode will be appreciably less than a microsecond. We need therefore make no significant amendment to our previous treatment.

3.3.3 *Proportional counter gases*

Many types of gases have been used successfully in proportional counters, the main exclusions being those exhibiting attachment of electrons to neutral molecules, which as we mentioned previously will lead to a dramatic drop in their mobility. So removal of any trace of oxygen, for example, from a fill gas is essential. Argon has been widely used for general-purpose counting, and xenon for a particular case we will come to shortly. However, for gas gains much above 100 it is necessary to add to the main gas an organic vapour, a common mixture being 90 per cent argon and 10 per cent methane (or ethanol). The necessity for the addition of these more complex molecules arises from the fact that many ultraviolet and visible photons are generated during the multiplication process from collisions producing excited atoms. These photons may generate photoelectrons when they strike the cathode, which in turn can produce, after a short delay, additional avalanches, and the process can continue on. We have here a form of 'positive feedback' and like all positive-feedback systems the gain is increased to a larger value M^*. It is not hard to show that M^* is given by $M^* = M/(1 - MF)$, the usual feedback formula. Here $F\,(\ll 1)$ is the number of electrons produced from the cathode per ion formed in the gas: it is of course equal to the number of photons formed per ion multiplied by the probability of a photon ejecting an electron from the wall. Clearly when MF approaches unity M^* becomes very large, which simply means we have moved to the Geiger region of operation. Even before that, however, and while still in the proportional region, if MF is not far from unity, the multiplication M^* becomes, as in all feedback systems, very sensitive to small changes in M (and M itself is already a strong function of voltage). The overall gas gain thus rises extremely rapidly with applied voltage, which is not a convenient regime to operate in. For stability then we add some of the complex molecules we mentioned earlier. These have a high cross-section for photon absorption but the photon energy goes, not to produce further ionization, but to harmlessly decomposing the molecule. The large molecules are also capable of de-exciting (or 'quenching') excited atoms directly by collision. The photon contribution to the process is thus eliminated, or at least sharply reduced. As a small delay due to electron drift time is associated with every avalanche, elimination of the photons also means a return to the optimum rise time associated with a single avalanche.

We make two final remarks before leaving the subject. The first is that the presence of methane gives us a further bonus in that the electron velocity is increased by an order of magnitude over that of pure argon. This arises because inelastic collisions between electrons and methane (or other large molecules) reduce the thermal agitation energy of the former, and from simple kinetic theory this increases the drift velocity in an electric field. We have already mentioned this fact in our discussion on signal rise-times. Finally, although the photons have so far appeared in an undesirable role, it is possible to obtain an output signal from them directly, using a photosensitive detector. The mechanism and advantages of such a 'gas proportional scintillation counter' will be discussed briefly in Chapter 8.

The pressure of gas in a proportional counter is usually about an atmosphere, which conveniently allows us to employ very thin entrance windows. Higher pressures may be used if we wish to increase the density and thus the stopping power of the gas in the device: lower values may be employed if we wish to operate at lower voltages. Counters can also be operated with a continuous flow of gas passing through them, which is either vented to the atmosphere, or recycled after repurification. These 'flow counters', though obviously awkward in some respects, have two advantages over their sealed off counterparts. Firstly they are immune to gas contamination from air leaks or slow gas desorption from the walls, and secondly they can be made demountable to allow a source of feebly penetrating radiation to be placed inside the counter volume.

3.3.4 *Applications of proportional counters*

Proportional counters may be used to detect all three basic radiations – alpha, beta, and gamma or X, as well as neutrons. Before referring briefly to a number of such applications, let us remind ourselves of the two main attributes of this type of counter. Firstly because of its built-in amplification we can conveniently detect low-energy radiations, and secondly because of its proportionality we can produce an energy spectrum of the incoming radiation, or if we do not need this information, at least discriminate against extraneous pulses whose amplitudes lie either above or below a selected range of interest.

We hardly need to say much about alpha detection except to mention that we can achieve 'windowless' operation without absorption by using a flow proportional counter as mentioned above. Beta particles, unless of very low energy, are not particularly suited to the proportional counter. Clearly we can not produce meaningful energy spectra if the range of the beta particles is much longer than the counter dimensions, and if we are only interested in detecting them without any energy information, then the Geiger counter is a simpler, more effective alternative.

For low-energy beta particles like those from ^{3}H and ^{14}C, proportional counters have been used to study the shape and end point of their spectra, the sample being introduced in gaseous form. Although largely superseded by mass spectrometric methods, an historically important application of beta counting by proportional counters was the dating of of old carbon-containing specimens by the radiocarbon method—that is by detecting the presence of small amounts of the ^{14}C isotope. This isotope is produced by the action of cosmic-ray neutrons on the nitrogen in the air, and is subsequently incorporated into all living material. However, when a tree say, dies, the ^{14}C present decays away with a 5500 year half-life, thus allowing the determination of the age of an old wood (or other carbon-bearing) specimen by its residual radioactivity. The snag is of course that this activity is very low, particularly for an older sample, and consequently for meaningful counting statistics we must reduce the background from ambient radioactivity and cosmic rays as much as possible. This is done in the first instance by shielding, but signals due to energetic cosmic rays which still enter the counter can be identified because their amplitudes are greater than that corresponding to the end point of the ^{14}C spectrum and thus can be eliminated from the count record. So although we are not using the proportional counter to indicate the energies of the betas themselves, but only to count them, we still have a significant statistical advantage over the corresponding Geiger counter arrangement—which was in fact used in earlier studies. In both these radiocarbon dating systems the radioactive ^{14}C material was introduced into the counter as part of the counting gas itself (usually carbon dioxide or acetylene): this guaranteed absorption-free counting at maximum efficiency.

It is much the same story for gamma rays. For moderate- and high-energy quanta interacting in the gas a count is all that can be recorded: exact information on the energy is unavailable, as only a small fraction of the energy of the fast secondary electron produced will be expended in the gas. However, proportional counters have found widespread application in the detection and energy measurement of low-energy gamma rays and X-rays in the region up to 100 keV. To increase the probability of photon absorption (which at these low energies is almost entirely due to the photoelectric effect) one often replaces the argon ($Z = 18$) with krypton ($Z = 36$) or xenon ($Z = 54$). As an added bonus these heavier gases also reduce the range of the photoelectron. For example, experimental results show that a 50 keV photon would have a 10 per cent probability of being absorbed in a 50 mm layer of krypton at a pressure of 1 atm. The resultant electron, whose energy would be 50 keV minus the K-shell binding energy of krypton (≈ 13 keV), or 37 keV, would have a range of only about 6 mm, which increases the likelihood that all its energy will be expended within the gas.

There is, however, a problem remaining, because to retain the complete energy of the incoming photon in the gas we must also capture the K-shell X-ray that for a high-Z material like krypton (or xenon) is the usual accompaniment of the rearrangement of the orbital electrons. Now an X-ray of about 13 keV, as we expect in the case of krypton, might have for a normal-sized counter a 50 per cent chance of escape. Thus for about half the number of events recorded we would have all the 50 keV deposited in the counter, and for the other half only 37 keV. A pulse-height spectrum then would exhibit not only the expected characteristic peak at 50 keV, but also an 'escape peak' at 37 keV of comparable size. If we are trying to unravel a complex spectrum with X-rays of different energies coming from a number of isotopes, the presence of pairs of peaks separated by the relevant K-shell binding energies would make interpretation difficult. Ironically this problem can be eased significantly by returning to an argon filling where the rearrangement X-ray is of much lower energy and more easily retained in the gas – but of course only at the expense of a much lower counting efficiency for the original incoming quantum! So it depends on what our priorities are.

The other question we must ask before leaving the topic of low-energy photon spectrometry is how well the counter can 'resolve' two X-rays of energies close to one another – where we use the word 'resolve' in the same sense as in our analogous discussion for the ionization chamber and alpha particles. Again we propose to discuss this matter in considerable detail in Chapter 7, so all we need to say at the moment is that moderate resolution for photons is available in the proportional counter – worse than for the semiconductor detector, but appreciably better than for a scintillation counter. In numerical terms we would have no difficulty in resolving the 50 keV X-ray we were discussing previously from another X-ray differing from it in energy by 3 keV.

A final important use of the proportional counter is for slow neutrons. As we mentioned in Chapter 2, slow neutrons must be detected via an ionizing particle coming from a nuclear reaction, and the $^{10}B(n, \alpha)^7Li$ reaction is a popular one. A bit more than 2 MeV energy is produced in this case of which about two-thirds is carried away by the alpha particle and the rest by the recoiling ^7Li nucleus. To produce a neutron detector, boron (usually enriched in the ^{10}B isotope over its normal 19 per cent level) is introduced into a proportional counter either as a coating on the counter walls or by using boron trifluoride (BF_3) as the counter gas. Both methods have their own advantages and disadvantages, but the latter is more commonly used, so we will restrict our discussion to it. If the counter is then irradiated with neutrons, each neutron that reacts with a boron nucleus in the gas ideally produces ionization equal to the total reaction energy. (We say 'ideally' because there clearly is a difficulty with the reactions which occur near the wall: however, this will not affect the general correctness of our argument). The output pulses then are quite large, and in particular very much larger than the smaller pulses produced even by energetic gammas which happen to enter the tube. This is important because in environments where neutrons are being produced — say near a nuclear reactor — considerable gamma radiation will usually be present also. The proportional counter thus has the ability to distinguish between the pulses due to the gammas and those due to the neutrons, and associated electronics can reject the smaller gamma pulses, to record only the neutron counts. The BF_3 counter can also detect fast neutrons if these are first moderated (in order to increase the cross-section for interaction) in a sheath of hydrogen-rich material such as paraffin wax.

3.4 Geiger counters

3.4.1 *Mechanism of pulse formation*

If we continue to increase the gas multiplication (either by increasing the applied voltage or more conveniently by reducing the gas pressure) then we pass from the proportional region through the region of limited proportionality to the Geiger region. In this case the electron cascade spreads (in a manner to be discussed in more detail in a moment) from the immediate neighbourhood of the initiating track, until it encompasses the whole length of the anode. The resultant output pulse is thus very much larger than that obtained from a proportional counter (which is one of the great advantages of the Geiger counter), but is obviously a standard size independent of the initiating particle — which is the price one pays for this large output. As we noted earlier, even one initial ion pair can produce the full output signal, which may be of the order of volts or even tens of volts depending on the counter dimensions. We can now estimate the number of charges collected in a moderate-sized counter. The capacitance of such a counter together with the associated capacitance of the cables and electronics attached might be around 15 pF. Since $Q = CV$, for $V = 10$ V the charge arriving is 1.5×10^{-10} C, which corresponds to about 10^9 electrons. This number, as just mentioned, is independent of how many initial ions were present, but does increase with the voltage applied to the tube, as we shall see shortly.

The mechanism of a Geiger counter filled with a monatomic gas, say argon, is quite understandable. Because of reduced gas pressure the multiplication can exceed 10^6 even for moderate voltages. Consequently the photons from the excited atoms associated with the first avalanche can produce a further avalanche, via photoelectric emission from the cathode, and a self-sustaining chain comes into being. The result is an increasingly thick sheath of positive ions around the wire and along its complete length. This sheath is highly conducting and eventually shields the wire sufficiently to reduce the maximum field in the gas below that required for multiplication to be possible. (We can think of it informally as having produced a conducting 'wire' surrounding the real wire and large enough in radius to reduce the field below the critical value for multiplication.) The further production of avalanches thus ceases. The dense sheath continues to move slowly outwards, and reaches the cathode after a time rather better than we calculated for the proportional counter (because of the lower pressure we usually use in Geiger counters) but still of the order of 1 ms. An unwanted sequence of events now occurs. When a positive ion reaches the wall it is neutralized there by an electron. Each neutralized ion is then left with an excitation energy (about 10 eV for argon and many wall materials) equal to its ionization potential minus the work function of the wall. This is more than sufficient to eject a further electron from the wall, and although the probability of this occurring is small, the large number of ions in the avalanche ensures that at least one fresh electron is almost certain to be produced. The complete multiplication cycle thus restarts, producing again a full Geiger pulse, and thus a whole sequence of them. (We add in parenthesis that this difficulty also arises in principle with a proportional counter using a monatomic gas. However, there the total number of ions arriving is very much less, and in any case a spurious repeat pulse will usually be much smaller than the first genuine signal.)

There are two ways in which we can avoid the production of these repeat pulses. The first is to ensure that the counter is inoperative when the ion sheath arrives at the cathode. This can be done by electronically reducing the counter voltage for an appropriate interval after the main pulse by a special 'quenching circuit'. The same reduction in counter voltage during the critical period can be more simply achieved by employing a load resistor R sufficiently large (say $\approx 10^8 \, \Omega$) to make the recovery time RC (where C is the detector plus stray capacitance) suitably long. Both these variations of 'external' quenching — but particularly the latter — have the disadvantage that as the counter is effectively 'dead' for about 1 ms the count rate capability is low. The other, and preferred, alternative is to use a 'self-quenching' tube, where 'quenching' here refers to some mechanism for harmlessly dissipating the excitation energy of atoms — in this case those produced from the ions neutralized at the cathode. The solution involves adding a gas or vapour with a complex molecule — ethanol is a common choice — to the monatomic gas, say argon, in about the same 10 to 90 per cent proportions we used with the proportional counter (although here the total pressure might be typically 100 Torr or 0.13 atm). In this case of course the primary aim is to prevent the ejection of electrons from the cathode when the ion sheath arrives there, and this is achieved as follows. During its passage to the wall an argon ion will collide with neutral ethanol molecules, and as the ionization potential of ethanol is less than that of argon the possibility exists of the ionization being transferred from the monatomic ion to the ethanol. Since a very large number ($\approx 10^5$) of such collisions will occur, only ethanol ions will arrive at the wall. Here the mechanism is very different from the previous case, because after being neutralized, the preferred mode for the ethanol molecule to dispose of its excitation energy is by dissociation — which it does very rapidly — and our aim is achieved.

We are now, however, left with another problem — how to explain the initial build-up of the avalanches in the presence of the ethanol, which as we know from our discussion on the proportional counter is a good absorber of ultraviolet photons. In fairness, however, the conditions in the two cases are not the same. In the Geiger counter the multiplication in the avalanches is very much larger, and in addition the actual number of complex molecules is smaller, since the overall counter pressure, as we noted, might be typically 100 Torr with the ethanol accounting for only one tenth of this. So it is not too hard to believe that the ejection of photoelectrons from the cathode can continue, albeit to a much lesser extent. It is necessary none the less also to invoke the production of electrons in the gas by the photons as a source of secondary avalanches. This is more difficult to understand. Photons from excited monatomic atoms clearly have not sufficient energy to ionize further monatomic atoms (although those from any excited ions present would have). Photons from excited neutral atoms may of course ionize the complex molecules, but dissociation, rather than ionization, has already been postulated as the major mechanism there. In any case the discharge does spread along the wire, though this appears to occur more slowly than in a counter with monatomic gas alone, and also to be more localized near the wire. As a small corollary to what we have said about the quenching process it is clear that the useful life of this type of Geiger counter is ended when an appreciable fraction of the ethanol (or other quench vapour) has been decomposed, and this will happen after about 10^8 counts typically. (In principle this also happens with proportional counters, but because of the smaller discharges it is not really a practical problem.)

3.4.2 *Geiger counters with rare gas–halogen fillings*

As we have just noted, a conventional argon–ethanol counter has a limited life of about 10^8 counts because of the disappearance of a small but finite portion of the ethanol at each discharge. In an effort to eliminate this problem, halogens (and particularly bromine) were tried as a replacement for the ethanol, on the grounds of their known ability to reassociate after decomposition. In fact counters made to the old recipe of 90 per cent argon (or other rare gas) and 10 per cent quenching vapour to a total pressure of

100 Torr, but with a halogen replacing the normal quenching agent, produced poor counters. The reason was not hard to find—the old problem of
electron attachment, in this case to the halogen molecules. However, when
the halogen content was reduced to a fraction of 1 per cent (and with the
same total pressure) new and important counter characteristics developed.

Firstly the starting voltage of the counter (that is the voltage at which
Geiger pulses appeared) dropped dramatically. For conventional counters
starting voltages of around 1000 V would be typical. For counters with
halogens, starting voltages of a few hundred volts were obtained. More significantly, early work showed that starting voltages—particularly with neon-
halogen mixtures—were almost independent of the anode diameter over an
order of magnitude variation in the latter. Such a result is completely at
variance with our previous idea that the multiplication takes place close to
an anode wire of small radius, and consequently must be quite sensitive to
changes in this parameter. To underline this difference one can add that
Geiger operation for a pair of flat plates is now possible at reasonable voltages. For example, for a counter with a 10 mm plate separation, using a filling of neon gas at 100 Torr pressure and a bromine admixture of 0.3 Torr,
a Geiger starting voltage of only about 450 V has been reported. The value
of $\mathscr{E}/p$ (where as usual $\mathscr{E}$ is the electric field and p the pressure) in this
counter is about 3.5×10^5 V m^{-1} atm^{-1}, which is an order of magnitude
smaller than the values we quoted earlier as being necessary even to begin
multiplication in conventional counters.

Obviously with these newer fillings the discharge must now be capable of
starting anywhere in the gas volume, and we are no longer restricted to the
geometry of a thin anode wire surrounded by a coaxial cathode cylinder. Not
only can anodes then be made of much larger diameter (particularly convenient for counters with a thin end window where the wire is attached at the
far end only) but even more exotic geometries can be used. For example, the
anode may be large and hollow to allow a radioactive liquid to flow through
it for measurement. Another application is shown in Figure 3.13(b). Here a
small end window counter for use with low-activity beta-emitting samples
is nested inside an outer guard counter. This latter, also shown in Figure
3.13(a), consists essentially of two coaxial hemispheres, insulated from one
another, and spaced 25 mm apart, forming the cathode and the anode, but
even so the device operating voltage is only 1000 V. This guard counter is
used in conjunction with conventional shielding to reduce further the background of the inner counter. Any penetrating particles coming through the
shielding and producing a count in the inner tube will normally have also
triggered the outer tube. This count can thus be removed (as it should be)
from the tally of counts of the inner counter by anticoincidence circuitry,
as will be discussed in Section 6.8. Typically the count rate of the inner
tube can be reduced by a factor of about four (compared with that in the

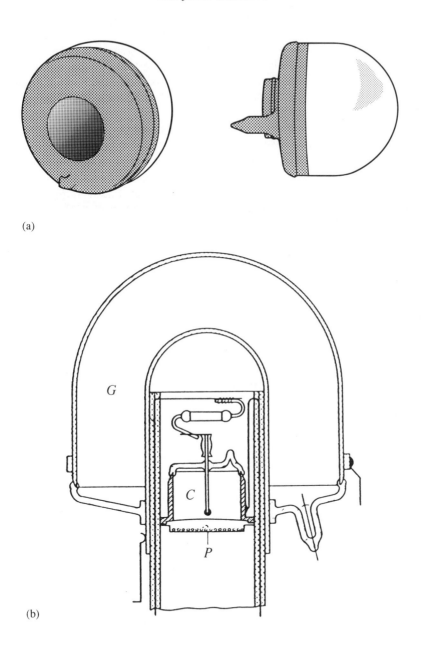

(a)

(b)

Figure 3.13 (a) Two views of guard counter. (b) Guard counter with nested beta counter: G, guard counter; C, beta counter; P, radioactive sample. (From Hofker, W.K. (1980), *Philips Technical Review*, **39**, 296–7, figure 2.)

same shielding but without a guard tube) thus enhancing its ability to count low-activity samples with small statistical errors.

Geiger counters with the small admixture of halogens that we have been discussing are only self-quenching near their starting voltage. Since it is necessary for a number of reasons (including reduction of electron attachment) to operate halogen counters at several hundred volts above their starting voltage, external quenching is necessary. Fortunately values of load resistor in the range 5–10 MΩ are sufficient in this case. Furthermore, because the pulses from these types of counter are very large, a tapped output from the device will suffice (see Figure 3.14). By accepting this smaller output one obtains an effective reduction, from the counter's point of view, of the stray capacitance between output and ground (coming from the recording circuit attached to the counter). This reduces the value of RC to quite a reasonable figure—a result we will refer to later when discussing counter 'dead times'.

We turn finally to look briefly at the actual discharge mechanisms in counters with rare gas–halogen fillings, and in particular at the neon–bromine case which has received most attention. Unfortunately the processes are more complex than in argon–ethanol systems, and depend strongly on the amount of bromine present. While a general understanding of what is happening exists, our knowledge of the basic parameters in the discharge does not seem to be sufficiently precise to allow numerical predictions on performance. For neon at 100 Torr and with bromine at 0.1 Torr the principal way in which the discharge is spread appears to be via resonance radiation from one excited metastable neon atom to another. The excitation energy (≈ 16.5 eV) finally goes by impact to the ionization of a bromine molecule

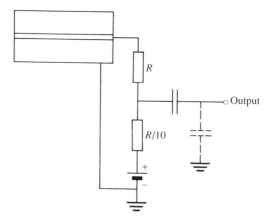

Figure 3.14 Output circuit for halogen tube.

(whose ionization potential is only ≈ 10.5 eV) and to the kinetic energy of the resulting electron. Because of the efficiency of such collisions for the conversion of excitation energy into electron production a substantial reduction in the counter starting voltage results. (This production of additional electrons by the introduction of small quantities of a gas or vapour whose ionization potential is less than the metastable excited state of the main inert gas is known as the Penning effect. Another well-known Penning mixture is neon with small amounts of argon, and this combination has been used to produce externally quenched counters with very low starting voltages. The addition of argon to neon–halogen counters has also been shown to improve their characteristics. Acetylene or xylene will similarly produce a Penning effect in argon. Such mixtures can also be used in proportional counters. They improve the energy resolution because, as we shall see in Chapter 7, this improves as the number of original ions created by the incident radiation increases. The operating voltage for a given gain is also reduced.)

In view of the various virtues of the rare gas–halogen mixtures we have been discussing, such fillings are, except for special applications, almost always employed for Geiger counters. For proportional counters traditional geometries and fillings are still required.

3.4.3 *The Geiger counter plateau*

We devote a short section to this topic because it is one that often leads to a certain amount of confusion. If, using an oscilloscope, we inspect the output pulses from a Geiger counter operating under normal conditions (and we beg the question for a moment of what 'normal conditions' means) then we will find they are all closely of the same size. However, unlike the proportional counter this size tells us nothing at all about the energy of the particles or photons producing the pulses: if for example we change over from 3 MeV alpha particles to 5 MeV alpha particles the pulse height will remain the same—as it will if we use 0.5 MeV betas or 1 MeV gammas. This is not to say that the pulse height for a given counter is independent of all parameters, and in particular if we increase the applied voltage, the height, although remaining substantially the same for all events, will increase. This is due to the physical fact that with a higher voltage on the wire, the number of ions produced can reach a larger value before the multiplication process terminates. In fact the pulse height is roughly proportional to the difference between the operating voltage and the Geiger threshold voltage.

We are now in a position to describe how to plot an operating curve for a Geiger counter and obtain its plateau, which is where we will run the device. What we do is to place a radiation source near the Geiger counter, and increase the counter voltage in steps while at the same time noting the corresponding count rates (in counts per minute, say) on an appropriate

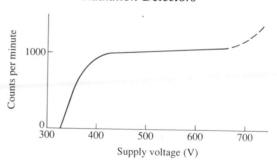

Figure 3.15 Typical plateau curve for halogen counter.

recording device ('scaler' or 'ratemeter') connected to it. Our results will typically appear as in Figure 3.15. (The numerical scale on the vertical axis will of course depend on the strength and position of the source used.) We emphasize the fact that this is not a curve giving us information about the energy of the radiation used, nor indeed is it directly related to that in Figure 3.10. We are here just recording the number of times the Geiger counter fires per minute at various voltages. For low voltages no counts are recorded because the output pulses are too small to trigger our scaler, which might typically need 1 V input to operate. The pulses are small because the counter is operating in the proportional region or in the early part of the Geiger region. Eventually the pulse heights reach the threshold level of the scaler and counts appear — at an applied voltage of 350 V in the case we have taken. Because all the pulses are, in proper Geiger fashion, almost exactly of the same size, the count rate rises very quickly to the 1000 counts per minute value which we have taken to be the number of particles entering the counter per minute from the particular source used. If we further increase the voltage then, as we have said, the height of the pulses will increase (as we can check with an oscilloscope), but the count rate naturally does not, being just dependent on the source strength. We are now on the flat part of the curve known as the plateau, which will typically be 100 V or more long. It is somewhere around the middle of this region that we normally operate the Geiger counter, because under these conditions the count rate is independent of the voltage (if we ignore the slight upward slope on the plateau, which arises for reasons we need not go into here). In practical terms this means that quite a simple high-voltage supply producing an output not particularly well stabilized is perfectly adequate for operating our counter. (This advantage only arises because the Geiger counter pulses carry no information on energy, and consequently, provided they are larger than the scaler threshold, correct recording of their number is assured. In contrast for a proportional counter the pulse heights, which carry energy information, depend strongly on the applied voltage, and as we noted earlier, a high-quality supply is required to prevent voltage fluctuations from smearing out this information.) Beyond the higher voltage end of the plateau, the curve starts to rise again sharply as the limiting mechanisms within the counter are unable to prevent spurious counts (that is ones without any initiating particle) occurring, and unless the voltage is reduced again quickly the counter will go into a continuous discharge and of course be useless as a detector.

3.4.4 *Count rate limitations with Geiger counters*

We have already noted the fundamental difference between the mechanisms of a proportional and a Geiger counter, in that for the former the discharge is restricted to one place on the anode, while in the latter it spreads along its length. This difference is reflected directly in their behaviour as regards count rate capability. If in a proportional counter a second event produces a discharge on the anode anywhere but in the same location as the preceding one, then the counter can respond with an output. (Of course if the count rate is very high an output pulse may occasionally occur sitting on the 'tail' of the preceding one. However, we can deal with this situation in the amplifier in a manner to be discussed in Chapter 6, provided the second pulse does not arrive at least until the rise of the first is substantially complete. We can expect the proportional counter then to be able to separate events only a few microseconds apart.)

For a Geiger counter, on the other hand, we must wait until the large number of positive ions produced along the anode length has at least partially cleared away before a further event can be recorded. We have already seen that for a Geiger counter containing a rare gas alone we must ensure (by means say of a large load resistor) that the counter voltage remains at a level insufficient for Geiger action until the positive ion sheath has reached the cathode. The resulting 'dead time' will typically be of the order of 1 ms, which implies a very poor count rate capability. For self-quenching counters we know that we do not need this large value of load, and thus the counter voltage can be restored quite quickly. However, the field in the counter will not return to its initial state until the ion sheath arrives and is neutralized at the cathode, and until this 'recovery time' has elapsed we can not have a full Geiger pulse. None the less when the sheath has moved out a certain distance from the anode the field is sufficiently restored to allow enough multiplication for small Geiger pulses to reappear, and if the associated counting equipment is sufficiently sensitive these will be recorded. The 'effective' dead time thus depends not only on the counter diameter, but also on the recording equipment, and might be typically about 200 μs. Halogen counters are of course not self-quenching under normal conditions, but only require, as we saw earlier, more modest values of load resistors than older externally quenched tubes. While low values of load (and stray capacitance) are certainly necessary for smaller dead times yet a completely satisfactory picture of the mechanisms controlling dead times in halogen counters does not yet appear to have emerged. However, experimentally determined times are in fact comparable with those in self-quenching tubes.

Note of course that if the effective dead time in a counter is known to be 200 μs, this does not mean that we can count events at anything like $1/(200 \times 10^{-6})$ or 5000 per second, because these events are arriving randomly distributed in time. So even at rates substantially below this value we may be missing significant numbers of counts due to the arrival of closely spaced pairs. For example at an observed count rate of only 50 per second we would still lose $50 \times (200 \times 10^{-6})$ second in dead time every second, that is 10^{-2} second per second. So our count rate would be low by 1 per cent. The losses with a Geiger counter are clearly not trivial even at moderate counting rates. Naturally it is important to know the exact value of the dead time in order to correct for counting losses when we are required to deal with sources giving moderately high count rates. This can be determined experimentally by taking two sources (which we have previously determined roughly to be active enough to give significant losses) and measuring them, first separately and then together. If our three experimental count rates are m_1, m_2, and m_3, respectively, then we do not expect m_3 to be equal to $m_1 + m_2$ because of the bigger losses at the higher count rate of m_3. However, from these three values we can easily work out the true values of m_1 and m_2 (if we need them), but more importantly for future reference, the value of T_D the effective dead time (see Section 6.5.4).

3.4.5 *Applications of Geiger counters*

We have already mentioned some of the practical advantages of the Geiger counter. The device itself is simple and cheap to manufacture compared with other detectors; it requires a relatively unsophisticated high-voltage supply, and little or no external amplification of the output pulses. In the many applications therefore where counting of moderately active radioactive samples without energy information is required, the Geiger counter is the first choice. It is particularly useful in counting high- and medium-energy beta particles such as arise in artificial radioisotope work. In these cases the counter usually takes the form of a short cylinder with a stiff anode wire supported at one end only. The other end of the tube is closed with a thin metal or mica window through which the betas (or even alphas) can pass. For any such particles entering the active volume, the counter is 100 per cent efficient. It is a different story with gamma radiation, where the probability of a high- or medium-energy gamma interacting in the counter gas is small. Indeed if one places a counter in the vicinity of a gamma-ray source most of the counts recorded will be due to quanta interacting in the counter walls and producing electrons which emerge into the gas. The efficiency for gamma counting can obviously be increased by increasing the wall thickness — but clearly when one has reached a value equal to the range in the wall material of the secondary electrons produced by the gammas, further increase is pointless. As a consequence efficiencies of only a per cent or two are about the best that can be obtained, even with high-Z wall materials, but even so the Geiger counter is still useful in circumstances where moderate gamma fluxes are encountered.

For the reasons enumerated at the beginning of this sub-section the Geiger counter is also widely used in portable instruments such as geological survey meters and health physics monitors. In short the Geiger counter has managed, despite the availability of more recent and more sophisticated devices, to retain an important role in the radiation detection field.

References

1. Knoll, G. F. (1989). *Radiation detection and measurement*, (2nd edn), pp. 142 ff. Wiley, New York.
2. Ibid., pp. 154 ff.
3. Ramo, S. (1939). Currents induced by electron motion. *Proceedings of the Institute of Radio Engineers*, **27**, 584–5.
4. Diethorn, W. S. (1956). *A methane proportional counter system for natural radiocarbon measurements*. Thesis. Carnegie Institute of Technology.

Further reading

Franzen, F. and Cochran, L. W. (1962). Pulse ionization chambers and proportional counters. In *Nuclear instruments and their uses*, Vol. 1, (ed. A. H. Snell), pp. 3–81. Wiley, New York.
Rice-Evans, P. (1974). *Spark, streamer, proportional and drift chambers*, pp. 247 ff. Richelieu Press, London.

4
Scintillation counters

4.1 Introduction

In the previous chapter on gas-filled detectors we discovered one area where these devices perform poorly — the detection and spectrometry of moderate- and high-energy gamma rays. Even for X-rays and low-energy gamma radiation where proportional counters are useful the counter efficiency is only moderate. Clearly what is needed, as indeed we indicated, is some form of solid (or liquid) detector, whose efficiency for stopping the original photon, and the resulting photo- or Compton electron will be much greater. Carriers can be produced by radiation in all sorts of materials besides gases, but the question is which type to choose. Should we go for a conductor, an insulator, or a semiconductor? Conductors are clearly ruled out because any carriers we produce and attempt to collect will be undetectable among the huge number of carriers normally present anyhow. Insulators would also appear to be unsuitable because of the difficulty of collecting the carriers we produce. This leaves us with semiconductors, and indeed these have proved to be, under the proper operating conditions, versatile and efficient radiation detectors: they will be treated at length in Chapter 5.

Let us return, however, to the insulators which we may have rejected a little prematurely. In a crystalline material for example, with a significant gap between the valence and conduction bands, incident radiation will lead to the formation of carriers of either sign — electrons raised from the valence to the conduction band, and corresponding positive 'holes' left in the valence band (Section 5.2). This is exactly the sort of mechanism the reader has probably met in the production of carriers in semiconductors, whether thermally or by radiation. In semiconductors, however, the carriers, as Chapter 5 will show, are more easily collected than in insulators. If instead then of attempting to collect the electrons and holes produced in the insulating crystal we allow them, or better still encourage them, to recombine with one another, then for almost every electron and hole recombining we get back a photon of energy of the same order as the band gap — at least for materials of interest in scintillation counting. For the sort of crystals we will be discussing this photon will usually be in the blue or near-ultraviolet region of the spectrum. Provided, as is normal, that the crystal is transparent to this radiation, then that part of the energy of say, an incident gamma initially used in the production of holes and electrons will reappear as a shower of thousands or even tens of thousands of photons, depending on the energy of the gamma ray (or other incident particle).

If this energy is large enough we can in certain cases see this flash, or 'scintillation' visually: however, it is more practical to convert the burst of photons by means of a photoelectric device into an electrical pulse. We then have our basic 'scintillation counter' for the detection of ionizing radiations.

So far in this introduction we have presented scintillation counting as one of the answers (and indeed the first satisfactory answer that became available) to the problem of gamma detection and spectrometry. Historically there is a twist to the story. Very early in the development of nuclear instrumentation, alpha or other heavy particles could be detected by the scintillations they produced in a thin layer of powdered zinc sulphide. This powder was normally spread on a glass disc and the scintillations were viewed through the glass by an observer using a low-power microscope, and with dark-adapted eyes. Such an arrangement was, for example, used by Walton and Cockcroft in their pioneering experiments in 1932 which demonstrated the first artificial disintegration of an element — as indeed we mentioned in Chapter 1. The alpha particles from the disintegration of lithium by protons were observed by the flashes they produced in a scintillator screen deposited on a window of the apparatus. Ironically this type of particle detection was superseded by methods using gas-filled detectors, and did not come into general use again until the early 1950s, with the development of appropriate photoelectric devices for the scintillation counting of gammas using large single crystals.

We now proceed to discuss in successive sections the crystals and other materials which are used as the primary detecting medium in scintillation counters (the 'scintillators' or 'phosphors' as they are called), and then the photoelectric devices (normally 'photomultipliers') which convert the scintillation into an electrical signal. Before doing so we show schematically in Figure 4.1 a typical scintillation counter assembly. An incident gamma ray, say, (1) enters the scintillator (2) where it is absorbed, let us imagine, by the photoelectric interaction we talked about in Chapter 2 to produce an energetic electron (3). The scintillation mechanism will convert a certain fraction of the electron's energy into into a shower of visible or near-ultraviolet photons, of which two are shown in the diagram (4). The photons pass through the glass top plate into the photomultiplier (which of course is highly evacuated) and eject electrons (5) from the thin photocathode deposited on its underside (6). These electrons are accelerated by an electric field (derived from the EHT supply and a resistor divider chain (7)) to the first electrode or 'dynode' (8) — in this case a 'venetian blind' type. Here each electron will eject typically three or four secondary electrons (9), depending on the applied voltage, and the process is repeated at ten or more dynodes. The

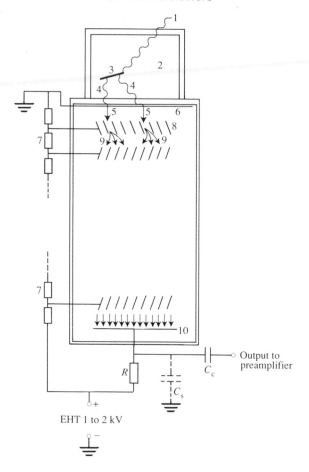

Figure 4.1 Scintillation counter (see text for key).

result is a multiplication of the electrons produced at the photocathode by a factor M of perhaps a million. These electrons arriving at the anode (10) alter the charge on its stray capacitance C_s to ground (C_s being the analogue of $C + C_i$ in Chapter 3 and C_{in} in Chapter 7) to produce an output signal of size $v = Q/C_s$. Here Q is the charge arriving, and we have assumed (again as we did with gas-filled detectors) that R (and whatever resistance in the preamplifier is in parallel with it) is large enough so that the time constant with which the signal decays is much larger than whatever time is required to effectively charge C_s. Because of the inherent multiplication in the photomultiplier the output voltage v will usually be of a size which requires only modest external amplification before being processed and recorded. Note that since the number of photons in the scintillator is closely proportional to the energy of the incident gamma, it follows that the number of photoelectrons released from the photocathode, and ultimately the charge arriving at the anode and hence the output voltage signal, is also proportional to this energy. The scintillation counter is thus one of the devices in which the output carries energy information, and with which therefore we can perform spectrometry of gamma rays. (One should note in passing that heavy particles like protons and alpha particles are generally less efficient in producing scintillation light than electrons — electrons of course being the means through which gamma rays interact. Depending on the type of scintillator in use, an alpha particle may produce anything from nearly 100 per cent down to only 10 per cent of the light output produced by an electron of similar energy. The linear relation between light output and energy is also less well obeyed for heavy particles than for electrons. Both these effects are attributed to the higher density of ionization associated with the heavy particles interfering with the scintillation process.)

The voltage between dynodes necessary to produce the appropriate multiplication is typically 100 V, thus requiring an overall supply of 1000 V or more. Naturally, as we are dealing with small scintillations and a highly sensitive photoelectric device, the whole assembly must be enclosed in a light-tight outer container, with a window opaque to light but thin enough to allow the entry of the radiation to be detected. The scintillator crystal itself must also have a reflective surround (usually white matt) to allow collection of those photons in the scintillator which move away from the photocathode initially. While crystals and photomultipliers can differ widely in their dimensions, a modest-sized assembly for general-purpose counting might have a cylindrical scintillator of say 50 mm diameter by 50 mm high, seated on a photomultiplier of similar diameter and 125 mm in height. Well-type crystals also exist. These have a cylindrical hole extending from the centre of the top face to just below the centre of the crystal, thus allowing a small source placed at the bottom of the well to be counted with high efficiency.

4.2 Inorganic scintillators

4.2.1 *Principles and materials*

Scintillators can be either inorganic or organic. Inorganic scintillators must be grown in a crystalline form and most must have small quantities of an 'activator' added to perform efficiently. Organic scintillators suffer from neither of these restrictions, but have their own peculiar disadvantages in the form of lower light output and poor suitability for gamma-ray spectrometry. We start with the inorganics. The most generally used scintillator is still sodium iodide activated with thallium — which we write NaI(Tl). Apart from activated zinc sulphide, it is the most efficient converter of radiation energy to visible light of all scintillators, and because of the relatively high atomic number of its iodine atom ($Z = 53$) is well suited for gamma-ray detection and spectrometry. Indeed until the advent of the semiconductor detector sodium iodide detectors reigned supreme in this field. However, it does suffer from the disadvantage that it is hygroscopic and must be encapsulated in a metal can (with a glass end window) to protect it.

Thallium-activated caesium iodide, CsI(Tl), on the other hand, does not require encapsulation, and its stopping power for gammas, because of the replacement of sodium by caesium, is somewhat better than NaI(Tl). It also produces more photons for a given gamma energy input, but unlike NaI(Tl) and most other scintillators, its light output is not at the blue end of the visible spectrum but more towards the centre, where the efficiency of photocathodes for converting photons into electrons is lower. Consequently the output pulse height under similar conditions is only about 60 per cent of that for the sodium iodide case. (Interestingly the caesium iodide light is much better matched to the response of silicon photodiodes, and this combination is now being used in high-energy physics where the large internal amplification of the photomultiplier is not required, and the cost saving by employing photodiodes in multidetector arrays is important.) Caesium iodide can also be used for pulse-shape discrimination, a technique which we will discuss in the applications Section 4.5.2 and Chapter 6. Sodium-activated caesium iodide, CsI(Na), has much in common with NaI(Tl), including the magnitude and wavelength of its light output, but rather strangely has never achieved a comparable popularity in spite of its better stopping power.

There is an obvious advantage in gamma-ray detection in looking for scintillators with high-atomic-number components, so that higher detection efficiencies may be obtained with the same scintillator size, or alternatively, as is often required in medical imaging where many closely packed detectors are used, so that the same efficiencies can be obtained with smaller scintillators. Bismuth germanate, $Bi_4Ge_3O_{12}$ — often referred to shortly as BGO — has an absorption coefficient more than twice as large as that of sodium iodide due to the high atomic number ($Z = 83$) of its bismuth atoms. It is also non-hygroscopic. The pulse height from a BGO–photomultiplier combination will be less than 20 per cent of that with sodium iodide, due both to the lower photon production and a light output, as with CsI(Tl), not well matched to high-efficiency photocathodes. However, this lower light output is not usually a serious disadvantage, and has not prevented BGO replacing NaI(Tl) in many of the imaging applications mentioned above.

Gadolinium silicate doped with cerium, $Gd_2SiO_5(Ce)$, or GSO for short, has a gamma absorption intermediate between sodium iodide and BGO due to its gadolinium component ($Z = 64$). Its light output is comparable to that of BGO, but appears at the blue end of the spectrum, and thus output pulse heights are appreciably better — but still less than one-third of sodium iodide. GSO has the added bonus of having a short decay time for the scintillation flash ($\tau = 60$ ns), much better than sodium iodide (230 ns) or BGO (300 ns), thus allowing higher counting rates to be achieved without the pulses 'piling up' on one another. Barium fluoride (BaF_2) is somewhat better than sodium iodide as regards gamma absorption (Z for barium is 56) and produces pulse heights comparable with or somewhat less than those from BGO. It has a main decay time of 620 ns, but the decay also contains a component (20 per cent) decaying very much faster (0.8 ns). This latter decay time, which is smaller than even those from the plastic phosphors we shall talk about later, means that BaF_2 can be used at extremely high gamma count rates and for fast timing. (There are, however, significant difficulties involved with such fast applications, inasmuch as the slow component must be removed or reduced electronically, and in addition special photomultipliers with quartz windows must be employed to pass the wavelengths around 225 nm where the fast component appears.)

We turn now to inorganic scintillators other than those orientated towards gamma-ray detection. Silver- or copper-activated zinc sulphide (the modern replacement for the naturally occurring material) has a light output appreciably better than NaI(Tl). It is only available, however, in powder form, and as this is not transparent in bulk, thin layers only can be employed. Its use is thus restricted to alpha or other heavy-particle detection, either in fixed installations or in portable monitors. Europium-activated lithium iodide, LiI(Eu), is important as a slow neutron detector. This arises because of the presence of the ^6Li isotope as one of the components (≈ 7 per cent) of ordinary lithium. As we mentioned in Chapter 2, slow neutron bombardment of ^6Li results in an (n, α) reaction, $^6Li(n, \alpha)^3H$ with a release of nearly 5 MeV energy shared between the alpha particle and the triton. A large scintillation thus results from the capture of a neutron. The efficiency of the device may be improved by using lithium enriched in the ^6Li isotope. At a ^6Li concentration of 96 per cent, which is commercially available, the absorption coefficient for thermal neutrons is so large (1.67 mm^{-1}) that a crystal slice of only a few millimetres thick is sufficient to effectively absorb a collimated beam of such neutrons. Glass scintillators activated with cerium also exist, and are useful in hostile environments. When doped with lithium they can also be used for neutron counting.

4.2.2 *Activators and decay times in scintillation crystals*

As not all transparent crystals will act as scintillators, there are obviously other interactions apart from light emission through which electrons and holes formed by incident radiation can de-excite. Indeed even for those which do scintillate, the dominant mode of de-excitation may in addition depend on temperature. For example, NaI without any added activators is not usable as a scintillator at room temperatures: at liquid nitrogen temperatures it performs better than activated NaI does in normal use at room temperatures. CsI shows a similar effect. So in the case of many crystal materials for use at room temperature we must provide activators to ensure that as large a fraction as possible of the energy available when the electrons and holes are formed will be channelled into light-producing transitions. A number of such activators have already been mentioned in our listing of scintillation crystals.

Let us look in a little more detail at the most common combination — thallium-activated sodium iodide. The light emission produced by radiation in cooled unactivated sodium iodide is in a fairly narrow peak centred around 300 nm, while that from the activated crystal is in a broader peak centred on 400 nm. This implies that the emission is no longer coming from the Na^+ ions but from the Tl^+ activator ions which have replaced some of these Na^+ ions in the crystalline lattice — even though the ratio of thallium to sodium ions is only of the order of one to a thousand. Clearly a very efficient transfer of energy to the activator ions is in operation. The actual mechanism of this transfer is complex and at least three modes of action exist. Details can be found in reference 1. It is sufficient here to note that an electron and a hole produced by radiation can be captured (in either order, or together as an exciton) by the thallium ion, to leave it in an excited state. This excited state then returns with its own characteristic decay time to the ground state, emitting a photon with an energy corresponding to the difference between the ground state and upper level (or band of levels). This is about 3 eV, corresponding to the 400 nm emission we noted earlier. As the sodium iodide energy gap is about 7 eV, clearly some of the initial energy of the electron and hole has been expended in non-radiative transitions (see Figure 4.2). This latter is in fact helpful in that the emitted photons have not now enough energy to produce further electron–hole pairs, and thus can escape fairly freely from the crystal — apart from the (admittedly small) chance of their being lost in a non-radiative process. We shall see later that a 3 eV photon, corresponding to a wavelength of 400 nm, has a much higher probability of producing an electron from the normal photomultiplier cathode than would one of 7 eV. So the thallium activator in fact produces a number of desirable effects.

We can make a simple but instructive calculation on the transformation of the energy of the original gamma ray into 400 nm light. It is a useful empirical relation for many insulators and semiconductors that about one third of the initial energy deposited in the crystal goes to produce electron–hole pairs, with the remainder taken up by the lattice. Now for NaI(Tl) the band gap is, as we mentioned, about 7 eV, so we expect to have to

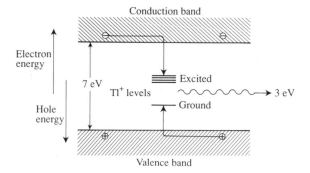

Figure 4.2 Energy transfer in NaI(Tl). (Energy band diagrams such as this are discussed more fully in Section 5.2.)

expend about three times this — or about 21 eV — to produce an electron–hole pair. Now direct measurement of the light output from the scintillator under gamma-ray bombardment establishes that just over 22 eV are needed on average to produce a photon — not much more than that for an electron–hole pair. Clearly then very few pairs are lost to processes other than photon production, and thus the thallium activator mechanism in NaI is confirmed as an excellent one for channelling the electron–hole energy into detectable light. Of course the calculation we have made must not be taken too literally because of the approximate nature of the one-third rule; however, our general conclusion about the excellence of the thallium activator mechanism is not in doubt.

The decay time of the excited state of the Tl$^+$ ion is about one-quarter of a microsecond, and because the time taken for the formation of the excited state is by comparison negligible, the light output decays exponentially, i.e. as $\exp(-t/\tau)$, with τ having this same one-quarter microsecond value. Indeed just about all of the inorganic phosphors have decay times of around a microsecond or an appreciable fraction of a microsecond. This means that we will have to wait in the case of NaI(Tl) say for a few microseconds to collect most of the light from the flash, and we should therefore select R (Figure 4.1) so that RC_s is at least this value — remembering to take account if necessary of the input resistance of the preamplifier. However, for $C_s = 20\,\text{pF}$ say, a value for R of 150 kΩ or more would give the appropriate value of RC_s, and as the preamplifier input resistance could be an order of magnitude or more larger, we can neglect it in this case. Note also that the fall time of the photomultiplier output pulse is now appreciably shorter than that of any gas-filled device, so the scintillation counter has the ability to deal with very fast count rates. Indeed when we come to discuss organic scintillators we shall find they have decay times more than two orders of magnitude shorter than those we are talking about here, so extraordinarily high count rates can be achieved with scintillation devices. We shall return again to the question of pulse shapes when we discuss photomultiplier action. In particular we shall deal with the response time of the photomultiplier itself, which in our discussion above we implicitly assumed was able to cope with the fast rise which precedes the exponential decay of the flash.

4.3 Organic scintillators

The general principles governing the production of light in an organic scintillator are the same as those for an inorganic crystal. In the organic case also a fraction of the energy of the incident radiation appears as excitation energy in the scintillator. Some of this excitation energy is removed in non-radiative transitions and the remainder appears as photons with energy less than that of the original excitation. The scintillator is again transparent to this fluorescent radiation. The actual mechanism of the light production is, however, very different for the two kinds of scintillator. In the organic case the molecule itself is the primary agent for the energy conversion, and neither activator nor crystal structure is required — the latter a most important attribute for some applications. So while organic materials are sometimes used in crystal form, they still perform well, as we shall see, when dissolved in a suitable solvent or even when the solution is polymerized into a solid solution or 'plastic phosphor'. Inorganic scintillator crystals, from what we have said earlier, will lose their ability to operate once their crystal structure is gone.

Figure 4.3 is the usual diagram showing how, in an organic scintillator molecule, the conversion to light of energy from the incident radiation is accomplished. The letters G and E refer to the ground and excited electronic states of the molecule. Vibrational energy levels of these states are indicated by the horizontal lines. The 'bond length' along the x-axis is intended to represent a relevant interatomic distance. The upper curve is shown displaced to the right to longer bond lengths because bonding in the excited state is usually weaker. The molecule will normally exist in the ground electronic state at the lowest vibrational level. When the molecule is excited with energy from the incident radiation a transition takes place as shown by the left-hand vertical arrow. The arrow is vertical because the Franck–Condon principle applies. This states that because of the very large mass of the nuclei compared with that of the electron, the transition takes place before the nuclei have time to move, that is at constant bond length. The molecule thus finds itself in an excited vibrational level of the upper state. It can quickly work its way down to the lowest vibrational level by thermal interaction with its

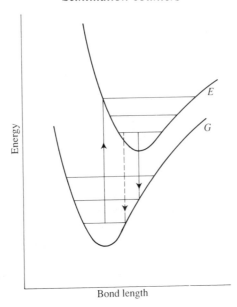

Figure 4.3 Excitation and de-excitation in an organic scintillator molecule.

neighbours. From there it can drop, with a characteristic decay time, to the ground electronic state, as shown by the right-hand full vertical arrow, emitting a visible or near-ultraviolet photon in the process. It then, as before, quickly discharges the remaining vibrational energy it finds itself with, and returns to its original condition. The photon emitted in the transition has insufficient energy (at least in our simple picture) to be reabsorbed in the scintillator, and so escapes. Other transitions are also possible such as that shown by the broken arrow, so we might expect an emission spectrum of the light to show a series of peaks, as is in fact experimentally found. Clearly the sequence we have just described will fail if there exists a raditionless transition which can compete successfully with the fluorescent photon emission. Such a transition could be indicated symbolically in Figure 4.3 if the G and E curves actually came in contact at some point on the right of the diagram. Dissociation of the molecule is another radiationless route for de-excitation, as we saw in another context with gas-filled detectors.

There are quite a few organic substances which can successfully convert incident radiation energy to light of wavelength suitable for efficient detection by photomultiplier cathodes: anthracene and stilbene are the most common for use in crystal form, while terphenyl is one of the most popular for liquid and plastic phosphor applications. The stable benzene ring structure with its shared electrons is a common feature of these compounds, and appears to be a very appropriate one for scintillator molecules. The decay time for anthracene is about 30 ns, that for stilbene about 5 ns, while for various liquid and plastic phosphors the values range from about 4 ns down to less than 2 ns. These are remarkably low values compared with most of the inorganic scintillators, and mean that liquid and plastic phosphors are excellent devices for fast counting and timing applications. Usually along with this fast decaying output, a smaller, more slowly decaying component also exists (which arises from states not shown in our simplified Figure 4.3). Normally this slower output is a nuisance, but it can be turned to good account in that the ratio of the amounts of the two components depends on the rate of energy loss of the incident particle. This allows us to distinguish between a pulse from, say, an electron and that from a proton, even though the actual pulse heights may be the same. The use of such 'pulse shape discrimination' is of obvious importance for example in attempting to detect fast neutrons (producing knock-on protons in the detector) in a situation where many gamma rays (producing electrons) are also present. We shall refer to this technique in a little more detail later.

The pulse height from anthracene is somewhat under half that from NaI(Tl) for a given energy deposited: that for stilbene is down by half again. Liquid and plastic phosphors have, depending on their constituents, outputs intermediate between stilbene and anthracene. Organic scintillators, because of their hydrogen content, are, as we implied above, often used for fast neutron detection via knock-on protons. However, because of the absence of any high-atomic-number atoms in their make up, they are not particularly efficient for gamma counting, and almost useless for gamma spectrometry. We shall be returning to a discussion of these and other applications later in the chapter.

Before we conclude this section on organic scintillators an interesting physical fact concerning liquid and plastic phosphors is worth noting. With the concentrations found experimentally to give the best light output, there may be only of the order of 1 per cent or less of the total number of molecules in the detector which are actual scintillator molecules. As the energy of the incident radiation is clearly absorbed almost entirely by the solvent (or plastic), while the final emitted light is characteristic of the solute, there must exist a very efficient mechanism (whether by photons or other means) for transferring the original excitation energy from the solvent or plastic molecules to the scintillator molecules. (This provides an interesting formal analogy with the transfer of the excitation in an inorganic crystal to the few activator ions, although the mechanisms are undoubtedly quite different in the two cases.) The active constituent in a liquid or plastic phosphor may perform satisfactorily in all respects except that the spectrum of its emitted light may be of too short a wavelength to match the response spectrum of the photomultiplier. In such cases a very small amount of a second solute may be added as a 'wavelength shifter' to absorb the light from the scintillator molecules and re-emit it at longer wavelengths. An overall gain in the number of electrons emitted by the cathode can thus be obtained with a corresponding increase in the output signal.

4.4 Photomultipliers

4.4.1 *Constructional details*

Although a number of photosensitive devices (including photodiodes and light-sensitive semiconductors) have been used to monitor the light from scintillation phosphors, it is the photomultiplier which is most frequently employed. We have already given a brief account of its mode of action and we now adduce some further information on construction and characteristics.

Photomultipliers for use in radiation detectors can have four main types of dynode arrangements as shown schematically in Figure 4.4. In the first two shown (Figure 4.4(a), (b)), electrons from a particular dynode are accelerated to the next grid and dynode assembly, pass through the grid and eject electrons from the dynode. The field at a particular dynode surface due to the following dynode is not very large, and so two successive electrons starting from the same point may follow very different paths depending on their initial directions and their (small) emission velocities. In this sense these types of dynodes are 'unfocused', and a significant spread in the arrival time of the electrons at the anode would result from this cause alone (even if the scintillation flash were infinitely short). So for work with fast pulses the linear focused type (c) is preferred. Here the fields at the dynode surfaces are higher, and in addition skilful design of the dynode shapes and positions allows control of the electron paths, and thus a more nearly constant travel time through the multiplier structure. These devices are thus the first choice for fast counting. However, a more complicated focusing system between cathode and first dynode (not shown in Figure 4.4) is needed to ensure efficient direction of the photoelectrons from the whole cathode into the relatively small aperture leading to the first dynode — a problem which is easier with the box and grid type, and particularly so with the venetian blind tube. Linear focused types are by their very nature the most sensitive to disturbance of the electron trajectories if the photomultiplier has to operate in an external magnetic field. The compact focused or circular cage type photomultiplier (Figure 4.4(d)) has a dynode structure rather like a 'rolled up' version of the linear focused dynodes, and as such shares many of the properties of the latter type. Although this compact structure was derived from that of a very early multiplier design it is still in use in a number of devices.

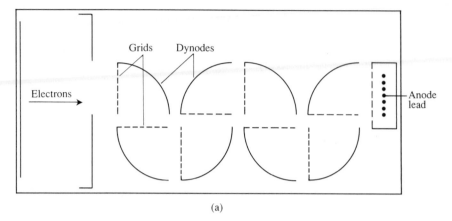

(a)

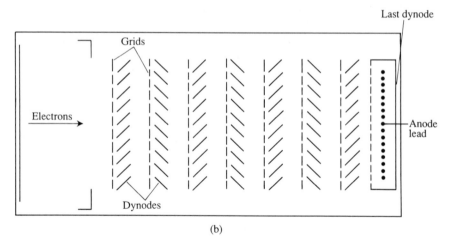

(b)

Figure 4.4 Photomultiplier dynode arrangements: (a) box and grid type; (b) venetian blind type;

It has the disadvantage that it is not possible to easily increase the number of dynodes (from the standard ten-dynode design shown) if a similar photomultiplier, but with higher gain, is required.

The design of the collector, or anode, or a photomultiplier is worth a moment's consideration. In our original simple picture (Figure 4.1) we showed it as a solid plate structure, but in practice this is not usually so. If, for example, one inspects, say, an actual venetian blind photomultiplier — or

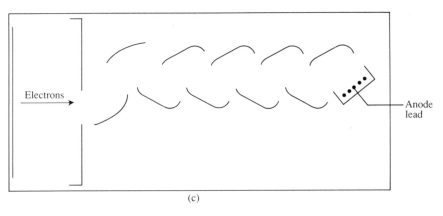

(c)

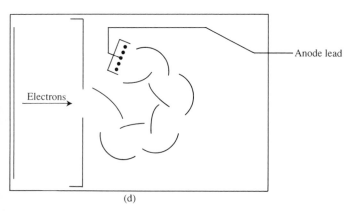

(d)

Figure 4.4 (c) linear focused type; (d) compact focused (or circular cage) type. (Dynode leads omitted for clarity.)

even looks at Figure 4.4(b) — one might at first sight identify the last solid plate electrode in the chain as the anode. In fact it is the last dynode, and the anode itself is a grid close to the last dynode, but between it and the penultimate dynode. The electrons from this penultimate dynode pass *through* the relatively coarse anode grid to strike the last dynode: the secondaries from there come *back* and are collected by the anode. The reason for this apparently curious arrangement is that it is the only logical way in which the anode and the last dynode can be placed sufficiently close together, and it is used, except in special circumstances, in all four types of photomultiplier.

There are two advantages in this close anode to last dynode spacing. Firstly, just as in the simple ionization chamber, where the rise-time of the output signal is determined by the transit time of the electrons and ions between the plates, so in the multiplier the electron transit time between the last dynode and anode could in principle have a major influence on the signal rise-time. However, because of the high velocity of the electrons it is not too difficult to provide spacings giving transit times below 1 ns, leaving other factors, to be discussed later in this chapter, to place a limit on the photo-multiplier rise-time. Secondly this close spacing is also advantageous in producing (for a given voltage) a large field between the last dynode and the anode, and thus preventing space charge effects at high anode currents, which would otherwise reduce the system gain. It might seem strange to be concerned with high currents in the context of an arriving charge at the anode which even after multiplication is only of moderate size. However, we must remember that this charge may arrive in a very short time interval, thus producing a very high *instantaneous* value of the current. We shall return to this point later when we discuss the current mode of photomultiplier operation.

4.4.2 *Photocathodes*

A photocathode for a photomultiplier must meet a number of criteria, some of which unfortunately are only partly compatible. These are:

1. The energy to produce an electron from the material of which it is made must be low enough for operation with visible light.
2. The photocathode thickness must be such that all, or at least a reasonable fraction, of the light falling on it is absorbed.
3. Despite requirement (2), there is nothing to be gained in making the photocathode thicker than that from which electrons can escape to the surface while still retaining enough energy for an appreciable fraction of them to surmount the barrier there. So a material with a large 'escape depth' is required.

While requirement (1) might point to the use of a metal with a low work function like caesium, it turns out that, primarily because of the large free-electron population in metals, the escape depth is much smaller than in semiconductors. Under criterion (3), then, semiconductors are preferred. Even with semiconductors the escape depth is still so small (a few tens of nanometres) that photomultiplier cathodes (at least in our 'head-on' photo-multipliers used in scintillation counting) are always thin enough to be semi-transparent. This small escape depth is an important factor in limiting our efforts under criterion (2) to convert as much as possible of the incoming light energy into electrons.

One of the commonest photocathode materials which has been used is, in line with what we have been saying, the semiconductor caesium antimonide (Cs_3Sb). A variant on this which is rather more efficient in electron production is the 'bialkali' photocathode (K_2CsSb), and Figure 4.5 shows the spectral response of these two types of cathode. (The code S11 indicates Cs_3Sb.) Notice how they both peak at around 400 nm, which makes them a good match for the light from most scintillators. The cut-off on the long-wavelength side is determined by the minimum energy to produce an electron: the cut-off on the low-wavelength side is usually due to the reduction in transmission of the glass window at these wavelengths. Note as a consequence how a quartz window (broken curves) can extend the response into the ultraviolet region if required. Incidentally it should be noted that while photocathodes are available with better sensitivity at the red end of the spectrum than in the case we have illustrated – the so-called 'trialkali' or 'multialkali' cathode – there is no necessity in scintillation counting to utilize such a cathode. Indeed they can be a positive liability in that good red sensitivity is usually associated with increased dark counts – that is the emission of electrons from the photocathode in the absence of incident light – as we shall discuss in a little more detail later.

The y-axis in Figure 4.5 is calibrated in terms of quantum efficiency (QE). This quantity is defined simply as the ratio (expressed either as a fraction or a percentage) of the number of electrons emerging from the cathode (on the dynode side of course!) to the number of incident photons. Alternatively it can be defined as the probability that any particular photon will produce a

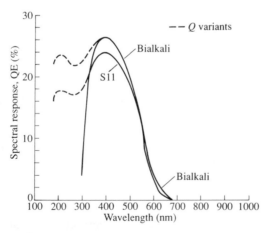

Figure 4.5 Spectral response of S11 and bialkali photocathodes. (Reproduced by courtesy of THORN EMI.)

photoelectron. (The *y*-axis is sometimes shown calibrated in mA W^{-1}, that is giving the output current from the cathode per watt of incident light, but the QE notation is more appropriate for our purposes.) The peak quantum efficiency shown in Figure 4.5 of around 25 per cent is typical of photocathodes used in scintillation counting, although values up to 30 per cent can be achieved.

4.4.3 *Dynodes*

One might expect that the active material on dynodes, which is required to produce secondary electrons under electron bombardment, could be the same as that for the photocathode, which produces electrons under photon bombardment. This is indeed true, and caesium antimonide (on a metal substrate) is widely used. Oxidized beryllium or beryllium copper, although giving rather less gain than caesium antimonide, is also a stable and popular dynode material.

The gain of a dynode, that is the average number of secondaries per incident primary (called variously *m* or δ) is at first an almost linear function of the primary electron energy, but the curve then flattens out and reaches a maximum as the primary electrons go deeper into the material and the secondaries have difficulty in escaping (see Figure 4.6). At a modest interdynode voltage of 100 V the gain *m* of a caesium antimonide dynode would be about four, that of a beryllium copper one about three. (The overall gain of the photomultiplier *M*, is clearly m^n, where *n* is the number of dynodes present. *M* of course, as we remarked earlier, can easily be made 10^6 or more, but note also how rapidly it changes with applied voltage. For very low

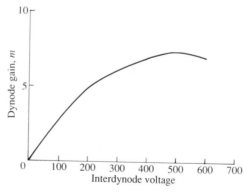

Figure 4.6 Gain of beryllium copper dynode versus interdynode voltage. (Reproduced by courtesy of THORN EMI.)

interdynode voltages m is a linear function of this voltage – and therefore of the overall photomultiplier voltage. Thus for a 12-stage tube ($n = 12$) we would expect the gain to vary as the twelfth power of the overall voltage. For a voltage of the size needed for normal photomultiplier operation the dynode gain, and thus the overall gain, varies more slowly, but even a tenth-power law means a 10 per cent change in gain for a 1 per cent change in applied voltage. A high-quality power supply is clearly required for stable operation and resulting reproducibility in pulse height measurements.)

There is an obvious advantage in a large dynode gain in that a given overall multiplication factor can be achieved with fewer dynodes and thus a more compact device. There is, however, a more subtle advantage. The figures we previously quoted for the dynode gain were only mean values. If, for example, the gain were three then a particular primary electron can produce any one of a whole statistical distribution of numbers of secondaries centred on mean three – including zero. This is one of the reasons why in a scintillation counter gamma rays of a given energy do not produce unique output pulses at the photomultiplier anode, but rather give a distribution of sizes. Clearly high-gain dynodes will help to reduce this spread, and in this connection the first dynode is of paramount importance, operating as it does at the place where the electron numbers are smallest and statistical effects greatest. For this reason it is usually sufficient with photomultipliers intended for high-resolution spectrometry to operate just the first dynode at a higher gain than all the rest. We shall be dealing with this in a more quantitative manner in Chapter 7.

The obvious way to achieve a higher first dynode gain is to increase the cathode to first dynode voltage – provided of course one does not pass the maximum in the curve of Figure 4.6. (Indeed a higher cathode to first dynode voltage is often specified for a quite different reason – to improve the collection of electrons from the cathode.) For caesium antimonide one can obtain a gain of about 12 at 600 V, and special processing can bring this figure up to nearly 20. If one wishes to push the matter further – as may be required for example when counting single photons – then with a special gallium phosphide (GaP) dynode a gain of 35 can be obtained at 600 V (and even higher figures if a greater voltage can conveniently be used).

The reason for the exceptional gains with GaP dynodes is the removal of the barrier which electrons normally encounter on emerging from the semiconducting material. This is achieved by evaporating caesium on the surface of gallium phosphide which has been heavily doped to be a p-type material (in the usual semiconductor sense), thus producing what is known as 'band bending' at the GaP–Cs interface. The mechanism can be understood simply as follows. Figure 4.7(a) shows a conventional semiconductor energy band diagram, with conduction and valence bands separated by a band gap E_G. The Fermi level E_F, which characterizes the occupancy of the energy levels

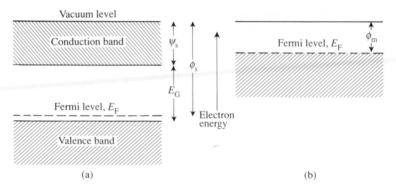

Figure 4.7 Band diagrams and Fermi levels for (a) a semiconductor and (b) a metal.

by electrons, can be anywhere in this band gap, depending on the semiconductor doping. We have shown it close to the valence band, where it would be in a well-doped p-type semiconductor, as this will be useful in the discussion to follow. The work function ϕ_s (where the suffix s is for semiconductor) is defined as shown, being the difference between the vacuum level and the Fermi level, while the electron affinity ψ_s is the difference between the vacuum level and the bottom of the conduction band. The latter is a measure of the additional energy an electron raised to the conduction band must have to reach the vacuum level. The diagram for a metal in Figure 4.7(b) is much simpler and only needs the work function, ϕ_m, to define it. The general scale of both diagrams is in the eV range: for example E_G for gallium phosphide is just over $2 \, eV$.

The end product from the entry of a primary electron into a semiconductor dynode is a large number of further electrons with a spread of energies centred typically a volt or so above the vacuum level. So an electron formed near the surface is likely to have enough energy to emerge through it into vacuum. Normally, however, one formed sufficiently deep in a dynode material will, through collisions, have lost energy and dropped below the vacuum level before it arrives at the surface. It thus can not emerge. The time available for it to get out has been estimated at about 1 ps. Before we consider the effect of bringing together the band structures of Figure 4.7(a), GaP say, and 4.7(b), Cs say, it might be helpful to consider briefly a diagram with which the reader may be familiar, and which will be treated in more detail in Chapter 5, that for the junction between p- and n-type materials to form a semiconductor diode. This is shown in Figure 4.8. Clearly the 'band bending' we mentioned earlier exists here, and indeed we shall find that the result of evaporating caesium on to p-type gallium phosphide is to produce an effect

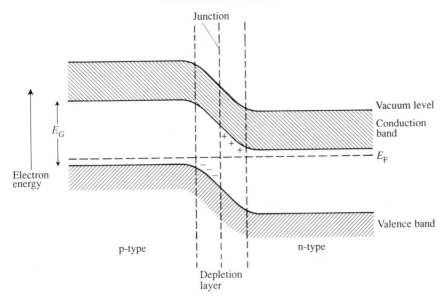

Figure 4.8 Band diagram for a semiconductor p–n junction.

on the GaP energy bands rather similar to that in Figure 4.8 to the left of the central vertical broken line. (Note also in Figure 4.8 the fixed positive and negative charges in the depletion layer as a result of charge redistribution when the Fermi levels equalized.)

Let us then perform the hypothetical experiment of bringing together the metal and semiconductor shown in Figures 4.7(a) and 4.7(b) (and we shall enquire shortly whether our simple procedures may need modification). To achieve the type of band bending we were talking about one needs to have the work function of the metal smaller than that of the semiconductor. However, to completely remove the barrier at the surface requires (as we shall see in a moment) the even stricter condition that the work function of the metal be smaller than the band gap of the heavily doped p-type semi-conductor employed. This latter condition is well fulfilled where caesium is used with gallium phosphide with values of $\phi_m = 1.2\,\text{eV}$ and $E_G = 2.2\,\text{eV}$, respectively. The alternative name for the Fermi level, the chemical poten-tial, indicates its role in junction theory, in that when equilibrium is estab-lished the Fermi levels on both sides of the junction must be coincident. This is achieved here by electrons spilling over from the metal (leaving it with a thin layer of positive charge) into the semiconductor, where the electrons and holes combine, leaving a fixed negatively charged depletion layer of finite width. This is shown in Figure 4.9, which is a composite of Figure 4.7(a) and

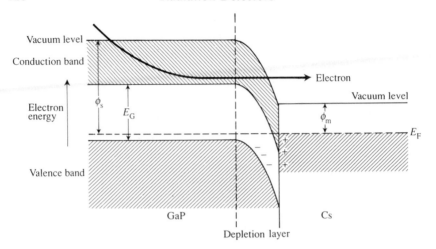

Figure 4.9 Band diagram for a GaP–Cs interface.

4.7(b), with the latter moved bodily downwards until the Fermi levels coincide. The vacuum levels on both sides of the junction can now be joined across the depletion layer with a curve as shown, and the other two levels also drawn in to give the completed picture with its characteristic band bending. The bending of course is just another way of saying an electric field now exists, preventing the movement of any further electrons from the metal into the semiconductor across the interface, but helping the movement of any minority electrons from the semiconductor into the metal. (The reader might like to check that if the work function of the metal is *greater* than that of the semiconductor, a very different type of diagram will result.)

Note what we have now achieved. Because of our extra condition that the work function of the metal be less even than the band gap in the semiconductor (and because the Fermi level in the heavily doped p-type GaP almost coincides with the bottom of the energy gap) the bottom of the conduction band now lies *above* the vacuum level in the caesium. Consequently electrons even at the bottom of the conduction band approaching the GaP/Cs interface find no barrier there, but instead can pass freely into the thin caesium layer and thence into the vacuum. For obvious reasons this gallium phosphide/caesium combination is said to have 'negative electron affinity' (defined numerically as the amount the bottom of the conduction band is above the vacuum level in the caesium), and this clearly leads to a much larger yield of secondary electrons than with conventional materials.

An electron arriving at the bottom of the conduction band in GaP has around 100 ps before it is lost by recombination with a hole. So compared with its counterpart in the conventional situation an electron in the negative affinity case has two orders of magnitude longer to diffuse around, with the continuing possibility of escaping, before being lost. Putting this another way, the depth from which an electron can escape (which is of course not the same thing as the total distance it travels in its random walk) can be several hundreds of nanometres when negative affinity exists, compared with only a tenth of this value for the conventional case. It is only fair to add that our discussion on band bending and negative electron affinity has only been at a very simplified level. For example we can not really assume, as we did implicitly in Figure 4.9(a), that the band structure in a semiconductor is the same at its surface as in the bulk of the material. In fact the very presence of the surface, with associated 'surface states' in the normally forbidden band gap, can produce band bending (and a reduction in the affinity) in a p-type material not unlike that described earlier when a thin layer of caesium is applied to GaP. Such an effect is obviously of general importance for both secondary electron emitters and photocathodes. For those readers wishing to delve further into these problems reference 2, for example, can be consulted.

4.4.4 *Dark current and dark count*

When a photomultiplier is operated without any scintillator present (and of course in complete darkness), a small residual anode current — typically a fraction of a nanoampere — is found to still exist, the so-called 'dark current'. This current, whose origin we will examine in a moment, places, because of the fluctuations occurring in it, a limit on the minimum genuine light flux which can be measured by the tube. The dark current may be quantified by giving its mean value at a stated photomultiplier gain, or by specifying the light flux on the cathode which would produce an anode current equal to the dark current observed. However, for photomultipliers used in counting applications it is more appropriate to specify the actual number of dark pulses appearing per second at the anode. This could be as low as several hundred per second for a premium tube with a 50 mm diameter photocathode operated at room temperature. One has to specify this last parameter because for a good-quality photomultiplier the bulk of the dark current in such circumstances is due to thermionic emission of single electrons from the cathode, which is of course a strongly temperature-dependent process.

These dark pulses present no problem if we are, for example, detecting high- or moderate-energy gamma rays with NaI(Tl) detectors, as even a 100 keV photon, fully absorbed, will result in over a thousand photoelectrons leaving the photocathode. The output pulse due to a single electron can thus be easily discriminated against electronically. Photomultipliers are, however, sometimes used (for example in astronomy) to monitor light sources so weak that the photons can be detected singly — in the so-called 'single-photon-counting' mode. In this case fluctuations in the dark count set a limit to the weakest source that can be meaningfully detected in a given counting time. A related problem arises in the scintillation counting of very-low-energy radiation (or with the Cherenkov detectors to be mentioned in Chapter 8). To be precise let us consider a flash in the detector which produces 12 photons. With normal photocathode efficiencies this would produce three photoelectrons on average. However, we can not afford to ignore the statistical fluctuations occurring in this average number of three. We deal with this matter in detail in Chapter 7: suffice it to say here that because of the fluctuations both in the number of photons produced and in their conversion to electrons at the photocathode, quite often only one or two electrons are produced (as well of course as larger numbers to bring the average to three!). The background of single electrons is thus of continuing importance. The problem is made even more complicated by the statistical fluctuations in the multiplication process in the multiplier which produces, as we noted earlier, not a unique size of output pulse for a given number of electrons emitted from the cathode, but a whole distribution of sizes. This spread is most marked for low initial numbers of electrons such as we are talking about here, and means that an appreciable number of the smaller pulses from two- and even three-electron events are less in size than the larger pulses from single electron events (see Figure 4.10). The upshot of all this is that we must either count all the output pulses (including those originating from a single electron) and accept the statistical implication of having to subtract off the dark pulses, or else set an electronic discriminator on the output to eliminate the bulk of these dark pulses, recognizing that in doing so we are eliminating also an appreciable fraction of the genuine counts with its corresponding statistical implication.

Clearly methods to remove or substantially reduce this dark count are needed. One way is to use two photomultipliers 'in coincidence', that is with

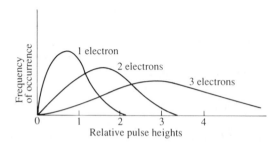

Figure 4.10 Computed pulse-height distributions for equal numbers of output pulses initiated respectively by one, two, and three electrons. (Note: because the distributions are asymmetrical, the means do not exactly coincide with the peaks.)

each looking at opposite faces of the scintillator, and counts only being recorded if one or more electrons are produced at each photocathode simultaneously. This will of course eliminate the vast majority of the dark counts (except for the very few cases where, fortuitously, dark pulses occur simultaneously in both tubes), although at the expense of dividing the photons from the flash into two groups. An alternative solution, if the experimental arrangement will permit it, is to use a single photomultiplier which is cooled, usually thermoelectrically. Thermionic emission is an exponential function of temperature, and cooling to around $-30°C$ will reduce the dark thermionic emission to negligible amounts.

The small number of counts remaining are found to be large multiphoton pulses. These are due to Cherenkov interactions (see Chapter 8) in the glass envelope of the tube, and particularly in the thick faceplate, caused by radiations from small amounts of radioactive materials (potassium, uranium, and thorium) occurring naturally in the glass. Cosmic rays and local radioactivity can produce similar Cherenkov effects. This concludes our listing of sources of counts produced in a photomultiplier even with no scintillator present. These represent an ultimate limiting factor in the counting of all low-level light sources, including scintillation counting of low-activity radioactive sources, particularly those emitting low-energy radiations.

Naturally with a scintillator *present* we will also get ordinary background counts from any cosmic radiations that have entered the scintillator in spite of whatever heavy shielding and anticoincidence arrangements (see Chapter 6) we have provided. Further background counts can also come from minute traces of radioactive contaminants in the shielding and in inorganic scintillators, while radiations from the photomultiplier faceplate and walls will be more efficiently counted by the scintillator than by the Cherenkov effects in the glass we mentioned previously. Just what in practice is the limiting factor depends very much on the actual experimental arrangements, type of photomultiplier, scintillation size, etc. To be fair one should conclude by emphasizing that our discussion in the preceding paragraphs has been most relevant to situations where the scintillation counting method is being pushed to its limits; for counting or spectrometry of moderate-activity sources emitting radiations of moderate energy, the background due to factors like dark counts and radioactivity in the multiplier envelope is not a critical consideration.

4.4.5 *Voltage and current modes*

In all our discussions on scintillation counting up to now, we have implicitly assumed that we were operating the photomultiplier in what is known as the 'voltage' or 'charge' mode. (We will only make use of the former term, as the latter is somewhat ambiguous in the context of the terminology in Chapter

6.) This was the arrangement we discussed in our introductory remarks in connection with Figure 4.1, and results in an output voltage proportional to Q, the arriving charge, and thus to the incident gamma energy. However, it is also possible to operate in the 'current' mode and a detailed analysis of both these arrangements will be given in Chapter 6. Here therefore we will treat the matter briefly and in a plausible fashion.

The form of the light output from a scintillator is assumed here to be a simple exponential as shown in Figure 4.11(a), with the decay time τ of the flash lying in the nanosecond to microsecond range depending on the nature of the scintillator. The electron current to the anode of the photomultiplier will follow the same pattern and can be written $i = I_0\exp(-t/\tau)$, where time is being measured from the leading edge of the signal. What happens next depends very much on the value of R the anode load (which we remember from Section 3.2.3 and Figure 3.3 must take account of the resistor at the input of the preamplifier effectively in parallel with it). If R

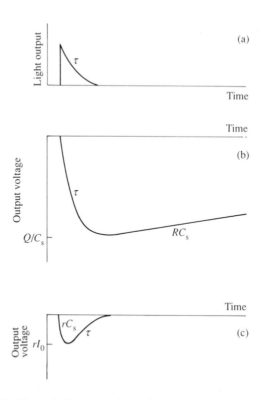

Figure 4.11 Scintillator light output (a) and alternative pulse outputs (b) and (c) at the anode.

is very large (and we will specify more precisely what we mean by this in a moment) then it is the stray capacitance C_s (see Figure 4.1) which plays the dominant role, with negligible signal current following through R. The current i flowing to C_s alters its charge quite rapidly and thus its voltage by an amount $v = Q/C_s$ where, as before, Q is the total arriving charge. (Q in fact can be seen to be $I_0\tau$ by integrating the expression $I_0\exp(-t/\tau)$ from $t = 0$ to $t = \infty$.) The voltage returns to its normal level on a longer time-scale with a characteristic time RC_s. The output signal to the preamplifier (which is the same as the voltage change across C_s) is shown in Figure 4.11(b), and is of course negative because we are collecting electrons at the anode. As well as the long return exponential with time constant RC_s, more detailed analysis in Chapter 6 also shows that the initial fast voltage change occurs with a characteristic time of τ as shown. (In fact its magnitude is initially proportional to $1 - \exp(-t/\tau)$.) It is clear now that to ensure the voltage on C_s effectively reaches a value of Q/C_s before starting to return to its original level, R should be chosen so that $RC_s \gg \tau$. For example if $C_s = 20\,\mathrm{pF}$ then a value of $R = 1\mathrm{M}\Omega$ will make $RC_s = 20\,\mu\mathrm{s}$, which amply fulfils the condition for a sodium iodide scintillator ($\tau = 0.23\,\mu\mathrm{s}$) and even more so for organic or liquid scintillators. (If the condition $RC_s \gg \tau$ is not too well obeyed, then the maximum value of the signal is less than Q/C_s; however, even for $RC_s = 10\tau$ the output reaches 77 per cent of its ideal value.)

If, on the other hand, we reverse our condition and make $RC_s \ll \tau$, then we are in what is called the 'current' mode of operation – the reason for the nomenclature being clear in a moment. Our new condition implies that we have chosen R to be very small so this component now plays the dominant role, with negligible current going to alter the charge on C_s. The current i then flows substantially through the resistor producing a signal voltage across it of magnitude ir or $rI_0\exp(-t/\tau)$ where r is the new reduced value of R. Again detailed analysis shows that the time constant rC_s, now very small, appears in the early part of the signal, as shown in Figure 4.11(c). Note that the maximum voltage reached in the current mode is (unless the photomultiplier gain is increased) very much less than in the voltage mode: in fact the signal heights are in the ratio of rC_s to τ. (We note for future reference in Chapter 6 that the current mode situation can be achieved by reducing the value either of the anode load itself or that of the preamplifier input resistance which is effectively in parallel with it.)

The reason for the name 'current' mode should now be obvious: the signal follows almost exactly the current in the photomultiplier, which in turn is a reproduction of the shape of the scintillation flash. In the 'voltage' mode, on the other hand, the output is 'integrated' on the capacitor to produce a maximum voltage directly proportional to the total arriving charge, and thus to the light emitted from the scintillator, and ultimately to the incoming particle or photon energy. Clearly if we are engaged, for example, in gamma-ray spectrometry—that is in determining the energies of a number of gamma rays coming from a radioactive sample—then the voltage mode with its greater output (and direct link to the gamma energy) is to be preferred. It has the additional advantage of a comparatively long signal fall time (RC_s) which can be adjusted to suit the input requirements of the commercially available amplifiers used in pulse-height analysis systems.

Current pulses on the other hand, although small in amplitude, have the advantage of providing the fastest possible arrangement—rising quickly, as we have pictured them, with the time constant rC_s, and decaying with time constant τ. They are therefore used in circumstances where very fast counting rates are being encountered, or where the accurate determination of a pulse arrival time is essential to an experiment. With such short pulses it is important that they are transmitted on from the photomultiplier down properly terminated coaxial cables, in order to avoid pulse reflections (Section 6.7.2). For that reason a load of $50\,\Omega$ (appropriate to the most commonly used fast signal cable) is normally employed. In that case the value of rC_s would, for our previous value of C_s, be $50 \times 20 \times 10^{-12}$ or 1 ns, which would handsomely satisfy the condition $rC_s \ll \tau$ for a sodium iodide crystal with $\tau = 230$ ns. However, for the fastest counting rates one would obviously use an organic or plastic phosphor with a decay time of a few nanoseconds. In this case it is not possible (unless we can reduce C_s substantially) to satisfactorily fulfil the requisite conditions using the chosen $50\,\Omega$ load, and this will lead to some loss in maximum pulse height.

In addition, in these extreme conditions a further contribution to the rise-time comes from the photomultiplier itself. This occurs because a group of electrons starting simultaneously from different places in the photocathode will not arrive exactly together at the anode, due to their slightly different paths from cathode to first dynode and different subsequent trajectories through the multiplier structure. Even electrons leaving simultaneously from a small area of the cathode will not arrive together because their emission in different directions with small intrinsic velocities will produce slightly different paths. Thus even if the value of rC_s was small enough to be negligible, the output pulse would not be able to follow the sharp initial rise in light output shown in Figure 4.11(a). As we mentioned earlier in Section 4.4.1, linear focused (and circular cage type) photomultipliers, particularly those with a cathode deposited on a faceplate with a concave inner surface to help equalise cathode to first dynode times, are the best performers as regards arrival-time spread. With these types of tubes and various design sophistications beyond the scope of the present book, intrinsic photomultiplier rise-times of around 1 ns can be achieved.

As a final postscript to this discussion one might perhaps imagine that in spite of the presence of a finite rise-time produced by a combination of the factors we have been talking about, it would be possible, by the methods to be mentioned in Chapter 6, to determine the moment of arrival of the peak of the pulse (or any related feature) extremely accurately. Such information would be of importance for example in time-of-flight measurements in nuclear physics. That this is not so is due to the same considerations of varying electron arrival times that we have just been discussing. A series of successive pulses which we would expect to arrive at a fixed time interval after a corresponding reference pulse, will have a 'time jitter' or 'transit time spread' between them because of the small random fluctuation in the moment of arrival of the first electron in each pulse. This jitter will be worst in the most unfavourable statistical case — that of a single electron leaving the cathode — and here it is in the region of half a nanosecond or less. This effect obviously places a limit on the most accurate timing we can provide.

The special design requirements for photomultipliers used for fast counting are beyond the scope of this book. Two practical changes needed in the arrangements shown in Figure 4.1 should, however, be mentioned. First of all we note that in that diagram the photocathode is at earth potential and the anode at the full supply voltage. This is the usual arrangement for the voltage mode, particularly for low-count-rate applications, as an ungrounded cathode may, in the presence of surrounding grounded regions, produce spurious counts due to leakage currents in the glass. However, for the current mode case other considerations are more important, and the grounding point of the voltage supply is reversed, with the bottom of the load resistor R (and the bottom of the dynode chain) being grounded, and the appropriate negative voltage applied to the cathode. As a result of the anode (in the absence of a signal) now being at ground rather than at high voltage, the coupling capacitor C_c of Figure 4.1 can be eliminated. This is desirable because with very fast pulses the small residual inductance unavoidably associated with capacitors makes good matching to the connecting cable difficult, and produces a characteristic distortion ('ringing') in the pulse shape. With C_c removed the anode resistor and that determining the input resistance of the preamplifier can be amalgamated. As we noted earlier, the coaxial cable connecting the photomultipier anode to the preamplifier input must, for fast pulses, be properly terminated (commonly with 50 Ω). This is usually achieved by eliminating the anode resistor and connecting the anode via the cable to a preamplifier having an input resistance of 50 Ω (as we shall see in Figure 4.13).

The other change arises from the fact that we often ask photomultipliers operating in the current mode to produce pulses of size large enough to perform whatever function (timing, coincidence, etc.) is required of them, without additional amplification. This avoids any degradation of the excellent photomultiplier rise-time by that of a following amplifier. As a consequence, relatively high anode currents may be demanded during the pulse. For example if we require a peak output of 1 V across a load of 50 Ω then 20 mA must be supplied, and an electron current of the same magnitude must leave the last dynode. Assuming a dynode gain of say four, only 5 mA of electron current need reach this last dynode from the preceding one, so the balance of 15 mA must be supplied through the resistor chain network. Now it is the usual practice in any voltage divider network, in order to avoid significant voltage changes at the outputs, to have the current through the network at least an order of magnitude greater than any current which it is anticipated will be drawn from it. So we might at first sight plan to have at least 150 mA in the network. However, if we assume an overall photomultiplier voltage of 2000 V (and that would be a modest figure in view of the fact that above average high voltages are commonly used in current pulse work to reduce rise-times) then the power dissipated in the resistor network would be 300 W — a totally unrealistic figure. We are saved of course by the fact that we only require the 15 mA current from the network at the leading edge of the pulse, and this value will decay exponentially with the short time constant of the fast scintillator. Put another way, what is required is some sort of reservoir capable of supplying a total charge of $\int_0^\infty I_0' \exp(-t/\tau)\,dt$ or $I_0'\tau$ (where $I_0' = 15$ mA in our case and τ is, say, 4 ns). This can conveniently be done by placing a capacitor C of appropriate size across the last dynode resistor (i.e. the bottom one in Figure 4.1), thus enabling the value of this, and of course the other, network resistors to be increased substantially over that previously implied: the total power dissipated will fall correspondingly. Let us attempt a numerical calculation on the matter.

Figure 4.12 shows how we would expect the dynode voltage to vary when a train of pulses (assumed in the first instance to be spaced regularly a time T apart) is arriving at the dynode. Every time a pulse occurs a negative charge of amount $Q' = I_0'\tau = 15 \times 10^{-3} \times 4 \times 10^{-9}$ or 60 pC in our example, must be provided by the capacitor C in a time effectively of a few times τ. If T is 10^{-5} s, corresponding to an arrival rate of 10^5 Hz, then with $\tau = 4$ ns,

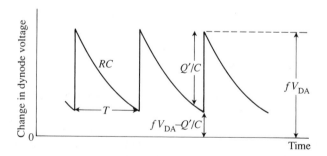

Figure 4.12 Dynode voltage variation with time.

a sharp voltage rise, almost instantaneous on the time-scale of the diagram, and of amount Q'/C occurs, as shown. During the time to the next pulse the dynode voltage returns exponentially towards its resting value with a time constant $R'C$ where R' is the value of the last dynode resistor and C the capacitance across it. (In fact if one draws the appropriate equivalent recharge circuit, the charging resistance is R' in parallel with the sum of all the other dynode resistors, or a little bit less than R' — but we'll just take it as R'.) Before the dynode actually reaches its resting voltage another pulse arrives and the cycle is repeated. In order for the photomultiplier to function correctly we clearly require the maximum excursion of the dynode to be limited to a small fraction f (say 1 per cent) of the dynode to anode voltage, V_{DA}, as we have shown in the figure. The distance between the lowest part of the waveform and the time axis is then clearly $fV_{DA} - (Q'/C)$. Writing down the usual decay equation then for one complete period T we have

$$fV_{DA} - (Q'/C) = fV_{DA} \exp(-T/R'C)$$

or

$$\exp(-T/R'C) = 1 - (Q'/CfV_{DA}). \qquad (4.1)$$

Clearly for a solution to exist we must choose C greater than Q'/fV_{DA} — as is intuitively obvious anyway — and we find as we increase C that R' also increases, but its value levels out for $C \gg Q'/fV_{DA}$. Choosing such a value for C and remembering that $\ln(1-x) \sim -x$ for small x, gives us $T/R'C = Q'/CfV_{DA}$ or $R' = TfV_{DA}/Q'$, independent of C provided the latter is large.

We can now put in some numbers from our example. As we indicated already $Q' = 60\,\text{pC}$, $T = 10^{-5}\text{s}$, $f = 1/100$, and we will take V_{DA} as 200 V. This gives the minimum value for $C = Q'/fV_{DA}$ as 30 pF. Choosing at least ten times this, i.e. 300 pF or more, gives us a value for $R' = TfV_{DA}/Q'$ of 330 kΩ. This allows us to count regularly arriving pulses at a rate of 10^5 Hz without breaking our conditions. For randomly arriving pulses we would have to restrict the mean rate to say 10^4 Hz, or else decrease R' to 33 kΩ. Even with the latter value the steady current flowing in the resistor (and in the chain) is 200 V/33 kΩ or 6 mA, which if the overall voltage is 2000 V gives a total power of 12 W, a manageable figure. Reservoir capacitors are usually added not just to the last dynode, but to the last three or four: clearly the value of C can be progressively reduced as we move back from the last dynode by a factor equal to the dynode gain (four as we have assumed) at each stage. The values of the other resistors in the chain can not of course be altered progressively upwards as our formula might suggest, because their primary role is to set the steady dynode voltages at equal intervals — or whatever variation on this may be required in special circumstances. For the earlier stages of a photomultiplier no external capacitors are required, the stray capacitance of each dynode to ground being quite adequate.

Incidentally all the previous remarks and calculations also apply in principle to a photomultiplier operating in the voltage mode, with say a NaI(Tl) crystal of $\tau = 230$ ns. However, both because the current is integrated on the stray capacitance to produce a charge, and because external amplification appropriate to the modest rise-times used is readily available, conditions are much less critical, as an example will show. Suppose we wish to produce an output voltage of 0.1 V on a stray capacitance of 20 pF, then we need only an integrated charge of 2 pC. So the total drawn from the network at the last dynode is 1.5 pC. The critical value for $C = Q'/fV_{DA}$ is, with a value of V_{DA} more appropriate to the present situation of 100 V, thus 1.5 pF, and we need to use at least ten times this, i.e. 15 pF or more for maximum value of R'. This latter turns out to be 6.7 MΩ for $T = 10^{-5}$ s (regular pulse arrival) or a mean rate of 10^4 Hz (random pulses). We would hardly be able to count faster than this anyhow, because of pulse overlap due to the fall time of a few microseconds which the 0.23 μs rise-time due to the sodium iodide demands. Even if we required a 1 V anode output we could achieve this with $C = 150$ pF or more and $R' = 0.67$ MΩ, or if we preferred we could with these values have a dynode shift of only 0.1 per cent of V_{DA} with an output of 0.1 V. So our problem of network dissipation is easy to solve in the voltage mode, as the network current for $R' = 0.67$ MΩ and $V_{DA} = 100$ V is only 0.15 mA, corresponding to a total power of 0.15 W for an overall voltage of 1000 V.

We close this sub-section by remembering again that for a variety of reasons beyond the scope of our present discussion, the simple potential divider network we have been assuming, consisting of a string of equal resistors, is not always used in practice. Indeed we already know for the reasons given in Section 4.4.3 that the cathode to dynode voltage is usually made considerably larger than the normal interdynode voltage, and thus the resistor between cathode and first dynode can be two or three times the value of the others. Above-average resistor values may also be used in certain circumstances towards the output end of the string. However, none of these types of change interfere with the principles we have been establishing for the prevention of excessive dynode voltage changes during pulses.

4.4.6 *Simultaneous voltage and current mode operation*

We have just discussed voltage and current operation of photomultipliers as two distinct modes, and this is generally true. However, there are occasions when a fast current signal for timing and a slower voltage signal for pulse-height analysis are desirable simultaneously. While there is more than one method of doing this, the option of using the anode to produce a current signal while taking a voltage signal from a dynode is a very simple and convenient one. This is achieved as follows. When we wished to obtain a voltage signal from the anode as in Figure 4.1 we placed a large resistor between the anode and the positive supply, and took off the resulting voltage signal on the stray capacitance via a coupling capacitor: now when we wish to obtain a voltage signal from a dynode we similarly place a large resistor, say 1 MΩ, between it and its normal point of voltage supply (point A in Figure 4.13) and take off the signal through a coupling capacitor C_c to block off the d.c. voltage. We remember when a pulse occurs that a dynode will be a net loser of electrons (rather than a net collector like the anode) so the stray capacitance C_s' *gains* positive charge, and a positive-going signal results. This decays slowly in the usual way with a time constant given by the product of the load resistor (1 MΩ here) and the total stray capacitance associated with the circuit and whatever circuitry is attached to it (say 20 pF in all). The fact that the signal is positive going presents no difficulties as it can always be inverted in the external amplifier if required. The reason why we pick an earlier dynode rather than the more obvious choice of the final one is that current signals are, as explained earlier, much smaller in amplitude than voltage signals. Consequently if we arrange for an appropriate signal at

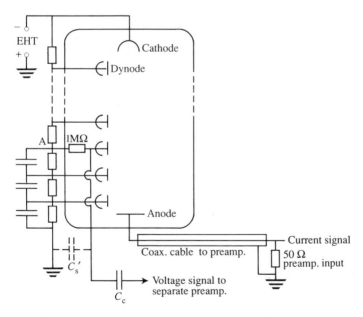

Figure 4.13 Arrangement for simultaneous voltage and current signals. (Note the conventional symbols used for the cathode and for any type of dynode.)

the anode then the change in final dynode voltage when the voltage signal occurred there could be an appreciable fraction of the dynode to anode voltage — the very thing that we were at so much pains to avoid in our previous discussion of dynode network design. By using an earlier dynode we can keep the signal there to say 0.1 V maximum — well below our 1 per cent limit — and use an external amplifier of gain say 100 to produce output signals suitable for pulse-height analysis.

4.4.7 *Continuous dynode multipliers*

So far we have discussed only photomultipliers equipped with a number of discrete dynodes, and indeed this arrangement is still the most common. However, multipliers with a continuous dynode structure have also been available for many years. In one such device secondary electrons emitted at one end of a long dynode strip travel, under the influence of crossed electric and magnetic fields, in a cycloidal path and are returned to the strip further along with enough energy to produce secondaries. The process is then repeated. An increasing cloud of electrons thus moves down to the far end of the strip where they are collected by an anode.

The most popular multiplier arrangement nowadays requires only an electric field. The idea is shown in Figure 4.14. A long narrow tube has a secondary emitting coating on its internal wall, and a potential of the order of 1000 V applied between its ends. An electron entering the left-hand end will strike the secondary-emitting surface and produce say two secondaries. These in turn will move diagonally across the tube to make further collisions

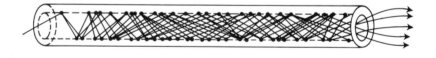

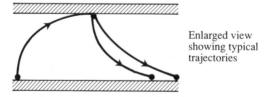

Enlarged view showing typical trajectories

Figure 4.14 Continuous channel multiplier. (From Goodrich, G.W. and Love, J.L. (1968), *IEEE Transactions on Nuclear Science*, **NS-15**, no. 3, 190–4, figure 1, © 1968 IEEE.)

with the walls and further secondaries. The multiplied pulse of electrons emerging from the right-hand end of the tube can be collected on an anode. These types of detectors, with diameters $\approx$ 1 mm (and some centimetres long) have been used for detecting not just electrons but ions, X-rays, etc.

Our interest here is more in a development of this device — the microchannel plate (MCP) — which, as its name implies, is a flat disc of glass typically some tens of millimetres in diameter and half to one millimetre thick containing an array of millions of closely packed pores (typically 12 μm in diameter spaced on 15 μm centres). The channels are coated with a secondary-emitting material, so in fact what we have is an assembly of millions of a miniaturized form of the single-channel detector of Figure 4.14. (A description of the fabrication method for such a microchannel plate makes interesting reading: an account can be found in reference 3.) A photomultiplier can obviously be formed by placing a photocathode close to one face to the MCP, a collecting anode close to the other face, and enclosing the lot in an evacuated envelope very similar to that of a conventional photomultiplier. Voltages from 500 to 1000 V are normally applied to the MCP itself, while voltages of a few hundred are required between cathode and MCP, and between MCP and anode. However, because the MCP itself is a fast device, higher voltages between cathode and MCP and between MCP and anode are often used to reduce electron transit times through these regions.

Let us look in detail at the internal operation of one of these channels. It is an important property of electron optical systems with electrostatic acceleration and focusing that for a given geometry the electron paths do not depend on the applied voltage. (This arises because if we, say, increase the applied voltage, then while we increase the velocity of the electrons and thus the rigidity of their tracks, we at the same time increase the electric field everywhere and thus its ability to bend the track. The actual paths are therefore the same.) Thus if we stick with the same ratio of channel length to diameter (often 40:1) then we have the same average number of collisions with the walls for an electron starting from the input end, independent of applied voltage. For the 40:1 ratio mentioned above this number of collisions (which corresponds to the number of stages in a conventional multiplier) comes out at about 19. Increasing the voltage does not, as we noted, change the average number of collisions, but it does increase the energy with which the electrons impinge on the walls, and thus the secondary emission ratio and the overall gain. Ignoring for a moment the first collision an electron makes on entering, as this is rather different from the others, let us assume a voltage difference between successive collisions for the remaining 18 of 50 V — requiring an overall voltage of 0.9 kV. The secondary emission ratio under these conditions will be about 1.5, giving an overall gain of $(1.5)^{18}$ or about 1500. The energy an electron has on making its first collision depends on the photocathode to MCP voltage used, which will certainly be much greater than 50 V. If we are calculating average current gain we may also want to take in to account the probability that an electron from the photocathode may strike the area between the channels, and all the secondary electrons produced may not be drawn into the pores. If we take a figure of three then for the effective secondary emission ratio at the first collision we can't be too far out, giving an overall gain of 3×1500 or around 4500, which agrees well with experiment.

If we look for very high current gains (say much greater than 10^4) with a simple MCP we run into difficulties due to ion feedback. Positive ions are produced in the residual gas by the dense cloud of electrons at the output end of a channel, and these are accelerated back towards the input end by the electric field. One (or more) of these may not strike the wall until it arrives back near the input, and by liberating an electron there can restart the whole multiplication process. Under high-gain conditions this can repeat itself many times, thus temporarily rendering the channel inoperative. There are two remedies for this problem: the first is to use a MCP fabricated with curved channels. More usual, however, is to employ two or more MCPs in series, with each operating below this critical gain. In this case the MCPs are fabricated so that the channels are at a small angle (of the order of 10°) to the axis of the disc. Placing two such discs as shown schematically in Figure 4.15(a) will ensure that positive ions can not find their way back to the input

(a)   (b)

Figure 4.15 Elimination of feedback in high-gain MCPs (schematic, not to scale). In practice channels in successive MCPs cannot be exactly lined up as shown because of their random positioning. However, it is worth noting that typically the open channel area on the face of an MCP represents 50 per cent or more of the total.

of the system. Such a two-stage arrangement is often known as a 'chevron plate' or a 'V plate', while for still higher gains a three stage arrangement in a 'Z plate' configuration, as shown in Figure 4.15(b) can be employed. While some losses obviously will occur at the interfaces, three stage MCPs with gains of greater than 10^6 are commercially available.

The main advantage of MCP photomultipliers over conventional devices is their superior performance in fast timing applications, basically due to their more compact construction. Transit times of about half a nanosecond, with rise-times of a quarter of a nanosecond are normal, while the transit time spread which controls the ultimate time resolution can be better than 100 ps (0.1 ns) even for a three-stage device. Considerable tolerance to magnetic fields is also a feature of MCP photomultipliers — again due to their compact structure. Information on the spatial distribution of the light falling on the cathode is also available if the anode is segmented and individual outputs taken from each segment (or if a resistive anode is used — see section 5.5.7).

What we can not expect a MCP photomultiplier to produce are large fast pulses at high repetition rates. This follows from an extension of our argument in the previous section regarding the current drawn through the resistor chain of a conventional photomultiplier. In the case of the MCP this resistor chain is just the resistive coating on the channel walls. Now for an MCP with 10^6 channels the overall resistance between the ends could typically be 500 MΩ. Each element therefore has a resistance of $5 \times 10^{14}\,\Omega$. Furthermore we obviously can not do anything analogous to what we did with a conventional device when, in order to improve pulse response, we paralleled the later resistors in the chain with reservoir capacitors. We are therefore compelled to depend only on the charge storage associated with the channel stray capacitance.

An extreme manifestation of this difficulty occurs at the output of a high-gain MCP photomultiplier (whether it be a multi-plate or a curved-channel device). The large amount of charge then being drawn from the stray capacitance in an active channel (and which can not be quickly restored) means that the accelerating potentials are substantially and progressively reduced, and with them the gain at successive 'dynodes'. The electron cloud thus passes through the final part of the channel with hardly any increase in electron numbers. Increasing the overall applied voltage will clearly not improve matters, and the channel is said to be 'saturated'. Saturation occurs for gains in the 10^6 to 10^7 region, depending on the MCP type, being higher for the multi-plate devices, as the charge originating in a single channel is in this case shared between a number of further channels at the interfaces. (An electron entering a single channel at the input will normally produce seven saturated channels at the output of a three-stage MCP photomultiplier.)

With such severe deformation of the potentials inside the MCP channel, it is clear that a considerable time will be required for the stray capacitance to recharge via the large wall resistance towards its original state. We can not apply our simple treatment at the end of Section 4.4.5 to this more complex case, but in fact a time of the order of 0.1 s could be needed. This seems extraordinarily long at first sight, but one must remember that multiplier gains of the sort of size we are talking about are employed for the case of relatively small numbers of electrons arriving at the MCP input (arising say from a weak scintillation). Consequently, since these electrons are arriving randomly at an input face with more than a million channels, any particular channel will almost certainly not be called on to provide a further output until a large number of scintillation flashes have occurred. A reasonable scintillation counting rate is thus possible. Nevertheless even in the most favourable case of single-photon counting, appreciable effects can be seen at only moderately high rates. In one test a variable frequency light source producing predominantly single-photon pulses was used with a three-stage 18 mm diameter MCP photomultiplier, whose output went to a discriminator (see Chapter 6) and a counter. A 5 per cent drop in count rate was found to occur when the expected value was 100 kHz, due to incomplete channel recovery affecting gain.

There is one important characteristic of the MCP photomultiplier which does improve substantially in the saturated mode, and that is the pulse-height distribution at the output due to a single electron at the MCP input. We have earlier noted briefly why, for a conventional photomultiplier, statistical processes in the multiplication produce, not a unique output pulse, but a range of pulses giving a pulse height distribution like that labelled '1 electron' in Figure 4.10. For a lower-gain MCP photomultiplier (a one-stage device, say) the distribution is even broader than for a conventional photomultiplier and has a quasi-exponential rather than a peaked shape. (This arises because of the statistical variation in the number and energy of wall collisions.) In the saturated condition, however, the spread in pulse size decreases and the distribution takes on a peaked shape which is now appreciably narrower than that associated with conventional photomultipliers. Such a narrow peaked distribution makes it easier to count the bulk of the single-electron pulses, while electronically discriminating against the smaller spurious pulses usually present (and whose existence is *not* indicated in the idealized sketch in Figure 4.10).

In summary then, although the MCP photomultiplier has obvious advantages in a number of special circumstances, the less complex (and cheaper) conventional photomultiplier — which of course has its own particular virtues — has not been displaced from the bulk of applications in radiation measurements.

4.5 Applications of scintillation counting

4.5.1 *Some general remarks on gamma-ray spectrometry*

One of the main applications of scintillation counting for many years was the spectrometry of gamma rays. By this we mean the detection of the presence of various radioactive elements in a source from their characteristic gamma rays, and with suitable calibration, the determination of the actual amounts of each element present. However, for a reason that will be apparent very shortly the better choice for many critical measurements in this area is now a semiconductor germanium detector. Nonetheless we will deal in some detail with the principles underlying the practical use of scintillation counters for gamma spectrometry, as most of these principles are applicable also to the semiconductor detector. We will thus be in a position to make comparisons between the two detector types. (A full account of semiconductor detectors and their operation will be given in Chapter 5. All we need to know for the present discussion is that a semiconductor detector is the solid state analogue of the gas-filled ionization chamber. Thus in a germanium detector used for gamma work, the gamma ray produces electrons and holes—the latter the analogue of positive ions—in the germanium, and these, moving under the action of an applied electric field to their respective electrodes produce the output signal. In size germanium detectors are comparable with the smaller NaI(Tl) crystals available.)

We restrict our discussion for the moment to physically 'small' detectors. We shall see the reason for this restriction, and what exactly we mean by 'small' detectors, very shortly. As we noted in Section 2.3, gamma rays can interact with a detector through the photoelectric, Compton and pair production mechanisms, of which the photoelectric interaction is the only one which is likely with our present restriction to deposit the full characteristic gamma energy. For simplicity let us consider at first gamma rays with energies of less than 1 MeV so that we will only have photoelectric and Compton interactions to take account of. Figure 4.16 shows what we should expect ideally if we produce a size analysis—a 'differential spectrum'—of the output pulses from a NaI(Tl) scintillation counter (or a germanium detector) when it is irradiated with monoenergetic gamma rays. (What we are doing, as will be explained more fully in Chapter 6, is simply producing a histogram of sizes, with the pulses sorted into typically many hundreds of boxes or 'channels', of which a few are shown. We remember that the pulse size is directly proportional to the energy deposited in the crystal.)

The spike on the right-hand side of the diagram represents, in the small-detector situation, the counts associated with the photoelectric absorption, and as we expect these to be all of the same size, they are confined to a single channel. We saw in Chapter 2 that with a Compton interaction the

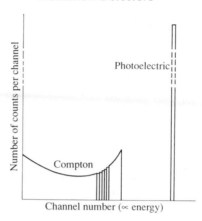

Figure 4.16 Idealized gamma spectrum. (Drawn for a gamma energy of 0.5 MeV.)

maximum energy which can be deposited in the absorber is less than the full gamma energy, and for example with a 0.5 MeV gamma this maximum energy was about two-thirds of the full amount. The vertical line, or 'Compton edge' to the left of the photoelectric spike represents this limit. The Compton edge corresponds to a collision between a gamma quantum and an electron where the quantum is backscattered through 180°: the rest of the curve in Figure 4.16 to the left of the edge relates to quanta scattered through various angles from 180° to 0°, each with its corresponding electron energy deposited in the detector. To derive this curve analytically would require us to compute the fraction of incoming quanta scattered through a particular angle – a calculation beyond the scope of the present work. We will be content to accept the result as shown: for the theory any standard textbook on nuclear physics can be consulted. We note in passing that as the Compton interactions are spread over a large number of channels, while the photoelectric interactions occupy only one channel, the total number of Compton counts will, as drawn, exceed the photoelectric counts, even though the photoelectric spike is the most prominent feature of the spectrum.

The curve we obtain experimentally is shown in Figure 4.17 and differs from the previous idealized one in that all the latter's sharp features have been smeared out. In particular the photoelectric peak extends over a number of channels, and no clear gap occurs between the Compton and photoelectric sections of the curve. Both scintillator and germanium detectors show the same basic pattern, but the departure from the ideal diagram is, as we shall see, much less for the germanium device. The spectrum shown in Figure 4.17 would be typical of the scintillator case. The reason for

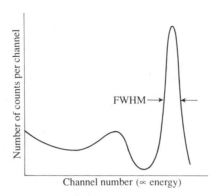

Figure 4.17 Experimental gamma spectrum.

this smearing out is that our assumption that a given energy deposited in the detector will produce a unique corresponding output pulse is not exactly true. The matter will be dealt with in detail in Chapter 7. All we need to say here is that for a scintillation counter, statistical fluctuations in the production of photons in the crystal, in their conversion to electrons at the photocathode, and in their multiplication in the photomultiplier imply that a series of identical energies deposited in the crystal will produce, not a series of identical outputs, but a narrow range of pulse sizes distributed around a mean value. Clearly the resulting spread in the width of the photoelectric peak, will for one thing, mean a significant drop in its height. For a germanium detector, statistical fluctuations in the production of the charge carriers, and limitations associated with the high-gain external amplifiers needed for these devices, play a similar role in degrading the ideal response — although as we have already noted, in a smaller measure. The full width of the peak at half its maximum height, labelled 'FWHM' in Figure 4.17, is clearly a measure of the detector's ability to 'resolve' two close peaks in a spectrum. We return to it in more detail in Chapter 7.

In the light of the previous considerations let us see how we can improve the visibility of the photoelectric peak — which after all is the feature characterizing the gamma ray — with respect to the Compton interaction. This will be particularly important if we are trying to analyse a spectrum with many different gamma rays present, where successive Compton spectra overlap each other to form a general background, with the peaks projecting from it. There are four factors which control the peak–Compton relationship: the energy of the gamma ray, the atomic number of the element or elements present in the detector, its efficiency for energy conversion, and finally its physical size, and these factors are relevant for both scintillation and semiconductor devices.

With regard to the first factor it is clear from our discussion in Chapter 2 and in particular from Figure 2.10, that for energies up to at least 100 keV with detectors even of moderate atomic number like germanium ($Z = 32$), and the more so for sodium iodide ($Z_{iodine} = 53$) there is no real problem, as the cross-section for photoelectric absorption is greater than that for Compton interaction. At higher energy values we must look to other favourable factors to help. As regards the second factor, the atomic numbers of the elements present in the detector, it is clear why we prefer say a NaI(Tl) crystal to any form of organic scintillator (crystal, liquid, or plastic) for gamma spectrometry. This follows from the fact that the photoelectric effect increases as Z^5 while the Compton interaction goes linearly with Z. Consequently, for an organic scintillator whose primary components are carbon ($Z = 6$) and hydrogen ($Z = 1$) the photoelectric peak is almost entirely absent except at extremely low energies.

By the detector's efficiency for energy conversion — the third factor — is meant the ability of the detector to convert as much as possible of the energy of the incoming radiation into photons, and thence via the photocathode into electrons (in the case of a scintillation counter), or directly into electrons for a semiconductor detector. For example if we compare the scintillators NaI(Tl) and BGO, then the latter, because of its heavy bismuth atoms ($Z = 83$) will capture a greater percentage of gamma rays passing through a crystal of given volume, yet because of its poorer conversion efficiency it will produce fewer photons, and thus fewer electrons in the photomultiplier per event. This means that not only will the output pulse be smaller, but, more importantly, that the fractional fluctuation in numbers from pulse to pulse will be much larger, thus producing broader peaks. For purposes of spectrometry therefore (but not for counting efficiency) sodium iodide is preferred. The most striking comparison under this heading is, however, between sodium iodide and germanium detectors. The energy to produce a photon in NaI(Tl) is, as we have noted previously, about 22 eV. With a photocathode efficiency of say 25 per cent this means nearly 90 eV on average to produce an electron. By comparison the corresponding energy in germanium is about 3 eV. Thus in spite of its worse situation as regards atomic number, germanium holds an overwhelming advantage as regards the visibility of its peaks. Not only that, but their extreme narrowness allows us, in say an environmental sample containing a very large number of radioactive elements, to resolve (i.e. distinguish clearly between) two peaks of energies very close to one another and which would, with the best scintillation counter, be either partially or totally merged, as we shall illustrate in Chapter 5.

The final factor in this discussion is the physical size of the detector, and at first sight this would hardly seem to have much bearing on the photoelectric–Compton ratio. In a direct sense this is true; however, we must consider a little more closely our assumptions about the deposition of energy by Compton interactions in a detector. Up to the present we have been assuming that we have been using a 'small' detector, and we now explain what we mean by this. A 'small' detector is one of such size that the gamma photon scattered in a typical Compton interaction will have little chance of interacting again before escaping from it. However, it is clear that there is always a finite probability that such a second interaction will take place, and this probability obviously increases with detector size. If this second interaction is by photoelectric absorption then the total energy of the incoming gamma will have been trapped (almost instantaneously) in the detector, and the output pulse will contribute to the 'photoelectric peak'—even though the first interaction was by the Compton process. For this reason what we originally called the photoelectric peak is more properly referred to as the 'full energy peak' as, particularly in large detectors (and especially for germanium with its only moderate value of Z) many of the counts under it can be attributed to interactions other than purely photoelectric ones. (Of course the second interaction can also be a Compton one, with the scattered gamma photon escaping. In such cases the output pulse lies between the Compton edge and the photopeak, and helps to fill up the gap which ideally exists there.) With a very large detector and a gamma source placed at its centre (at the bottom of a very narrow well, say) it is clear that the total gamma energy will almost always be trapped. This can be the result of a single photoelectric interaction, or a chain of Compton interactions terminated at some point by a photoelectric interaction—the latter becoming increasingly probable as the energy of the scattered photon decreases. In fact even a plastic scintillator can produce a significant full energy peak in such circumstances.

The words 'small' and 'large' in all our previous discussion must of course be on a scale determined by the relevant half thickness for the gamma energy and detector material in question. We remember from Chapter 2 that the half thickness $d_{\frac{1}{2}}$ is the thickness of material which reduces the number of photons per second in a beam to half its original value: alternatively it could be defined as the thickness of material in which a single gamma has a 50 per cent chance of being absorbed. It is given by $\mu d_{\frac{1}{2}} = \ln 2 = 0.69$, and for sodium iodide and a gamma ray of 0.5 MeV the Compton absorption coefficient is about $0.27 \, \text{cm}^{-1}$, giving $d_{\frac{1}{2}}$ as about 2.5 cm or 25 mm. So making a crude calculation, if we had incoming gamma rays of 1 MeV giving Compton-scattered photons of around 0.5 MeV, then with a crystal of 25 mm dimensions we might expect to be beginning to get some appreciable contribution from them to a full energy peak. A 'large' crystal would have to be say at least five times these sorts of dimensions. (The same sort of rough calculation can of course give us an idea of the overall efficiency of detection – say again for a sodium iodide detector. For a 1 MeV incoming gamma the total absorption coefficient for sodium iodide is about $0.2 \, cm^{-1}$ which corresponds to a half thickness of about 35 mm. Therefore for a source placed at the centre of a sphere of this radius we would expect 50 per cent of the gammas to interact. For a cylindrical well type crystal of diameter and height both 70 mm we might, because of the extra material, do a little better.) Note a final but somewhat academic point in this connection. Because the total cross-section for gamma absorption rises again at higher energies – due to pair production – so will the efficiencies of detectors. For sodium iodide the minimum efficiency, which is of course at the same place as the minimum cross section, occurs around 5 MeV.

It is hardly necessary to elaborate on features produced in gamma-ray spectra by high-energy gammas when pair formation occurs, as these are analogous to what we have already discussed. For an ideal 'large' detector – and it really would have to be large because of the greater energies involved – we expect not just the electron but also both gammas from the annihilation of the positron to be absorbed, thus contributing to the full energy peak at E_γ, the gamma energy. For 'small' detectors, both annihilation gammas will escape, taking with them $2m_0c^2$ of the incoming energy and leaving a detector peak at $E_\gamma - 2m_0c^2$ (where $m_0c^2 = 0.511 \, MeV$). For intermediate-sized detectors one could have three peaks, at E_γ, at $E_\gamma - m_0c^2$, and at $E_\gamma - 2m_0c^2$, the second of these being due to the escape of one annihilation quantum alone.

Clearly the analysis of a spectrum from say an environmental sample containing a large number of gamma emitters is a difficult business, particularly as in our necessarily brief survey we have not included additional features that occur, such as X-ray peaks and the fact that a particular nucleus may emit more than one gamma ray. In the light of this complexity, automatic computer analysis of spectra with identification using a stored library of peaks from various isotopes, is widely used. For more detailed information on this topic manufacturers' literature will often prove helpful.

4.5.2 *Scintillation counting with solid scintillators*

From our remarks in the previous sub-section it is clear that the role of scintillation counters in gamma detection and analysis will be restricted to particular areas. For example if we are monitoring just a single gamma-emitting element, say in a biological or medical experiment on radio-isotope take-up, than a sodium iodide detector is perfectly satisfactory and much cheaper than its germanium alternative. For very-large-volume gamma detectors, such as might be required in whole-body monitoring there is no alternative to scintillation counters. Sodium iodide crystals of dimensions two to three hundred millimetres (and viewed by a number of photomultipliers) are commercially available, but a germanium detector of dimensions around 50 mm would be considered a large device.

A large annular sodium iodide crystal can also be used to surround a smaller detector which is looking at weak sources, and thus reduce the latter's background due to cosmic rays and local radioactivity. Using an anticoincidence arrangement (see Section 6.8) pulses occurring in the inner detector are cancelled if they have also deposited energy in the outer detector. Thus the counts due to radiation which has penetrated the conventional outer lead or steel shielding are drastically reduced, as are those due to any residual radioactivity in the shielding itself. If the source is a gamma emitter and is viewed by say a germanium detector, then Compton events in this detector where the scattered photon enters the annulus may also be suppressed, thus improving the visibility of the full energy peaks. Plastic phosphors are a cheaper alternative for these guard detectors, and they can be machined easily into complicated shapes. However, their efficiency for gamma rays is appreciably less than sodium iodide. They also can be used for general-purpose gamma work in their own right.

We already noted earlier that plastic phosphors are excellent for use in fast timing, and they are also employed as inexpensive neutron detectors. For fast neutrons their action depends on the detection of knock-on protons produced from the hydrogen in the scintillator itself. Plastic phosphors (in common with some other organic scintillators) can also be used with pulse-shape discriminator (PSD) circuitry to distinguish between neutron counts and counts from the gamma radiation which is often present at the same time. The method (mentioned briefly earlier) depends on the fact that as well as the fast decay component we noted in our simple picture as being characteristic of plastic phosphors, there is also a much slower component, and the helpful fact is that the amount of this slow component present will be greater for a heavily ionizing particle like a proton than for an electron. The different amounts of slow component present (which correspond to slightly different pulse shapes) can be detected electronically, and in the case of neutrons and gammas those pulses arising from gamma interactions can be discriminated against in this fashion, even if it is not possible to do so by conventional pulse height methods (see Section 6.7.6). We noted earlier in this chapter that the inorganic scintillator CsI(Tl) can also be used for PSD. Plastic phosphors loaded with ^{10}B can be used for slow neutron detectors, although as we saw earlier LiI(Eu) is also available for this purpose and is a superior though more expensive performer.

Scintillation counters using a thin sodium (or caesium) iodide crystal or a thin layer of zinc sulphide are widely used as general-purpose alpha-particle counters, particularly in portable monitors. However, for critical applications semiconductor detectors are preferred, again because of their superior performance in resolving groups of alpha particles close together in energy. Thin slices of nearly any of the scintillators we have been discussing, organic or inorganic, can be used in beta detection, but plastic phosphors are usually preferred, both for economy, and because their low atomic number reduces backscattering out of the detector. However, the main role for scintillation counting of betas involves liquid scintillators, to which the next sub-section is devoted.

4.5.3 *Liquid scintillation counting*

Liquid scintillators share many of the properties of their plastic counterparts, and have in addition the ability to provide detectors of extremely large volumes. They have therefore been used in experiments like neutrino detection where interaction between the incoming particle and the detector material is extremely small. However, the main application for liquid scintillation counting — and probably now the largest application of scintillation counting generally — is more prosaic, but still a key element in medical and biological work. This is its use in tracer studies which employ beta-particle emitters, particularly carbon-14 (^{14}C) and tritium (^{3}H). Tritium is the only radioactive isotope of hydrogen, and carbon-14 the most widely available radioisotope of carbon, and as such they are vital in the study of living and organic material. That they are beta emitters (rather than say gamma emitters) in principle presents no difficulty, as we saw that the Geiger counter for example is a simple and effective device for beta detection in general. However, both these isotopes emit beta particles with very low energies: the end point for the ^{14}C spectrum is at 156 keV, and that for tritium at 18.6 keV, and of course many of the betas will be emitted with energies very much less than the end point.

The thinnest windows available in a Geiger counter are unlikely to transmit more than 50 per cent of the arriving betas even from a ^{14}C source, and while of course windowless proportional counters do exist, liquid scintillation counting systems, with their additional ability to deal directly with liquid samples, are now the norm. The radioactive samples to be counted must be in a solvent which can be intimately mixed with the scintillator, and which does not substantially 'quench' (i.e. degrade) its light output. The beta particles can thus be counted in what is effectively a 4π windowless situation. Fully automatic sample changing and data recording allow large amounts of material to be routinely dealt with.

Liquid scintillators are down by roughly a factor of four as regards photon production compared with sodium iodide. So if in the latter we need 22 eV per photon then for the liquid scintillator we need about 90 eV per photon or 360 eV per electron, assuming a photocathode efficiency of 25 per cent. Thus even making an allowance for the quenching we mentioned above, we should still get at least two electrons from the photocathode for every 1 keV of beta energy in the scintillator. So if in order to eliminate the bulk of single-photon tube background, we set a discriminator to cut out all pulses less than the mean value corresponding to two electrons at the photocathode, we still should have no difficulty in detecting beta particles of say 2 keV energy. So very high efficiencies approaching 100 per cent for tritium can be achieved. For very-low-activity sources two photomultipliers viewing the opposite sides of the sample container, and operating in coincidence (Section 6.8), can further reduce background.

4.5.4 *Fast neutron spectrometry with scintillators*

As well as merely detecting the arrival of a fast neutron, the knock-on proton in a solid or liquid scintillator containing hydrogen must carry information on the energy of the initial neutron–as indeed would be true in other detectors such as a proportional counter using a hydrogen or hydrogen compound gas filling. Indeed even if we use detectors involving nuclear reactions with fast neutrons, the energies of the outgoing components should be capable of indicating the incoming neutron energy, provided that we know the energy available in the reaction itself. However, the actual spectra produced in both these cases are not at all as easy to interpret as the corresponding ones for gammas, and consequently we shall not treat them further here. Some of the principles involved are discussed in Section 5.10 in the context of solid-state devices, and references for further reading on this topic are provided at the end of Chapter 5 and in the general bibliography.

References

1. Murray, R. B. (1975). Energy transfer in alkali halide scintillators by electron–hole diffusion. *IEEE Transactions on Nuclear Science*, **NS-22**, 54-7.
2. Bell, R. L. (1973). *Negative electron affinity devices*. Clarendon Press, Oxford.
3. Dhawan, S. (1981). Introduction to microchannel plate photomultipliers. *IEEE Transactions on Nuclear Science*, **NS-28**, 672-6.

Further reading

Birks, J. B. (1964). *The theory and practice of scintillation counting*. Pergamon Press, Oxford.
Murray, R. B. (1962). Scintillation counters. In *Nuclear instruments and their uses*, Vol. 1, (ed. A. H. Snell), pp. 82-165. Wiley, New York.

5
Semiconductor detectors

5.1 Introduction

The development since the 1940s and 1950s of ionization chambers which use semiconductors rather than gases as the stopping medium has had a profound impact on the detection of all types of nuclear radiation. Primarily this arises because the quantity of ionization produced by an event is much greater when the stopping medium is a semiconductor. We shall see in Section 5.2 that because of the different electronic energy level structure in a semiconductor only about 3 eV is required for every charge pair produced, rather than 20 or 30 eV as in a gas. In comparison with scintillation counters, where the order of 100 eV is required to produce one 'information carrier' (a photoelectron), semiconductor detectors are in this respect even more outstanding. This follows because the larger number of carriers for a given incident radiation energy leads to smaller effects due to the statistical fluctuations in this number. The net result is that greatly improved energy resolution is possible, as we shall discuss in more detail in Chapter 7.

Furthermore, much smaller active volumes are possible when charged-particle detectors are made from semiconductors rather than gases because of the much higher stopping power (energy loss per unit path length) shown by solids. Apart from the intrinsic convenience offered by this reduction in size, it is one of the factors leading to the possibility of making very fast timing measurements with nanosecond resolution. In the case of gamma ray detectors, much better detection efficiencies for a given detector volume are obtained with semiconductors than with gases. (Here, of course, scintillators also possess these advantages.)

There are, however, some less desirable features with semiconductors, compared with gases, which have to be overcome or at least appreciated. Efficient charge collection requires a proper understanding and control of the materials properties of the semiconductors used. In this respect silicon and germanium have become pre-eminent. The detectors have also a relatively high susceptibility to performance degradation through damage induced by radiation. But, most fundamentally of all, no solid useful for detectors is as good an insulator as a gas, and so there is a standing current when a voltage is applied. Even with solids the quantity of charge released by radiation is in macroscopic electrical terms very small, and it would normally be obscured by 'noise' (see Chapter 7) unless special measures were taken. We shall see in this chapter that this will usually mean cooling the detector and/or constructing a reverse-biased junction to act as a barrier to this current flow.

5.2 Band structure of semiconductors

In order to understand the physics of semiconductor detectors and their advantages and drawbacks we must consider the energy levels that electrons can occupy. We shall do this only in outline; some of the material to be covered has already been mentioned in Section 4.4.3, and a full description of the theory is contained in most textbooks on solid-state physics (see 'Further reading' at the end of this chapter).

From what we said in Section 5.1, our main concern will be with the elemental semiconductors silicon and germanium. These belong to group IV of the periodic table, which means that they are tetravalent, with four valence electrons per atom. These electrons bind the atoms together in the crystal lattice of the semiconductor, with every atom having four 'covalent' bonds with each of its four nearest neighbours.

In a gas the locations of the electron energy levels for an isolated atom basically set the energy required to produce an electron–ion pair, which we saw was typically 25 eV. In a solid these levels are broadened into bands of many closely spaced levels because of the proximity of neighbouring atoms, such that the bands are brought close together. Figure 5.1 shows that in a semiconductor such as silicon or germanium there is below the vacuum level (corresponding to zero electron kinetic energy outside the semiconductor) a 'conduction band' of energy levels located above a 'valence band'. There is an energy gap between the bands such that in a perfect semiconductor an electron cannot possess an energy which occurs within this gap. By 'perfect' we mean that effects due to impurities, surfaces, and crystallographic defects are neglected.

The bands can accommodate only a finite number of electrons, since each level can (according to the Pauli exclusion principle) be occupied by at most two electrons. For a perfect semiconductor crystal at the temperature of absolute zero the valence band, in the absence of ionizing radiation, can be shown to be full of the valence electrons described above, and the conduction band to be empty. No current flows when an electric field is applied because the valence electrons are in an already full band of levels which has an energy gap above it. They cannot therefore absorb any net energy from the field.

The process of ionization, by an alpha particle, say, is able to raise negative electrons to the conduction band. This leaves in the valence band

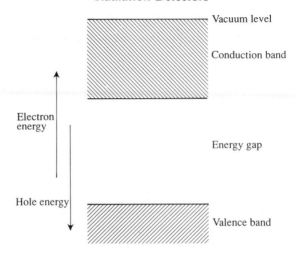

Figure 5.1 Energy band diagram for a perfect semiconductor, showing an empty conduction band and a valence band full of electrons, as would be the case at absolute zero.

an equal number of 'holes'; these behave like positively charged particles, and are the counterpart of the electrons raised into the conduction band. Being now free to change their energies within their respective bands, the electrons and holes can undergo a net absorption of energy from an applied electric field, and as a result they drift in opposite directions. Hence in a semiconductor detector they can, while thus drifting in the field between two electrodes attached to, say, each side of the semiconductor, induce a flow of current in an external circuit. This, of course, is not dissimilar to what happens in a gas-filled ionization chamber. The energy gap between the valence and conduction bands is only about 0.7 eV in germanium, but on average 2.96 eV (at what will be seen to be a typical operating temperature of 77 K) is required to be expended per electron–hole pair produced. For silicon the corresponding figures are 1.1 eV and 3.62 eV (at 300 K). The energies per pair are higher than the band gaps themselves because some of the energy from the radiation is dissipated in the creation of lattice vibrations.

In a metal the conduction band, which as in a semiconductor is the highest band below the vacuum level, is already partly filled with electrons even at absolute zero. (The level in a metal up to which the electrons fill is known as the Fermi level, as mentioned in Section 4.4.3.) There are therefore always electrons which are able to absorb energy from a field and hence give a large conduction current which would completely mask any ionization produced by incident radiation. Insulators, on the other hand, possess similar band structures to semiconductors, but with band gaps of around 4 or 5 eV or more rather than about 1 eV. The number of ion pairs created would therefore be less than in semiconductors. In practice, however (as we shall see in Section 5.3), it is charge-collection difficulties that unfortunately tend to rule out insulators, and indeed many semiconductors other than silicon and germanium. Nevertheless there are exceptions, which are considered in Section 5.12, to this general principle of using only silicon and germanium.

Detectors, of course, cannot in practice be cooled to absolute zero. Consequently a significant number of electrons together with an equal number of holes are produced by thermal excitation across the relatively small band gaps of silicon and germanium. If this is the main source of carriers (and we shall see in a moment that impurities or other features of the material may contribute), then the semiconductor is described as being 'intrinsic'. For this case it can be shown that the hole or electron density p_i or n_i at an absolute temperature T is given by

$$n_i = p_i = AT^{3/2} \exp\left(-E_G/2kT\right) \qquad (5.1)$$

where A is a constant for a given material, k is Boltzmann's constant and E_G is the band gap.

At room temperature these densities are 1.5×10^{16} m^{-3} in silicon and 2.4×10^{19} m^{-3} in germanium. Even in intrinsic material there would therefore be a 'background' of thermally generated carriers on which any ionization induced by radiation would be superimposed. With a field applied to any semiconductor detector the current obtained is always the sum of that from the signal pulse and the contribution from thermally generated carriers. We shall consider this point further in Section 5.3.

In practice most semiconductors have impurities deliberately added which considerably modify the carrier densities from those in intrinsic material. An extrinsic semiconductor is one in which many more charge carriers are generated at the temperature concerned from impurities than by the intrinsic excitation mentioned above.

The impurities which are most often added in extrinsic silicon or germanium are those from groups III or V of the periodic table. Such impurities are respectively trivalent (three valence or bonding electrons per atom) or pentavalent (five valence electrons per atom). 'Doping' a semiconductor refers to the addition during crystal growth (or later) of very small quantities (perhaps only a few parts per million) of impurities such as these. The reasons for doping will become clear in Section 5.4

When a pentavalent impurity from group V, such as phosphorus, is introduced at these low concentrations, the impurity atoms substitute themselves at the lattice sites which would have otherwise been occupied by the tetravalent silicon or germanium atoms. Four of the five electrons per impurity atom are used for covalent bonding to the four nearest neighbours. It can be shown that the fifth is easily excited thermally to become free to move through the semiconductor, just like the few conduction electrons already thermally excited from the valence band. The impurity atom is then left as a fixed positive ion. Such a semiconductor is called 'n-type', after the negative carriers which have been introduced.

The pentavalent 'donor' atoms, so called because they can each donate an electron for conduction, can be shown on an energy band diagram as in Figure 5.2. The donors are shown in levels lying in the energy gap just below the conduction band edge at a depth corresponding to the energy required for ionization. This ionization energy (given by the distance between the donor levels and the conduction band edge) is small, and therefore at room temperature (and indeed often down to 150 K and lower) almost all of the donors are positively ionized. The Fermi level, which in a semiconductor characterizes the occupancy of the energy levels by electrons, lies in this case near the top of the band gap. We shall not, however, attempt a full definition and description of the Fermi level — books on solid-state physics cover this topic more fully.

Even dopant concentrations of only a few parts per million generate an electron density in the conduction band at room temperature of about 10^{23} m^{-3}, which completely swamps the density of intrinsic electrons already present. We can therefore conclude that in such material the electron den-

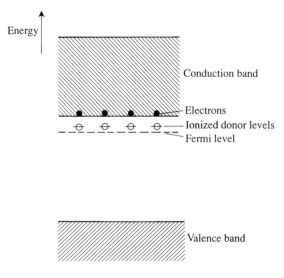

Figure 5.2 Energy band diagram for an n-type semiconductor, showing fully ionized donors.

sity, n_n, (where the suffix n refers to n-type material) is normally essentially equal to N_D, the donor density, and that both n_n and N_D are much greater than the intrinsic electron density, n_i. This is certainly true in silicon doped to make transistors and integrated circuits, for which N_D can be 10^{25} m^{-3} or even greater, and is still true even for the lighter doping densities, perhaps 10^{18} m^{-3}, typical of silicon radiation detectors. (In germanium we can see that a donor density of 10^{18} m^{-3} would, by contrast, from the figures given previously, still leave the material intrinsic at room temperature, though we shall later have to discuss what happens at the lower temperatures at which germanium detectors are actually operated.)

Such n-type impurities, however, do more than increase the electron density. There is a dynamic equilibrium between the generation and the recombination (the reverse of generation) of charge carriers (that is, electrons in the conduction band and holes in the valence band). Increasing the electron density n_n increases the rate of recombination between electrons and holes until a new equilibrium is established, in which the hole density p_n is much reduced. It can be shown that the product of the two densities remains constant, that is,

$$n_n p_n = n_i p_i = n_i^2 = p_i^2 \quad \text{where } n_n \approx N_D.$$

The electrons in n-type material are therefore called the majority carriers, and the holes, the density of which is (from what we have just said) many orders of magnitude lower, are called the minority carriers.

If we now consider trivalent impurities such as boron, their atoms at low concentrations again substitute themselves into the crystal lattice in place of silicon or germanium atoms. One of the four bonds around the trivalent impurity, however, now lacks an electron. An electron from a neighbouring bond can fill this vacancy, leaving a hole in this second position, and a continuation of this process enables a hole to move freely through the semiconductor just like the few intrinsic conduction holes already present. This is an exactly complementary picture to that of pentavalent impurities donating electrons to the conduction band. This time we have trivalent 'acceptors' accepting electrons, thus becoming fixed negative ions and creating holes. This process gives a 'p-type' semiconductor, so called because of the positive carriers introduced. Normally at room temperature almost all the acceptors are negatively ionized and lie in the energy gap just above the valence band edge, as Figure 5.3 shows. The Fermi level in this case lies near the bottom of the band gap, as shown.

The hole density p_p (the suffix p referring to p-type material) is essentially equal to N_A, the acceptor density, with both much greater than p_i. The rate of electron–hole recombination increases until a new dynamic equilibrium is reached with a much reduced electron density n_p. This time

$$n_p p_p = n_i p_i = n_i^2 = p_i^2 \quad \text{where } p_p \approx N_A.$$

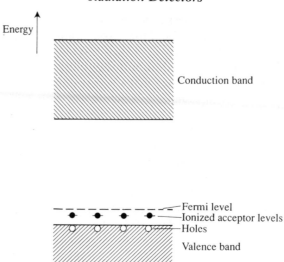

Figure 5.3 Energy band diagram for a p-type semiconductor, showing fully ionized acceptors.

The holes in p-type material are therefore the majority carriers, and the electrons the minority carriers.

When both donors and acceptors are present at the same place in a semiconductor, they partially 'compensate' for, i.e. cancel, the effects of each other. For example, if the donor concentration, N_D, exceeds that of the acceptors, N_A, then we may imagine that, of the initial N_D electrons from the donors, N_A of them recombine with this number of holes from the acceptors. (We neglect minority carriers and assume that the donors and acceptors are fully ionized.) The density of remaining electrons is thus $N_D - N_A$, and this is the effective donor density of this partially compensated n-type material. Similarly if N_A exceeds N_D, $N_A - N_D$ is the effective acceptor density of this partially compensated p-type material. If N_D equals N_A, then full compensation is achieved, and in many ways the properties of this compensated material resemble those of intrinsic material.

The way in which compensation can be most nearly achieved in practice and its application to radiation detection are considered in Section 5.6.

5.3 Current flow in bulk semiconductors

We are now in a position to analyse what happens when an electric field is applied to a piece of bulk semiconductor. The field causes the electrons and holes to move, not with constant accelerations, as would happen for electrons in a vacuum, but with constant average terminal velocities limited by scattering events in the material, just as for the charge carriers in a metal or a gas. Except at very high fields the velocity is found to be proportional to the electric field $\mathscr{E}$ at the point concerned. (This, incidentally, shows that the effective resistive force caused by the scattering is proportional to the velocity u.) We write for the electron and hole velocities u_n and u_p

$$u_n = \mu_n \mathscr{E} \tag{5.2}$$

and

$$u_p = \mu_p \mathscr{E} \tag{5.3}$$

respectively, where the constants of proportionality, μ_n and μ_p, are called the electron mobility and the hole mobility for the semiconductor concerned. (Note the analogy with carrier mobility in gases, as considered in Section 3.2.3, although here for semiconductors we have of course no pressure dependence in our equations.) In silicon the values at room temperature are 0.14 and $0.05 \, \text{m}^2 \, \text{V}^{-1} \, \text{s}^{-1}$, respectively, while for germanium they are 0.39 and $0.19 \, \text{m}^2 \, \text{V}^{-1} \, \text{s}^{-1}$. At lower temperatures the scattering is reduced, and at the temperature of liquid nitrogen, 77 K, for example, the mobilities are an order of magnitude higher.

As $\mathscr{E}$ is increased the velocities remain proportional to it until the velocity approaches $10^5 \, \text{m s}^{-1}$, at which value it saturates. At room temperature this corresponds to fields approaching $1 \, \text{V} \, \mu\text{m}^{-1}$ ($10^6 \, \text{V m}^{-1}$). These saturation conditions can sometimes occur in detectors at room temperature; at low temperatures, where with the higher mobilities the fields for saturation are lower, it is even more likely to happen.

It is important to appreciate the outstanding carrier transport properties possessed by silicon and germanium. In gases we saw that only the mobility of electrons is high; ion mobilities are much lower. Similarly in many semiconductors other than silicon and germanium, one mobility, usually that of electrons, is high, but the other (of holes) is rather low. The high mobility of both types of carriers in silicon and germanium greatly aids efficient charge collection in a detector. Furthermore, the lifetime of excess carriers before they are lost through trapping or recombination can sometimes in silicon and germanium be as long as milliseconds. This again ensures that carriers are able to travel easily all the way, through distances of up to 10 mm or more, to the collecting electrodes of a detector. These long lifetimes are a consequence not only of the high crystal perfection attainable in silicon and germanium, but also more fundamentally of the detailed nature of the structures of their electron energy bands. In other semiconductors carrier lifetimes are often limited to microseconds or less, and this can place design constraints on detectors made from them.

The drift of carriers in an electric field constitutes an electric current. Of course, being oppositely charged, electrons and holes move in opposite directions; the positive holes drift in the direction of the field and the negative electrons against it. The electric current generated at any point by the motion of electrons, however, adds to that produced by the holes. For one-dimensional drift motion of the charge carriers the steady current density j flowing (that is, the current per unit area) is in general

$$j = neu_n + peu_p \qquad (5.4)$$

where the electrons and holes have densities n and p. (For the moment we attach no suffixes to n and p as we have not yet specified whether the material is intrinsic, n-type or p-type.) Substituting for u_n and u_p from equations (5.2) and (5.3) into equation (5.4) we get

$$j = ne\mu_n \mathscr{E} + pe\mu_p \mathscr{E}.$$

The semiconductor conductivity, σ, is given by $j/\mathscr{E}$, and hence the resistivity ρ, given by $1/\sigma$, is

$$\rho = 1/(ne\mu_n + pe\mu_p). \qquad (5.5)$$

For intrinsic germanium and silicon at room temperature the values of $n_i (= p_i)$ can be calculated to be 2.4×10^{19} and $1.5 \times 10^{16}\,\mathrm{m}^{-3}$, as we saw in section 5.2. Thus equation (5.5) gives about 0.45 and 2200 Ω m, respectively, for the resistivities of germanium and silicon. These represent the highest possible values for the room-temperature resistivities, and no amount of materials purification can make them higher. We shall see that this has important consequences for the design of silicon and germanium radiation detectors.

In order to obtain germanium or silicon which is intrinsic at room temperatures (i.e. have values of N_D – or N_A – much less than the value of n_i given above), we need impurity concentrations of the order of 1 in 10^9 and 1 in 10^{12}, respectively. The former level of purity is technically possible, but not the latter, although impurity levels in silicon of at least as low as about $10^{17}\,\mathrm{m}^{-3}$ or better than 1 in 10^{11} have been attained. Suppose we have, for example, a sample of n-type silicon with a realizable resistivity of 50 Ω m. Since in n-type material the minority hole density is negligible (Section 5.2), the resistivity is now given simply by

$$\rho_n = 1/n_n e\mu_n. \qquad (5.6)$$

from which we deduce that the donor density would be about $10^{18}\,\mathrm{m}^{-3}$, that is, less than one donor per 10^{10} silicon atoms. Although a value of 50 Ω m is much higher than in electronic devices like transistors, it is, as noted, realizable in practice and is typical of values encountered in some high-resistivity detectors for charged particles.

Let us now try to design a 'conduction counter' made of this material. This is a detector made of a homogeneous slice of the material with both contacts of a simple, 'ohmic' nature to be discussed later (Section 5.4) and one of them thin enough to admit the radiation. Suppose that the applied voltage V is $10\,$V, the detector thickness d is $200\,\mu$m, and the area A of each face is $10\,$mm $\times\ 10\,$mm, or $10^{-4}\,$m^2. Consider an alpha particle, whose energy is taken for convenience as $3.62\,$MeV, incident on one face, say the negative one. Its range in silicon is only just over $10\,\mu$m, so it will generate by ionization electrons and holes immediately adjacent to this face. The number N of each will be 10^6, since $3.62\,$eV is required to create an electron–hole pair.

The electrons drifting to the positive face about $200\,\mu$m away will induce most of the signal in an external circuit (Section 3.2.3 discusses signal formation in the context of gaseous ionization chambers; the same principles apply here, and are considered in more detail in Chapter 6). The electron drift velocity is $u = \mu_n \mathscr{E} = \mu_n V/d = 7 \times 10^3\,$m s^{-1} (which, we note in passing, is much less than the saturation velocity). They therefore take a time $t = d/u = 2.9 \times 10^{-8}\,$s to drift to the positive electrode. Note that this is a very short time, so there is indeed, as we remarked earlier, no difficulty in collecting the electrons, their lifetime in silicon being orders of magnitude longer.

During this drift time, however, the electrons generate in an external low-resistance circuit a current $Ne/t = 6 \times 10^{-6}\,$A. This is a very small current in comparison with, for example, the steady leakage current $VA/\rho_n d$ across the detector. For the resistivity ρ_n of $50\,\Omega$ m which we assumed earlier this leakage current comes to $100\,$mA.

This obviously presents a difficult measurement situation. One might think that it would nevertheless still be possible to see the pulse of current produced by the alpha particle as a 'blip' on top of the steady leakage current. However, in Chapter 7 we shall see that there are unavoidable fluctuations or 'noise' on the leakage current which will tend to obscure the signal pulse. From the account of such noise in that chapter we can estimate that the noise inherent in measuring a signal pulse of duration $3 \times 10^{-8}\,$s in the present circumstances could be of order $10^{-7}\,$A, which is not negligible in this context compared with the signal current. For example, if we attempted to perform energy spectrometry the resulting peak from these alpha particles would be very broad. For lower-energy radiation such as X-rays the signal would be completely swamped by the noise. Even the use of silicon of still higher resistivity would fail to make a satisfactory conduction counter. With germanium, which has a smaller band gap, the situation is much worse; at room temperature the highest resistivity possible is the value of $0.45\,\Omega$ m for intrinsic material.

True conduction counters can normally be made only from materials with wide band gaps, such as diamond and mercuric iodide, as described in

Section 5.12. These do not, however, possess to the same degree the out-standing carrier transport properties of silicon and germanium. Fortunately there are ways to reduce the leakage current in silicon and germanium detectors, and we shall describe these in the following sections.

5.4 Junction detectors: principles

The creation of a rectifying junction drastically reduces the leakage current in a semiconductor detector, as we shall now see. For example, we mentioned briefly in Section 4.4.3 what happens in a single crystal of semiconductor containing a region of n-type material and a region of p-type material. Let us recall what happens in this situation if we imagine two separate pieces of the same semiconductor (say, silicon), one n-type and the other p-type (Figure 5.4), being brought together into perfect electrical contact (if it were possible), such that charge carriers are as a result able to migrate across this p–n junction. Majority holes in the p region see a much smaller concentration of minority holes in the n region. Thus those in the p region can diffuse down the hole concentration gradient to the n region and recombine there

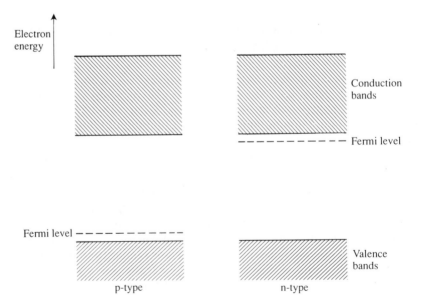

Figure 5.4 Energy band diagram for a p- and an n-type semiconductor before being brought into contact.

with electrons. Similarly electrons can diffuse from the n to the p region and recombine there with holes.

This hole diffusion leaves a region of fixed acceptors depleted of its holes in the p material next to the junction. This produces a layer of immobile negative charge. Similarly the electron diffusion leaves in the n region next to the junction a layer of immobile, positively charged, depleted donors equal in number to the depleted acceptors. These layers of space charge constitute the 'depletion layer' (or 'depletion region'), as shown in Figure 5.5 (which is a redrawn version of Figure 4.8). These charge layers set up a small contact potential, or 'diffusion potential', and an electric field across the region. As Figure 5.5 shows, the Fermi levels in the p and n regions are now at the same height, as must be the case for equilibrium.

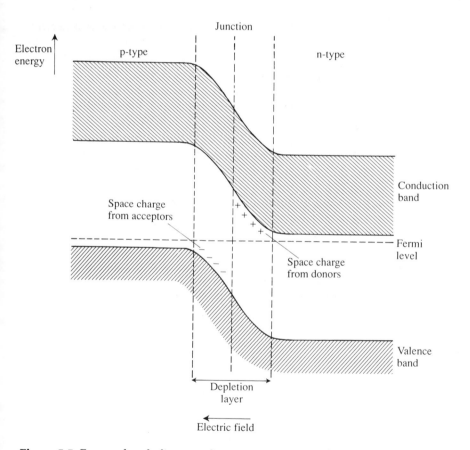

Figure 5.5 Energy band diagram for a p–n junction, showing donors and acceptors forming layers of immobile space charge in the depletion region.

No current flows in equilibrium once the diffusion potential has been set up, since the potential acts basically as a barrier to further diffusion of majority carriers. A few of the more energetic ones do in fact manage to diffuse across — holes to the n region and electrons to the p region. Their associated currents are, however, each counterbalanced by drift currents, through the depletion layer field, of minority carriers, such the minority holes drifting in the electric field from the n region balance the majority holes diffusing from the p region, and minority electrons similarly drifting from the p region balance the majority electrons diffusing from the n region. The hole and electron currents are thus each separately zero as well as the overall current.

Even with no external applied voltage, the depletion region with its associated diffusion potential across it can function as a detector of any ionization deposited in it by radiation. The electrons and holes which are created are separated by and drift in the built-in field so as to induce an observable current pulse in an external circuit. However, for a number of reasons (one of which is the necessity to increase the size of the depletion layer and thus the sensitive volume of the detector), an applied voltage to bias the detector in a reverse sense is always used. This reinforces the built-in potential and field across the high-resistance depletion region, and, as we shall see later in this section, increases its width, as desired. It also provides more efficient collection of the carriers produced by radiation. Note that the minority carrier flows are now no longer counterbalanced by those of the majority carriers because the latter are almost completely held back by the increased potential barrier. Thus a small reverse current now exists; it consists of the minority carriers from either side drifting, as before, through the electric field in the depletion layer. The 'noise' due to this reverse current will be taken account of in our discussions in Chapter 7.

As the potential and field are increased, so are the amounts of space charge on either side of the junction (these amounts remaining equal to each other, as before). Since the space-charge density is fixed by the depleted acceptor and donor density, the width of the depletion layer on either side of the junction increases, as we mentioned a moment ago. We therefore have very good conditions for radiation detection — the sensitive volume (that of the depletion layer) is increased, and charge collection becomes very efficient because of the increased field. Furthermore, as we shall also see in Chapter 7, the electrical noise is reduced because of the decreased junction capacitance resulting from the increase in depletion layer width.

The standard 'planar' geometrical configuration for a junction detector is a thin disc or slice of silicon, with the junction parallel to the faces, on one of which the radiation is incident. We must remember, though, that the electric field in any low-resistance undepleted semiconductor on one or other side of the junction is small compared with that in the high-resistance depletion layer. It follows that any ionization created by radiation outside the depletion layer cannot be effectively collected before it recombines or is trapped. Such a region would act as a thick 'dead layer' if the radiation had to pass through it before entering the depletion region. Any such dead layer or 'entrance window', caused by this or any other factor (a metal electrode forming a front contact, for example) must therefore be thin.

We shall therefore consider only the situation where one side of the junction, the p side, say, is thin. For various reasons which become clear in Section 5.5, the p side will also be taken as heavily doped, giving us a so-called 'p$^+$-n' junction. These assumptions will also simplify the subsequent mathematics, and will enable us to extend the results to the types of junctions other than p-n ones which we shall see are also used. The configuration is thus as shown in Figure 5.6(a). The back contact, which may be a second metal electrode, is ohmic (non-rectifying) in character.

We now want to calculate the lateral extent of the depletion layer for a given reverse bias. Poisson's equation relates the potential V_x at any point x in the depletion layer with the volume charge density at that point, ρ_e, by (in its one-dimensional form)

$$\frac{d^2 V_x}{dx^2} = -\frac{\rho_e}{\kappa \epsilon_0}$$

where κ is the relative permittivity (or dieletric constant) and ϵ_0 the permittivity of free space. For a p$^+$-n junction, that is, one with the p side heavily doped, we need to solve this equation for the n side only, since we shall see that this is the region into which the depletion layer most penetrates. Here therefore

$$\frac{d^2 V_x}{dx^2} = -\frac{e N_D}{\kappa \epsilon_0} \tag{5.7}$$

where N_D is the density of the positively charged ionized donors in the n region.

To determine the electric field $\mathscr{E} = -dV_x/dx$ at any position x we integrate this equation once. We take the p$^+$-n junction itself to be at $x = 0$, and the edge of the depletion layer to be at $x = d$, as Figure 5.6 shows. At the edge the electric field is near zero, since it must match up with the low electric field existing in the undepleted bulk, and this is a boundary condition which the integration must satisfy. The result is then

$$\frac{dV_x}{dx} = \frac{e N_D}{\kappa \epsilon_0}(d - x) = -\mathscr{E}. \tag{5.8}$$

The electric field profile thus varies linearly through the detector, and is here negative, as shown in Figure 5.6(b).

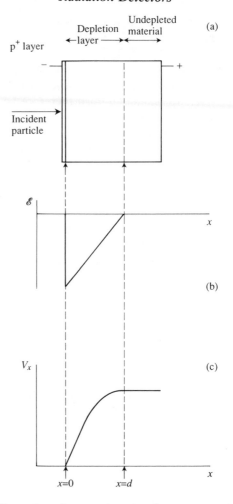

Figure 5.6 (a) Configuration of a p^+–n junction detector. (For clarity the detector thickness is exaggerated.) The front contact forms part or all of the entrance window, and the bulk of the detector is formed from n-type material. (b) The variation of the electric field $\mathcal{E}$ through the detector. Note $\mathcal{E}$ is negative through the depletion layer. (c) The variation of the potential V_x through the detector.

A second integration yields the potential V_x as a quadratic function of x:

$$V_x = \frac{eN_D}{\kappa\epsilon_0} x \left(d - \frac{x}{2} \right). \tag{5.9}$$

where V_x is taken to be zero at $x = 0$, as shown in Figure 5.6(c).

A complete derivation would indicate that in the p^+ region the electric field and potential also vary linearly and quadratically, respectively, out to the edge of the depletion layer. Here, however, the immobile charge density is high, and so therefore in order to achieve equal numbers of depleted acceptors and donors on either side of the junction, the lateral extent of the depletion layer into the p^+ region (not shown in Figure 5.6) is small. In addition, because the area under the $\mathscr{E}$ versus x line is small, the potential drop (equal to $-\int \mathscr{E} dx$) across the depletion layer in the p^+ region is also small.

It therefore follows that the applied reverse bias voltage V, if the small diffusion potential described earlier is negligible, is given from equation (5.9) by the value of V_x at $x = d$:

$$V = \frac{eN_D d^2}{2\kappa\epsilon_0}. \tag{5.10}$$

The depletion layer thickness is hence

$$d = \left(\frac{2\kappa\epsilon_0 V}{eN_D}\right)^{1/2}. \tag{5.11}$$

Since (from equation (5.6)) the resistivity ρ_n of the n-type region is given by $1/n_n e\mu_n = 1/N_D e\mu_n$, equation (5.11) may be rewritten as

$$d = (2\kappa\epsilon_0 \mu_n \rho_n V)^{1/2}. \tag{5.12}$$

For an n^+–p junction, in which the depletion layer extends largely into the lightly doped p region, the thickness of the depletion layer is

$$d = (2\kappa\epsilon_0 \mu_p \rho_p V)^{1/2} \tag{5.13}$$

ρ_p being the resistivity of the p region and μ_p the hole mobility. For silicon, since $\kappa = 12$ and μ_n and μ_p at room temperature are 0.14 and 0.05 $m^2 V^{-1}s^{-1}$, respectively, equations (5.12) and (5.13) give

$$d = 5.5 (\rho_n V)^{1/2} \tag{5.14}$$

for n-type material and

$$d = 3.3 (\rho_p V)^{1/2} \tag{5.15}$$

for p-type material, where d is in μm, ρ in Ωm and V in V. These last two expressions assume of course that the mobilities are constant; in fact they tend to decrease somewhat in general with increasing field, doping density, and temperature. Nevertheless we can see that in order to achieve a deep depletion layer, that is, a thick sensitive region for junction radiation detectors, ρ and V must be high. Particular importance is therefore placed on obtaining high-purity material. For n-type material of resistivity 50 Ωm, d is about 550 μm (0.55 mm) at 200 V reverse bias. Depletion depths up to about 5 mm in silicon are commercially available. (These resistivities and depletion depths are incidentally much higher than in electronic devices such as transistors and integrated circuits, where large depletion layers are not required.)

The maximum electric field, $\mathscr{E}_m$, in the depletion layer is at the junction itself, at $x = 0$. From equations (5.8) and (5.10) its numerical value (omitting a negative sign) is

$$\mathscr{E}_m = \frac{2V}{d} \tag{5.16}$$

which for a p^+-n structure becomes, on substitution for d from equation (5.11),

$$\mathscr{E}_m = \left(\frac{2eN_D V}{\kappa\epsilon_0}\right)^{1/2} \tag{5.17}$$

or, if instead we substitute for d from equation (5.12),

$$\mathscr{E}_m = \left(\frac{2V}{\kappa\epsilon_0\mu_n\rho_n}\right)^{1/2} \tag{5.18}$$

with corresponding expressions for an n^+-p structure. (Compare, by the way, the expression $2V/d$ for $\mathscr{E}_m$ — equation (5.16) — with the corresponding result for a gas ionization chamber, where the field is everywhere constant at a value of V/d. The former is a consequence of the variation of $\mathscr{E}$ with x as given by equation (5.8).)

In silicon the values of $\mathscr{E}_m$ are hence given, in $V \mu m^{-1}$, from equation (5.18) by

$$\mathscr{E}_m = 0.37 \left(\frac{V}{\rho_n}\right)^{1/2} \tag{5.19}$$

for an n^+-p structure. For a p^+-n structure the corresponding equation is

$$\mathscr{E}_m = 0.61 \left(\frac{V}{\rho_p}\right)^{1/2}. \tag{5.20}$$

These fields can easily exceed $1 \ V \mu m^{-1} (10^6 V \ m^{-1})$ with moderate applied voltages, with consequent possibilities of excellent charge collection in detectors.

It is, however, important to limit the voltage applied to below the breakdown voltage in order to avoid irreversible damage to a detector. In principle bulk silicon can withstand fields of well over $10 \ V \mu m^{-1}$. In a practical detector configuration, however, the maximum permissible voltage specified by the manufacturer will be much lower than that calculated with equation (5.19) or (5.20) from this field value because of surface leakage and detector edge effects.

Because a junction detector possesses two regions of space charge of opposite sign facing each other, the structure possesses capacitance, the value of which can be readily calculated. Per unit area it is

$$C_A = \frac{\kappa \epsilon_0}{d}$$

which for a p^+-n structure gives, from equation (5.11),

$$C_A = \left(\frac{\kappa \epsilon_0 e N_D}{2V} \right)^{1/2} \tag{5.21}$$

or, from equation (5.21),

$$C_A = \left(\frac{\kappa \epsilon_0}{2\mu_n \rho_n V} \right)^{1/2} \tag{5.22}$$

again with corresponding expressions for an n^+-p structure. These formulae give incremental or small-signal a.c. capacitance values, namely, the increase in charge per unit area for a small increment in voltage. This is the appropriate parameter to consider when a small signal induced by radiation is derived from a detector in which the capacitance varies with applied voltage. Equation (5.22) gives in $pF\,mm^{-2}$

$$C_A = \frac{19}{(\rho_n V)^{1/2}} \tag{5.23}$$

for silicon p^+-n detectors. For n^+-p detectors the corresponding expression is

$$C_A = \frac{33}{(\rho_p V)^{1/2}}. \tag{5.24}$$

Thus for a $50\,\Omega\,m$ resistivity $100\,mm^2$ area p^+-n detector biased at $200\,V$ the capacitance is $19\,pF$. In Chapter 7 we shall see that the amplifier noise is always reduced and the resolution improved as this capacitance is reduced. As well as low areas this implies thick depletion layers and thus high resistivities and voltages.

There is a further implication in the variation of capacitance with applied voltage. In Chapter 6 we shall see that this variation has a decisive influence on the type of preamplifier normally used with junction detectors.

5.5 Silicon junction detectors: practice

5.5.1 *General considerations*

Most simple junction detectors are made of silicon. Compared with types we shall meet later they have relatively small depletion depths, which are typically between 60 μm and 5 mm. They are particularly suitable for alpha and beta particles and other energetic charged ions above about 10^4 eV. If the thickness of the depletion layer exceeds the particle range, the energy of the particle can be measured from the amount of ionization collected. This condition is easily satisfied for alpha particles, for instance, since their ranges are short, usually less than 50 μm in silicon. The range of a 1 MeV beta particle is about 2 mm, so, unless the energy is much greater, it is also possible to meet this condition for betas. However, junction detectors are relatively insensitive to gammas and energetic X-rays because of their comparatively shallow depletion depths and the low absorption coefficients shown by silicon (Figure 2.10(b)). Other types of semiconductor detectors are better suited for these purposes, and these are discussed later in this chapter.

Silicon junction detectors for charged particles possess many attractive features, most of which will be discussed further in this section. These include freedom from drift, good timing characteristics, entrance windows which in some cases are very thin indeed, simplicity of operation, and flexibility of size and shape. This last feature refers, for example, to the range of detector areas available, from under 10 to at least 2000 mm^2, although large-area detectors with their larger capacitances show poorer energy resolution because of increased noise. In addition, as well as discs, annular-shaped detectors are made for use in backscattering measurements. Here a target is irradiated by a particle beam passing through the central hole in such a detector, which itself is looking at the particles backscattered from the target. Detectors totally depleted throughout their volume are also available, and we shall consider this further in Section 5.5.

Historically, the first junction detectors, which were made during the 1950s, used germanium rather than silicon. This mirrored the development of germanium diodes and transistors in electronics at the time. With its smaller energy gap, however, germanium has inherently higher carrier concentrations, and cooling is essential for the successful operation of detectors made from it. Following their introduction in 1959 silicon junction detectors, which could operate with no cooling, in general rapidly superseded those made from germanium. However, germanium junction detectors later reappeared for X- and gamma-ray detection, as will be discussed in Sections 5.8 and 5.9.

Silicon charged-particle detectors can be made by a number of different methods which we shall consider in the next three sub-sections.

5.5.2 *Diffused junction detectors*

These detectors are made by various techniques similar to the manufacture of electronic diodes by planar diffusion of impurities. A layer of n-type dopant, usually phosphorus, is diffused at fairly high temperatures, up to 800 °C, into a slice of p-type silicon. Diffusion from the gaseous phase can be used, or alternatively a phosphorus-based paint can be initially applied to one face. The depth of the phosphorus diffusion is typically about 1 μm. An n^+-p structure therefore results, and when a reverse bias is applied to suitable contacts the depletion layer extends almost entirely into the p region. The n^+ layer forms an entrance window which is relatively thin compared with the depletion layer, but which is fairly rugged in use.

The real difficulty with diffused junction detectors is that their entrance windows, which are about 1 μm, are still thick compared with those achieved by other methods described below, such as surface barriers or ion implantation. A 5 MeV alpha particle loses around 0.1 MeV energy in such a thickness. A 100 MeV, 100 u fission fragment, for which the stopping power is highest not at the end but at the beginning of its range (Section 2.2.2), would lose as much as 10 MeV before the active depletion layer is reached. A second problem is that of increased charge trapping and recombination as a result of the reduction in carrier lifetime which high-temperature diffusion tends to cause.

On the positive side, specially designed diffused junction detectors have in fact shown satisfactory charge collection with alpha particles at temperatures right down to 1.4 K. Reference 1 gives details as well as describing how they could be used in conjunction with searches for the 'dark matter' particles which will be mentioned in Section 8.5.

5.5.3 *Surface barrier detectors*

The earliest junction detector in 1951 used a metal–semiconductor rectifying contact at the entrance window rather than a p–n junction, and this is the basic principle of the surface barrier detector. It remains to this day one of the most widely used detector types for charged particles.

The potential barrier which is referred to in the name of the detector is a type of 'Schottky' barrier acting at the entrance window as a rectifying barrier to current flow. We saw in Section 4.4.3 that the theories which have been developed to describe the formation of such barriers consider the work function in the metal and electron affinity in the semiconductor, together with the effect of any states acting as electron traps which may exist at the interface. In the present cast the most usual combination is a gold to n-type silicon junction, which is the complementary case to that which we discussed in Chapter 4. Here, in order for the Fermi levels to be equalized, a layer of the n-type silicon is depleted of majority electrons through their transfer either to the surface states, which are believed to be of particular importance in our case, or to the metal. (We have seen that in an analogous manner the majority holes in the p region of a p–n junction deplete a layer of n-type material of electrons by capturing them through recombination.)

The gold is in the form of a very thin layer which is evaporated on to the silicon surface. For our purposes the resulting rectifying contact behaves rather like a p^+–n junction, with the metal playing the role of a very thin p^+ region. The various equations derived in Section 5.4 for the depletion layer thickness, electric field, and capacitance of a p^+–n junction thus also hold for a surface barrier detector.

The procedures for making surface barriers have evolved empirically. The surface is etched chemically and then allowed to stand in air to form a very thin layer of silicon oxide. This oxide layer is perhaps only a few nanometres thick (around $10\,\text{mg m}^{-2}$ or $1\,\mu\text{m cm}^{-2}$). This is even thinner than the gold layer, which is typically about 20 nm thick ($0.4\,\text{g m}^{-2}$ or $40\,\mu\text{g cm}^{-2}$). As a result a rectifying contact with a very thin entrance window is produced. The contact on the other face is an evaporated aluminium contact also of thickness around $0.4\,\text{g m}^{-2}$ or $40\,\mu\text{g cm}^{-2}$. In this case the contact between the materials is such that the barrier to the flow of current is relatively low, with no rectification occurring, and so an 'ohmic' contact results.

Surface barrier detectors have very good charge-collection properties because, unlike the thermal diffusion of dopants for diffused junctions, the barrier manufacture introduces no substantial degradation of carrier lifetime in the silicon from which a detector is made. The very thin entrance window is another major advantage of a surface barrier detector. A gold window of thickness as given above absorbs only about 10 keV for 5 MeV alphas and 1 MeV for 100 MeV fission fragments. This is at least ten times better than for the entrance window of a diffused junction. With these thin dead layers, variations in window energy loss due to straggling or variable angles of incidence are minimized, a feature particularly important in high-resolution studies.

It should be borne in mind that care is needed with such thin gold windows. Being optically partially transparent, they can make detectors light sensitive, as visible photon energies are greater than the energy band gap of the semiconductor. Often, of course, they are operated in light-tight vacuum chambers, in which case this disadvantage disappears. Surface barriers are also susceptible to contamination from, for example, vacuum-pump oil vapours. Furthermore, since the front surfaces are easily damaged, they must never be touched.

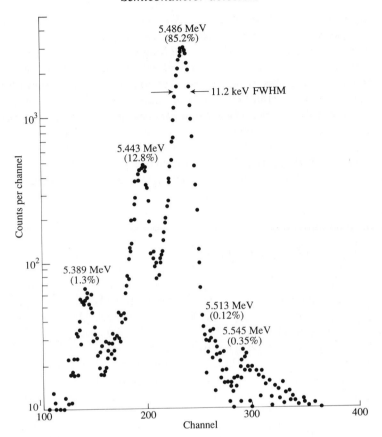

Figure 5.7 A spectrum of alpha particles from ^{241}Am using a surface barrier detector. The spectrum is 300 channels wide. Note that the channel number is here not proportional to the alpha-particle energy because the spectrum has been expanded around these peaks by a technique to be discussed in Section 6.6.1. (Reproduced by courtesy of EG&G ORTEC.)

Figure 5.7 shows an example of a pulse-height spectrum and the FWHM resolution (as defined in Section 4.5.1) obtained with alpha particles stopped in the depletion layer of a surface barrier detector. Here (as in Figure 4.17) the pulse heights produced (which, as usual, are closely proportional to the kinetic energies of the particles) are shown as a differential spectrum, with the heights grouped into consecutively numbered channels. FWHMs of less than 12 keV can be achieved, so we can easily distinguish peaks separated by, say, 20 keV. Similar results are obtainable with other light ions. With their thin windows and excellent charge-collection properties, silicon surface barriers make very good energy spectrometers for such ions.

Electrons have for a given energy much longer ranges than those of ions. Nevertheless, except for the highest energies, silicon junction detectors make very effective energy spectrometers for electrons. Figure 5.8 shows a spectrum from a surface barrier detector of secondary electrons electrostatically accelerated to only 30 keV. These originate from a caesium iodide dynode surface (as has been used in image intensifiers) which is bombarded with electrons. The higher peaks at integral multiples of 30 keV correspond to groups of 2, 3, 4, ... secondary electrons simultaneously arriving at the detector, where each group always originates from a single primary electron event.

5.5.4 *Ion-implanted detectors*

It has been known for a long time that a rectifying junction can be made in silicon by bombarding the surface with accelerated dopant ions. Junction detectors with dopants introduced by this technique of ion implantation are now routinely available. Either n^+ or p^+ layers can be formed by accelerating, for example, phosphorus or boron ions, respectively. By control of the energy, usually around 10 keV, the ion range can be closely defined, and an accurately 'tailored' impurity profile can be built up. Consequently the entrance window can be very thin. Radiation damage (Section 5.11) introduced by the ions has to be subsequently annealed out, but the temperature required (500 °C) is much lower than that needed to diffuse impurities to

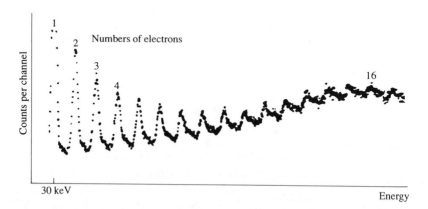

Figure 5.8 A surface barrier detector spectrum of groups of 1, 2, 3, 4, ... secondary electrons, in which each electron has been electrostatically accelerated to an energy of 30 keV. (C.F.G. Delaney, unpublished data.)

form a diffused junction. Carrier trapping is thus quite low, and, with the low reverse currents which occur, resolutions available are comparable with those using surface barriers. Ion-implanted detectors are also much more rugged than surface barriers, since in the former case the surface can be protected with a very thin layer of thermally grown silicon dioxide.

5.5.5 *Totally depleted detectors*

We have now considered the three main techniques used for making a rectifying junction in a semiconductor detector. In this and the next two subsections we discuss three situations in which special conditions apply for the collection of the charge deposited.

In a totally depleted detector the depletion layer is made to extend throughout the volume up to the back, ohmic contact. Figure 5.9 shows field profiles for a partially depleted detector, a just fully depleted detector, and a slightly and a strongly overdepleted detector. In all four cases the slope $d\mathscr{E}/dx$ of the field $\mathscr{E}$ is the same, since from Poisson's equation (as given by equation (5.7)) the slope (which is equal to $-d^2 V_x/dx^2$) is fixed by the density of the space charge due to the ionized donors (or acceptors) in the depletion layer. In other words the slope is fixed by the level of doping. The area between the field profile and the x axis gives the magnitude of $\int \mathscr{E}\, dx$, which is of course (apart from a minus sign) just the applied voltage. Thus line (d), for example, corresponds to the highest value of the voltage.

Totally depleted surface barrier detectors are made with thicknesses from 10 μm (or even less) up to about 5 mm. The thinner detectors, some of which

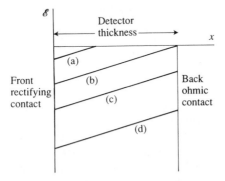

Figure 5.9 The variation of the electric field $\mathscr{E}$ for a detector (made from n-type silicon) which is (a) partially depleted, (b) just fully depleted, (c) slightly overdepleted, (d) strongly overdepleted.

are very fragile, can be made by the techniques of epitaxial growth. These involve the deposition of a thin layer of uniformly doped semiconductor from the vapour phase on to a semiconductor substrate. This substrate is then etched away to leave only the epitaxial layer as a thin wafer of highly uniform thickness. A rectifying and an ohmic junction are then made on the layer in the usual way. Thin epitaxial detectors are particularly valuable for heavy ions of high stopping power and short range, such as fission fragments.

A very common application is as a 'ΔE detector'. In this mode the detector is overdepleted so that the field extends all the way through, and the detector thickness is chosen such that the particles being detected pass right through, rather than stop in, the detector. Thus only the energy ΔE lost in the detector is measured, rather than the total energy. If the detector is thin compared with what the particle range would be in thick material, a signal is observed which is proportional to the stopping power $-dE/dx$ at the energy concerned.

Normally the detector is in a transmission mount such that the emergent ion can strike a second detector behind the first in a so-called 'telescope' arrangement. If the particles are arranged to stop within the depletion layer of the second detector, then, in addition to the measurement of ΔE, the responses of the two detectors can be summed to give the total particle energy. This assumes that the back ohmic contact of the ΔE detector is thin, which is normally the case in any such detector. It also assumes that the detector is sufficiently strongly overdepleted that the field is high enough even at the ohmic contact (as shown by line (d) in Figure 5.9) to ensure efficient charge collection along the complete path of the particle.

The importance of the combined measurements of E and ΔE is that they can be used to help to identify the incident particles. This is because for a given E the value of ΔE depends on the particle mass and atomic number. For example, for a completely ionized particle of mass M, atomic number z, velocity v, and kinetic energy E, Bethe's formula predicts for the stopping power $-dE/dx$ that (equation (2.1))

$$- \frac{dE}{dx} \propto \frac{z^2}{v^2} \propto \frac{Mz^2}{E}$$

apart from a factor only weakly dependent on v or E. The product of E and dE/dx is therefore proportional to Mz^2 and nearly independent of E. This information can be used to help to identify the particle. This or similar schemes are widely used in particle-identifier telescopes.

Totally depleted detectors are useful elsewhere as well as in telescope arrangements. The absence of an undepleted region for ordinary energy measurements means that the series resistance which exists between the depletion layer and the back contact is eliminated. We shall see in Section 5.5.9 that there can be advantages with such detectors in timing measurements.

5.5.6 *Avalanche detectors*

Avalanche detectors are the solid-state analogue of gas proportional counters. In such a detector the electric field in the depletion region of a p–n junction is raised so high that the drifting electrons create secondary ionization through an avalanche process. The internal gain is obviously attractive for the detection, for example, of low-energy radiation, because this gain amplifies only the signal and none of the subsequent noise generated in the electronics. However, this is at the expense of introducing statistical noise from the multiplication process, just as for proportional and scintillation counters (Chapter 7). With the high electric fields the carriers are quickly collected, so timing properties can be quite good. Care is needed with the design of the detector geometry in order to minimize the chance of unwanted breakdown due to high surface fields, and also to achieve a reasonably uniform gain throughout the detector volume. They have been used for low-energy X-ray detection, where their resolution is comparable with that of the best proportional counters.

5.5.7 *Position-sensitive detectors*

Many applications require the measurements of the position of arrival of radiation, for example, in the focal plane of a mass spectrometer using a magnetic field, in medical imaging applications, or in the measurement of particle tracks in high-energy collisions. Semiconductor junction detectors can and have been assembled in suitable arrays or mosaics of separate detectors. Information about both position and energy is available in this way. Clearly the handling of data from many detectors can get quite complex. Of course, such a position-sensitive array does not have to use semiconductor junction detectors; any detector type can in principle be used.

Much better spatial resolution in one dimension can be achieved with a detector which has separately addressed electrodes on a single slice of semiconductor. For example, one of the earliest techniques used was to evaporate gold through a comb of wires onto a silicon slice to produce a multi-electrode surface barrier detector. Modern techniques for the fabrication of electronic devices such as integrated circuits can be adapted with considerable success. Silicon strip detectors, used in recent work in high-energy physics, typically consist of p-type strips of implanted boron on an n-type substrate, with a single ohmic contact at the back. This forms an array of multiple detectors whose spatial resolution is set by the separation of the strips. This can be as good as a few micrometres. The use of these silicon fabrication techniques also introduces the possibility of integrating the preamplifiers for each strip into the same silicon slice.

The use of many electrodes on a single substrate does not easily remove the problem of complex electronics which we saw also existed with arrays of separate detectors. An electronically much simpler technique, which, however, does not possess the position or energy resolution of strip detectors, is to construct an ordinary surface barrier or p–n structure on a narrow rectangular strip of substrate, on which is a back ohmic contact of finite resistance as measured along the strip. Separate signals are taken between ground and the front contact and between ground and the back contact, as shown in Figure 5.10.

The first signal measures the particle energy E deposited in the usual way. The second is, however, affected by the place of entry of the radiation. The resistive strip can be shown to act as a charge divider, as indeed is intuitively obvious. If the strip is of uniform resistance per unit length, the second signal is proportional to Ex/L, where x is the distance of the incident radiation from the grounded end and L is the strip length. Values for E and x can be

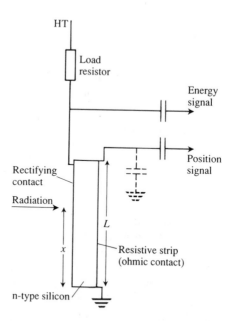

Figure 5.10 A resistive strip contact position-sensitive detector. The strip length L is much greater than the detector dimension perpendicular to the diagram, and for clarity the detector thickness has been exaggerated.

derived from the two signals with resolutions of 30 keV (for alphas) and 100 μm, respectively, for $L = 10$ mm. Lengths up to 50 mm are easily available. Variations include partially and totally depleted constructions and resistive front rather than back contacts. A variation in mode of use of this detector is to sense the position from the times of occurrence of the signals.

A newer type of position-sensitive detector with a resolution of a few micrometres is the so-called 'drift detector' or 'drift chamber' (first described in reference 2). This has a completely new geometry for the field which collects the charge, as can be seen from Figure 5.11. In essence we see that an n-type slice of silicon has p^+ ion-implanted junctions constructed on both faces in the form of parallel strips, and an n^+ anode strip at one side. If the p^+ junctions are reverse biased by the same voltage the depletion layers formed at each junction will, as the bias is increased, eventually meet in the middle of the slice. Now suppose an ionizing particle is incident on the face shown. Holes will be quickly collected by the p^+ electrodes. The electrons, however, will collect at what is for them the central potential minimum where the depletion layers meet: we thus refer sometimes to a potential 'gutter' or 'valley' for electrons between the p^+ electrodes.

These electrons are then made to drift down the gutter to the edge of the depletion layer at the anode, and are there collected. This drift can be achieved if progressively higher potentials are applied to successive strips as the anode is approached, such that the gutter is made to slope downwards, as Figure 5.11 also shows. Alternatively, continuous p^+ electrodes can be constructed, one on each side of a detector and each with a finite resistance, so that they can sustain potential differences along their lengths. Other techniques which can be used to make the gutter slope downwards include making either the dopant concentration or the slice thickness vary along the length of the gutter. Whichever method is used, complete depletion has to be maintained along the slice so that everywhere an electric field exists in which the carriers can drift to the electrodes.

The position of the incident radiation with respect to the anode can be calculated from the transit time of the electrons as they drift in the electric field down the potential valley towards the anode. The field in practice is quite low, about 10^{-2} to 10^{-1} V μm^{-1}, so it follows from the known mobility in silicon (Section 5.3) that this time is around a few hundred nanoseconds per millimetre drift distance. To measure the drift time we need to know when the radiation struck the detector, together with when the electrons reached the anode. The former can be measured either from the signal induced by the holes in the p^+ contacts as they drift towards them, or from a signal from a second detector also triggered by the radiation. As for the electron arrival time at the anode, it can be shown that because of the shielding effect of the p^+ electrodes the pulse induced in the anode circuit essentially appears only as the electrons arrive at the anode. (This, of course,

(c)

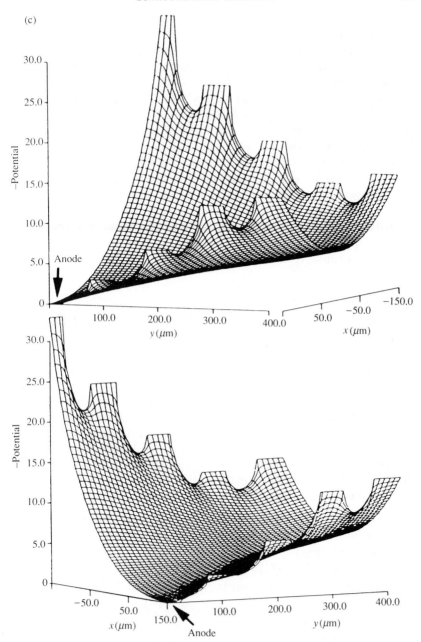

is unlike that for a simple two-electrode detector, in which the induced pulse starts as soon as carriers begin to move in the detector field.)

An important property of these detectors is that once full depletion has been achieved the capacitance between the n^+ anode and the p^+ electrodes drops to a very low value of 2 pF or less. This is because the capacitance appears across the small area of the depletion layer edge facing the anode, rather than across the large areas of the edges parallel to the p^+ electrodes prior to full depletion. We have said already that the energy resolution of a detector is better if the overall capacitance is low, this being because the associated noise in the preamplifier is reduced. Drift detectors can therefore be also used simply as large-area low-capacitance spectrometers. For this, since timing information is no longer required, the drift field is made much higher in order to minimize any carrier trapping effects.

Drift detectors can be made from germanium, rather than silicon, if the types of applications described in Section 5.8 and 5.9 are envisaged. An example is given in reference 3.

5.5.8 *Response to heavy ions*

We recall from Chapter 2 that when an alpha particle is slowed down in, say, a detector its ionic charge drops to zero at the very end of its range as it picks up the two electrons required to neutralize it. For a heavy ion such as a fission fragment being slowed down, it is found that electrons are picked up all along its range so that its ionic charge steadily decreases. Towards the end of its range the ionic charge falls to near zero, so it then ionizes less efficiently even though the energy may be still perhaps 1 MeV or more. However, it is found that it experiences at the end of its range competing elastic collisions with the atoms of the detector material. The resulting slow-moving recoils themselves ionize very inefficiently for similar reasons. Consequently two significant effects occur in a detector in which such an ion is completely stopped. These are:

1. A small percentage of the deposited energy that would be expected to appear as ionization does not in fact do so, but is instead lost to these 'atomic' collisions.
2. Since the number of these collisions per ion is quite low, the statistical fluctuations in the energy lost to them are relatively high.

As a result of the second effect the resolution obtained in heavy-ion spectra is often limited not only by detector and amplifier noise (as considered in Chapter 7) but also by broadening due to atomic collisions. Even for alpha particles the FWHM due to collision broadening is 7 keV at 5 MeV energy, which thus contributes a substantial part of the total observed resolution (perhaps 12 keV FWHM) with a good detector. (Equation (7.42), for example, will show how to combine the effects of broadening due to different causes.) For 100 u light fission fragments the broadening due to collisions is about 1.5 MeV, and for 140 u heavy fragments 2 MeV, which in both cases essentially determines the observed FWHM. This latter figure is nearly 3 per cent of the kinetic energy of a heavy fragment, which on average is about 70 MeV.

This situation is in sharp contrast to that with gamma-ray spectra, which, as we saw in Chapter 4, are detected by the electrons they produce. Because of its low mass an electron can transfer almost no energy in collisions with a heavy object like an atom, as momentum conservation shows, and so the resolution is limited almost entirely by the statistical noise effects described in Chapter 7, and not by atomic collisions. Similarly, electron energy spectra themselves are unaffected by atomic collision effects.

The ionization loss from (1) above is a major contribution to the 'pulse-height defect', which we call Δ. We define this for a given ion of given energy E_H as the difference $E_H - E_L$ between the ion energy and the energy E_L that a light particle needs in order to produce the same pulse height (Figure 5.12). When a defect occurs the pulse height is thus normally not quite proportional to the energy deposited. Some values of the observed defect are 12 keV for 5 MeV alphas, several MeV for heavy fission fragments, and, of course, zero for electrons (and thus gammas).

In general the total defect is a sum of the effects due to atomic collisions and any entrance window losses together with charge losses due to trapping and recombination. Charge losses are especially important for high-energy heavy ions, which are highly ionizing. In this case recombination

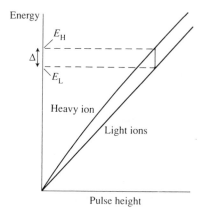

Figure 5.12 Diagram for definition of Δ.

almost inevitably occurs in the high-density 'plasma' of electrons and holes initially formed along the ion track. Losses occur even with the high fields in the special heavy-ion surface barrier detectors which are available. In these the resistivities and depletion depths are typically as low as $3\,\Omega$ m and $60\,\mu$m, respectively, with maximum fields of as high as 2 V μm^{-1}, even with applied voltages of well under 100 V (see equations (5.14) and (5.19)).

The defect can in special circumstances be reduced essentially to zero if highly collimated heavy ions are 'channelled' down one of the major crystallographic directions in the crystal lattice of the detector silicon. In this way the ions avoid close encounters with the silicon atoms (rather like a person walking through a forest plantation parallel to the lines of trees instead of pushing straight through the forest). This eliminates atomic collision losses; in addition, because the channelled ions sample a lower electron density, the range is increased, the ionization density is reduced, and charge recombination can be made negligible. Consequently the defect (Figure 5.12) is removed, that is, the calibration line for heavy ions coincides with that for light ions, and the peak broadening due to atomic collisions disappears. These dramatic features, unfortunately, cannot in general be made use of, because of the high collimation, much better than $1°$, which is required. In fact a detector is often fabricated from a slice cut so that the perpendicular to the front face is tilted well away from any of these crystallographic directions where channelling may confuse a spectrum.

5.5.9 *Timing with junction detectors*

Semiconductor junction detectors can show responses which are very fast. They therefore find widespread use, when used with suitable electronics, in fast timing applications. We shall discuss the pulse of current resulting during the time it takes for the carriers to drift from their point of release to the collecting electrodes. (Reference 4 may also be consulted.) In energy spectrometry this current is integrated on an external capacitor to give the charge collected, as described further in Chapter 6. We shall see that although the charge pulse can be used for fast timing, the unintegrated current pulse should be used if possible for the best timing resolution. Observation without distortion of such a pulse is a demanding requirement, as we shall see below, but we shall assume for the moment that it can be done.

Let us consider for simplicity a quantity Q of charge liberated close to the junction at the front face of a partially depleted detector. This will be the situation for a particle whose range is much less than the depletion depth. In a p$^+$-n detector, say, the holes are then immediately collected at the negatively biased front, rectifying, contact, whereas the electrons drift into the depletion layer, eventually to be collected at the back, ohmic, contact. A complementary picture holds for an n$^+$-p detector. We neglect for the moment circuit loading effects due to any undepleted material. Let us first imagine the field to be constant right through to the opposite side of the depletion layer depth, d (rather than, as is in fact the case, to decrease linearly with distance in the way described in Section 5.4). In this situation the current would be constant with time at a value of $i_s = Qu/d$, where u is the velocity of the carriers which are drifting in the depletion layer. Now d/u gives the carrier transit time across the layer and hence the duration of the current pulse. This is the same result as that for the conduction counter in Section 5.3, where $Q = Ne$, and also for a single charge q crossing the ionization chamber in Chapter 3, with $d = L_0$.

In reality, of course, a partially depleted detector possesses a non-uniform field profile due to the fixed space charge of the uncompensated donors in the depletion layer. This can be shown, however, to be important only in so far as it affects the motion of the mobile charge as it is being collected. The current is still given by the expression $i_s = Qu/d$, where now the velocity u of the carriers constituting the charge Q decreases because of the decrease in field experienced by the carriers as they drift further into the depletion layer; the induced external current must now also decrease with time.

We saw in Section 5.4 that the numerical value of the electric field in the depletion layer of a partially depleted detector varies linearly from a maximum value of $\mathcal{E}_m = 2V/d$ (equation (5.16)) at the junction itself at a position $x = 0$, to zero at the depletion layer edge at $x = d$ (Figure 5.6). Consequently the electric field $\mathcal{E}$ at any position x is given by the expression

$$\mathcal{E} = \mathcal{E}_m \left(1 - \frac{x}{d} \right)$$

and so the carrier velocity u as a function of x is

$$u = \mu \mathcal{E} = \mu \mathcal{E}_m \left(1 - \frac{x}{d} \right)$$

where μ is the mobility of the drifting carriers. Writing u as $\mathrm{d}x/\mathrm{d}t$ and integrating we obtain

$$x = d \left[1 - \exp \left(-\frac{\mu \mathcal{E}_m}{d} t \right) \right]$$

and therefore, substituting this back into the previous expression for u, we get

$$u = \frac{d}{\tau} \exp \left(\frac{-t}{\tau} \right)$$

where τ is defined by

$$\tau = \frac{d}{\mu \mathcal{E}_m}. \tag{5.25}$$

The current as a function of time is thus given by

$$i_s = \frac{Qu}{d} = \frac{Q}{\tau} \exp\left(\frac{-t}{\tau}\right). \tag{5.26}$$

Mathematically this expression holds for all time since the carrier slows down in the decreasing field as it penetrates more deeply into the depletion layer, never quite reaching the edge of the layer. Clearly for fast timing purposes the current pulse described by equation (5.26) is very satisfactory, since the pulse rises immediately to a maximum value at the instant at which ionization is released. If, however, this pulse is integrated on a capacitor, the resulting charge pulse has a finite rise-time of τ (as we consider further in Section 6.3), which of course for fast timing is inferior.

The time constant τ characterizing the decay is more fundamental in character than this derivation appears to indicate. For n-type material, for example, d and $\mathscr{E}_m$ are given by $(2\kappa\epsilon_0\mu_n\rho_n V)^{1/2}$ and $(2V/\kappa\epsilon_0\mu_n\rho_n)^{1/2}$, from equations (5.12) and (5.18), respectively, where the symbols are as previously defined. Thus equation (5.25) gives for τ the expression $\kappa\epsilon_0\rho_n$. In a similar way τ is given for p-type material by $\kappa\epsilon_0\rho_p$. Alternatively, dropping the subscripts denoting n- or p-type, we can write in general that

$$\tau = \kappa\epsilon_0\rho. \tag{5.27}$$

This is a constant called the dielectric relaxation time (called τ_d) for the semiconductor from which the detector is constructed. For silicon the value of τ_d in nanoseconds is about 0.1ρ when ρ is in Ω m. For $\rho = 50\,\Omega$ m this is 5 ns; for $5\,\Omega$ m, as in a detector with a high field, it is only 0.5 ns.

So far we have neglected the effect of any undepleted material in our partially depleted detector. If the thickness of the whole detector is w ($> d$) it turns out that the time constant τ in equation (5.26) is not the dielectric relaxation time τ_d but has a larger value equal to $\tau_d w/d$. The fastest exponential decay is therefore obtained if a detector is just fully depleted. On the other hand, if a detector is overdepleted, the waveform loses its purely exponential character and eventually approaches, as the voltage is increased and the field becomes more uniform (see Figure 5.9), the situation in which a rectangular current pulse is obtained. In all cases, however, the current pulse has in principle an infinitely fast rise-time (as we would like), and a characteristic time usually on a nanosecond scale.

These small fast current pulses always need to be amplified, and here there is a difficulty; no standard amplifier exists which has a low enough input impedance to do this perfectly. The effect of this input impedance makes it impossible for an amplifier to 'see' at its own input the exact form of the current pulse shape given by, say, equation (5.26) at the fast speeds concerned; in particular its sharp rise is appreciably slowed down. This is considered further in Chapter 6. In general, however, despite this difficulty, timing resolutions in the nanosecond region can be achieved with junction detectors.

With highly ionizing heavy ions the possibility always arises of an additional contribution to the rise-time. The electron–hole plasma formed along an ion track, which was referred to in Section 5.5.8, can be so dense that the collecting electric field cannot immediately penetrate it. Consequently the release of the charge from the plasma takes place over a finite period which is usually characterized by a 'plasma decay time'. For many practical detectors it can be shown that the rise-time is slowed down and dominated by the plasma decay, with consequent possible problems for fast timing measurements. For heavy ions such as fission fragments the value of the decay time is typically a few nanoseconds; for alpha particles it is around 1 ns.

An associated difficulty due to plasma effects that can arise for heavy ions is an actual delay which can occur between the incidence of an ion on a detector and the subsequent detectable appearance of an electrical pulse. This 'plasma delay time' is typically of order 1 ns for fission fragments. The possibility of one or both of these plasma effects being present must always be borne in mind when attempting to use semiconductor detectors for such measurements.

5.6 Lithium-drifted detectors: principles

The lithium drift technique, which was developed from 1960 onwards, was designed to produce silicon and germanium detectors having large volumes as well as high resolution. These are used for X-rays and for penetrating radiation like gamma rays and energetic electrons because they can be made to have deep depletion layers. This is unlike the situation with simple silicon junction detectors; these, as we said in Sections 5.4 and 5.5, cannot be obtained with depletion depths greater than a few millimetres. This is because silicon impurity concentrations, which for large depths need to be as low as possible, cannot be reduced much below about 10^{17} m^{-3}, as we saw in Section 5.3. This corresponds to a maximum obtainable resistivity approaching 1000 Ω m. In practice silicon resistivities above about 100 Ω m have been relatively difficult to obtain. Of course, a far greater demand for silicon comes from manufacturers of electronic devices like integrated circuits. For them deep depletion layers are a hindrance, not a help, and normally they need material of much lower resistivity. Reference 5 gives more details.

Material of the highest purity in practice tends to be p-type, due to the different metallurgy of residual impurities. We can see from Section 5.2 that compensation of these acceptors by an equal density of donors would in principle provide material of effectively zero dopant concentration. This is not a trivial task; for example, sufficiently exact equality of donor and acceptor concentrations can never be achieved by, for example, adding an impurity to the melt at the crystal growth stage.

The process of drifting donors at elevated temperatures by an electric field into the bulk of a p-type crystal has provided the method by which exact balance between donor and acceptor concentrations can most nearly be achieved. This remarkable process has led to detectors with very large depletion regions. These (when cooled to eliminate thermally generated carriers) provide us with large-volume detectors of high resolution—the latter resulting not only from the virtual absence of carriers but also from the wide depletion region with its low capacitance (see Chapter 7).

The donors considered so far, such as phosphorus and arsenic, can never drift in an electric field because they are substitutional impurities. They occupy lattice sites in the crystal into which they are firmly bound, that is, they substitute themselves for the semiconductor atoms. The alkali metals like lithium, sodium, and potassium (which are from group I of the periodic table rather than group V) can also be easily ionized. They thus tend also to form donors but this time they take interstitial positions between the lattice sites. In this case the donors can, at raised temperatures, drift by 'slipping through' the lattice. For lithium, the smallest in physical size of these donors, concentrations large enough to compensate for acceptor impurities can be introduced. The important feature of the drifting process is that a nearly exact compensation automatically takes place throughout the volume being drifted. This is because the total space-charge density in a conductor at equilibrium is required from electrostatics to be zero at every point, and the mobile charges move so as to ensure this. Thus, if the compensation were perfect, the field across a 'planar' drifted detector (one with parallel electrodes on either side of a semiconductor slice) would be spatially uniform when an operating voltage is applied.

To construct such a detector, an n-type surface layer rich in lithium is first made by thermal diffusion of the lithium into the p-type slice of silicon or germanium. The resulting p–n junction is then reverse biased and held slightly above room temperature. Over a period of days or weeks the positively charged lithium ions drift into the negatively biased p-type bulk until a layer around 10–20 mm thick becomes compensated. Contacts to the compensated region are made through the remaining uncompensated p region on one side and through the n^+ region still doped with lithium on the other. This thin n^+ region forms the entrance window. Figure 5.13 indicates for the case of an ideal planar detector the structure which results and the electric field profile through its thickness. Note how this device shares with a gas ionization chamber the property of a uniform collecting field. The lithium-drifted region is often loosely described as 'intrinsic' (although at the low

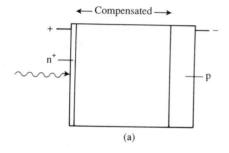

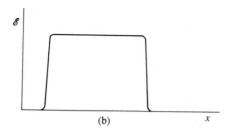

Figure 5.13 (a) Structure of a planar lithium-drifted detector, showing material which is (from left to right) n^+ (doped with lithium), compensated, and uncompensated p-type. (b) Variation of electric field $\mathscr{E}$ with distance through the detector.

operating temperatures used — normally the 77 K of liquid nitrogen — there is insufficient thermal energy to produce any intrinsic carriers), and the complete detector is often referred to as a 'p–i–n' device.

In practice perfectly compensated material can never be made, even with further sophistications in the drift process. Hence appreciable net space charge remains which, by distorting the field in the detector, can degrade the detector charge collection properties below what might otherwise have been expected. It is essential therefore in detector operation that we apply a high enough reverse bias not only to deplete fully the lithium-drifted region of majority carriers arising from the residual uncompensated acceptors, but also to produce as uniform a field as possible. If the compensation were perfect the field would, as we said, be always uniform. In practice the level of uncompensated acceptors can with care be kept below $10^{15} \, \mathrm{m}^{-3}$; consequently deep depletion depths and large volumes are possible.

In addition the charge carriers, as always, must be collected in a time shorter than their recombination or trapping lifetimes. Consideration of this criterion is here particularly important because of the large depletion depths. However, since the maximum detector voltage is usually limited by surface leakage or breakdown effects, which can be particularly troublesome in lithium-drifted devices, the maximum fields obtainable tend to be lower than those in junction detectors. Very often 1 or 2 kV are applied across compensated regions of perhaps 5 to 20 mm thicknesses. Carrier velocities at 77 K tend to saturate at around 10^5 m s^{-1} at the fields concerned, giving carrier transit times of order 100 ns. Not only is this longer than the characteristic times in the junction detectors which we considered previously, but also the carrier lifetimes tend to be shortened by the compensation process. Pulses of reduced amplitude and lengthened rise-times, if they occur, are an indication of charge-collection problems.

In the ideal planar detector, with its uniform field and without any charge-collection problems, the current pulse shape formed as the ionization is collected can be calculated somewhat as for the case given in Section 3.2.4 for an alpha particle stopped in a gas ionization chamber. Suppose a penetrating gamma ray interacts at a point well inside the depth of a drifted detector (unlike the case in Section 5.5.9 of a junction detector used with short-range radiation, where the ionization could be considered to be created close to the entrance window). Suppose that in addition the ionization created lies entirely within a volume small compared with that of the detector. This could be the case with a gamma ray of moderate energy which produces a short-range electron through the photoelectric effect. We would thus find first that a high current flows while both electrons and holes are in motion, followed by a lower current in the interval when one set of carriers (which may be either electrons or holes) has been collected but the other set is still in transit. These two phases correspond to a rise in the charge pulse which is steep at first and slower later. This is like the behaviour of a gas ionization chamber, although there are significant differences in detail because the hole mobility is only about one-third of that for electrons, instead of the much wider factor in gases separating ions from electrons.

The exact shape naturally depends on the point at which ionization occurs in the detector. Even for a monoenergetic, collimated beam of gammas, this point may well occur at very different places from event to event, as we saw in Chapter 2. A whole range of shapes can thus be produced, with the variation in effective rise-time of charge pulses amounting to a factor of perhaps about two. Field non-uniformities and charge trapping during the long transit times can increase still further the complexity of the situation. Even though the energy resolution can be very good, we see that fast timing measurements present considerably greater problems than with charged particles of short range in junction detectors. We shall have to consider these problems again in Chapter 6.

5.7 Si(Li) detectors

Successful lithium-drifted detectors were first constructed with silicon rather than germanium. Si(Li) (colloquially pronounced 'silly') detectors, as they are known, find widespread application with fast electrons such as beta particles and with photons, like X-rays, of low energy, whereas, as we shall see, Ge(Li) detectors (referred to, with equal stylistic acumen, as 'jelly') have most often been used for gamma rays with higher energies.

It is possible to use Si(Li) detectors at room temperature if high resolution is not required. For example, high-energy betas can be detected and indeed completely stopped in the depletion layer thickness of such a detector even when the beta energy is many MeV. The large depletion depths, sometimes 10 mm or more, mean, however, that carriers generated thermally in the depletion layer contribute more significantly to the noise in a signal than with surface barrier and other junction detectors with thinner depths. As a result most low-noise applications (as in gamma, X-ray, and beta energy spectrometry) require the detector to be cooled. To prevent surface

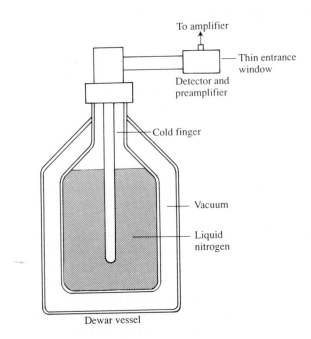

Figure 5.14 Typical configuration for a Si(Li) (or similar) cooled detector. Many configurations are available from manufacturers.

condensation and contamination it is kept under vacuum, cooled usually by fixing to a cold finger in a liquid nitrogen dewar, as shown in Figure 5.14. To minimize electronic noise the preamplifier is normally also attached to the assembly to enable its first stage to be cooled (Chapter 7).

In order to admit low-energy radiations into the vacuum chamber some form of thin window is required. Beryllium is a particularly suitable material, as its low atomic number (four) means that the intensity loss for X-rays is low. Thicknesses down to about 8 μm are possible, although such foils are very fragile. However, 'windowless' detectors, with the source and detector crystal in the same vacuum, have been used down to below 200 eV (for example, with K_α X-rays from boron).

Differential spectra of the pulse heights produced by X-rays of well-defined energies show peaks with FWHMs down to 100 eV at these low energies, while for 5.9 keV X-rays from ^{55}Fe, 150 eV FWHM is routinely available. This latter resolution, for example, is several times better than with proportional counters, and around an order of magnitude better than with scintillators. This is because of (among other factors) the low energy required in silicon to produce an electron–hole pair. Details of the electronics required to achieve these resolutions with Si(Li) detectors are given in Chapter 7. An example of an X-ray spectrum from a Si(Li) detector is shown in Figure 5.15. The shapes of the peaks are approximated by Gaussian distributions (Section 1.4.3). (The 'escape peak' is discussed in a moment.) We note in passing that X-ray energy spectra can also be measured with crystal diffraction instruments. The resolution obtainable is indeed better than with Si(Li) detectors, but their construction can be very complex. Sometimes only one wavelength (that is energy) at a time can be counted by the instrument, and this is a very inefficient and slow process.

The main response mechanism by which an X-ray is detected in these low-energy cases is the photoelectric process (Section 2.3.2). In this the energy of the X-ray, minus the binding energy of the electron in its original shell, is transferred inside the sensitive volume of the detector to a photoelectron, which is then stopped entirely within this region. Any characteristic X-rays (or Auger electrons) emitted from a silicon atom as the resulting electron vacancy is filled are likewise normally stopped within the depletion layer, so that the total pulse height from all the electron–hole pairs produced is proportional to the energy of the incident X-ray. An exception can occur for X-rays with lower energies, in which most of the photoelectric absorptions will occur near the detector surface. In this case some of the characteristic X-rays can escape. The pulse-height spectrum in Figure 5.15 is of iron K_α and K_β X-rays. The large K_α peak (note the logarithmic scale) clearly shows an 'escape peak' at just under 2 keV lower in energy, corresponding

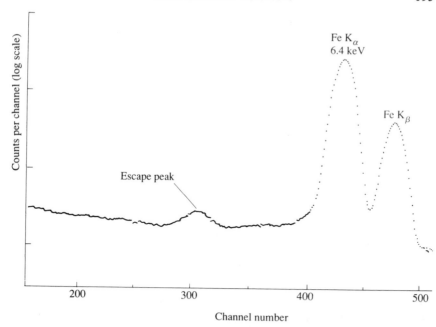

Figure 5.15 A spectrum from a Si(Li) detector of low-energy X-rays from iron following the decay of ^{57}Co. The energy of the iron K_β peak is 7.1 keV. (From Wood, R.E., Venugopala Rao, P., Puckett, O.H. and Palms, J.M. (1971), *Nuclear Instruments and Methods*, **94**, 245–52, Figure 2.)

to the energies of the unresolved silicon K_α and K_β X-rays (1.8 and 1.9 keV, respectively). We have already discussed this and other types of escape peaks in Chapters 3 and 4.

Figure 5.16 shows that, for the detector thicknesses shown, near 100 per cent X-ray detection efficiency can be achieved in this low-energy region down to 2 or 3 keV, below which the radiation is increasingly absorbed in the window. Here we are showing the 'intrinsic full energy efficiency', which is the probability that a photon of given energy impinging on the detector front contact will give a full energy peak. (We can also define an 'absolute efficiency', which in addition takes account of the fraction of photons coming from the source which reaches the detector, but this is not of direct interest to us here.) Figure 5.16 shows that above about 20 keV the intrinsic efficiency starts to fall with energy as the half thickness for absorption, $d_{1/2}\,(=1/\mu)$, becomes comparable with the silicon thickness.

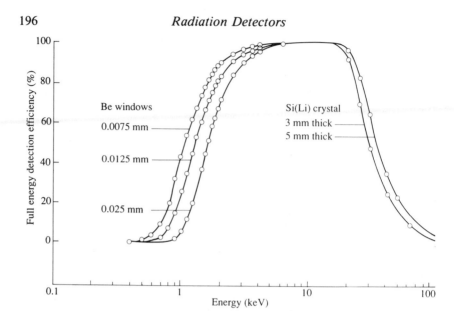

Figure 5.16 Intrinsic full energy efficiency versus photon energy for Si(Li) detectors for various detector thicknesses and beryllium window thicknesses. (Reproduced by courtesy of EG&G ORTEC.)

As the photon energy is raised, the photoelectric absorption cross-section falls rapidly until, as described in Section 2.3.5, above about 55 keV it becomes less than the more slowly decreasing Compton cross-section. At these higher energies in Si(Li) detectors an X- or gamma ray will often fail to deposit all its energy in a detector, since any scattered Compton gamma ray usually escapes. Just as for scintillators in an energy spectrum (Figures 4.16 and 4.17) a Compton continuum appears up to the Compton edge. This is distant below the full energy peak by up to 0.25 MeV, corresponding, as we saw in Section 2.3.3, to the minimum energy of an escaping photon produced in the Compton scattering process. This continuum can obscure lower full energy peaks which may also be present.

As the photon energy rises further the peak-to-Compton ratio (that is, the height of the full energy peak relative to that of the flat part of the Compton continuum) rapidly falls, and above 100–150 keV Si(Li) detectors are seldom used for spectrometry. This is the main limitation of such detectors, and it normally prevents their being used for all gamma rays except those of very low energy. An exception occasionally occurs for gammas above 3 MeV, for which Si(Li) detectors have an advantage over their germanium counterparts. We shall see why this is so when we discuss below, in Section 5.8, the response mechanism involved.

5.8 Ge(Li) detectors

Lithium-drifted germanium detectors appeared in 1962, only a couple of years after their silicon counterparts. As expected, it was found that because of their higher atomic number they could be used with higher-energy gamma rays as well as with those of lower energies measurable with Si(Li) detectors. This follows because the likelihood of a gamma ray depositing its full energy in a germanium detector and hence adding to the full energy peak is much greater. This is primarily because germanium, with its higher atomic number of $Z = 32$, has a photoelectric absorption cross-section more than an order of magnitude greater than in silicon, for which $Z = 14$. Much higher peak-to-Compton ratios with germanium can be obtained even at quite high gamma energies; for example up to 60:1 at 1.33 MeV can be obtained with certain detector configurations (as described below).

More recently gamma detectors have been fabricated without the use of lithium drifting. Many of the remarks that can be made about the operation and use of Ge(Li) detectors apply equally well to the newer devices; a description of the features unique to the latter is deferred to Section 5.9.

One property which is common to all germanium detectors is that because of the small band gap (0.7 eV as against 1.1 eV in silicon) cooling is always essential during operation in order to reduce thermally generated currents.

The outstanding feature of germanium detectors, whether drifted or not, is their very good energy resolution for high-energy gamma rays. At 1 MeV the FWHM of a spectral peak (as usual, of approximately Gaussian shape) is often better than 0.2 per cent of the gamma-ray energy (compared with about 5 per cent for scintillators), as we shall see in Chapter 7. Indeed in this energy region the resolution can be better than from crystal diffraction spectrometers, since at these higher energies gamma-ray wavelengths, and therefore diffraction effects, become very small indeed. Germanium detectors are thus especially suitable for the analysis of complex gamma spectra.

One of many examples in which germanium detectors have revolutionized radiation measurements is the clear separation available between the lines from ^{137}Cs and ^{134}Cs. Satisfactory resolution of these lines, which is important for environmental measurements involving fission products, is much harder to achieve with a scintillation counter. This is shown in Figures 5.17(a) and 5.17(b), respectively. Thus in (b), the scintillation counter spectrum, the ^{137}Cs 662 keV line is just separated from the ^{134}Cs 796 keV line, and overlaps the ^{134}Cs 605 keV line, whereas in (a), the Ge(Li) detector spectrum, despite the very low ^{134}Cs concentration in the sample concerned, the lines are all clearly visible. In addition spectrum (a) shows (among other features) a 609 keV line close to but resolved from the 605 keV

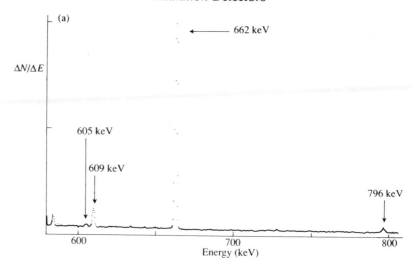

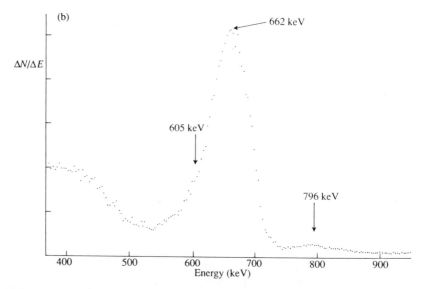

Figure 5.17 Radiocaesium gamma-ray spectra from two Irish Sea fish samples, one taken with a Ge(Li) semiconductor detector (figure 5.17(a)), the other with a NaI(Tl) scintillation counter (figure 5.17(b)). (Spectrum (a) by courtesy of I.R. McAulay, Trinity College, Dublin. Spectrum (b) reproduced from Doyle, C. (1982), *An investigation of radiocaesium levels in the Irish Sea*, M.Sc. thesis, Trinity College, Dublin, Figure 2.8.)

line. This additional line is associated with the decay of ^{214}Bi, a member of the uranium series, and must also be present in spectrum (b).

An additional advantage, particularly where weak sources of radiation are concerned, is that the much narrower line widths with germanium detectors make peak heights stand higher out of any Compton continuum or other background. This usually more than offsets their lower efficiencies compared with scintillators of higher atomic number and larger volume, such as NaI(Tl).

Nevertheless, one of the important aims in germanium detector manufacture is often to design configurations so as to increase efficiency and reduce losses due to escaping radiation as much as possible. Figure 5.18 shows how the intrinsic full energy efficiency of a planar germanium detector (constructed, if a Ge(Li) detector, as in Figure 5.13) varies with photon energy. For example, in the energy range above around 100 keV the full energy efficiencies are falling rapidly with energy, but it is also the region where very many gamma-ray energies of interest lie. As with scintillators, this efficiency can be increased by making the active volume of the detector as large as possible. This not only increases the chance of photoelectric absorption in the

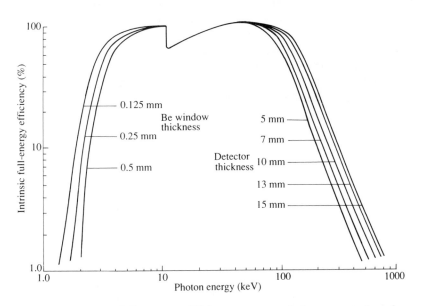

Figure 5.18 Intrinsic full energy efficiency versus photon energy for planar germanium detectors for various detector thicknesses and beryllium window thicknesses. Note logarithmic scale on efficiency axis. (Reproduced by courtesy of EG&G ORTEC.)

detector, but also makes it possible for a significant number of Compton events likewise to lead to a full deposition of the energy, as we have seen in Section 4.5.1. In silicon detectors, by comparison, full-energy Compton events are much less probable in this energy range because the scattered Compton photon has only a low chance of being reabsorbed even in a large detector.

Germanium detectors with planar geometry are available with dimensions typically up to about 15 mm active depth and 36 mm diameter, corresponding to an active volume of around 15 cm^3. Good intrinsic full-energy efficiencies for the detection of full energies can be maintained up to 100 keV or more. In this region the planar geometry offers (especially with a small detector) the best energy resolution available.

Germanium detectors of larger volumes and with usable efficiencies over much wider energy ranges can be made with coaxial electrode structures. One procedure, in the case of a Ge(Li) detector, is to drift in lithium from the outer surface of a cylindrically shaped crystal until only a p-type core is left along the cylinder axis. An optional additional drift can be performed into the end not required for a contact. Figure 5.19 shows the arrangement which can result. The collecting field is now no longer uniform but increases towards the central core. For cylindrically symmetric geometry and perfect compensation the field would vary as $1/r$, r being the radial distance from the central axis. Active volumes of up to 150 cm^3 are possible. Such detectors have the highest peak efficiencies and peak-to-Compton ratios, although there is an increased risk of pulse-shape and charge collection problems due to the non-uniform field.

Provided that a satisfactory electrical contact can be made, there is in fact an advantage in having the core radius (r_1 say) small despite the

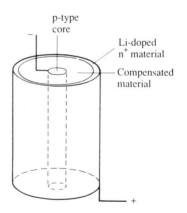

Figure 5.19 A coaxial Ge(Li) detector.

resulting increased non-uniformity of the field. This is because the detector capacitance, given per unit length for a fully depleted structure by

$$C = \frac{2\pi\kappa\epsilon_0}{\ln(r_2/r_1)}$$

can be seen to be thereby reduced. In the formula κ is the dielectric constant of the germanium, $r_2 (\gg r_1)$ is the outer radius of the detector, and end effects are neglected. We mentioned in Section 5.4 that a low capacitance improves the energy resolution of a detector system, as will be shown in Chapter 7.

A variant on this coaxial design is a well structure, where part of the core of uncompensated silicon is removed over most (but not all) of its length. Just as with well-type scintillation counters this gives a very good geometrical efficiency for detection of radiation from a source placed inside the well. Actual intrinsic efficiencies in the 1 MeV region are then typically up to 10^{-1} or more.

Manufacturers by convention frequently quote the efficiency of a large-volume detector as a percentage of that of a standard 3 inch × 3 inch cylindrical NaI(Tl) scintillation crystal for ^{60}Co 1.33 MeV gammas and a source-to-detector distance of 250 mm. Values of this 'efficiency ratio' can for the largest germanium detectors now approach or even exceed 100 per cent.

At high gamma energies above about 1.5–2 MeV, when (as with scintillators) pair production becomes significant, deposition of the full energy through this process is more likely to occur with large-volume detectors. With pair production, however, the increased escape probability with small detectors leads to less serious problems than with the Compton continuum for the following reasons. As described in Section 4.5.1 and 2.3.4, the electron and positron produce, as they slow down, ionization corresponding to their combined kinetic energy of $E_\gamma - 2m_0c^2$, where E_γ is the gamma energy and m_0 is the electron or positron rest mass. The subsequent 0.511 MeV annihilation photons may, especially for detectors with large volumes, be stopped in the detector, adding to the full-energy peak. Alternatively one of them may escape, giving a single-escape peak 0.511 MeV below the full peak. If both escape a double-escape peak appears at 1.02 MeV below the full peak. Figure 5.20 shows an example.

Normally all three peaks are equally suitable for energy measurements, provided of course that they are properly identified. In the highest-resolution work, however, the single-escape peak is found to show slight broadenings and shifts of a fundamental nature, which in this case detract from its use. The added spectral complexity introduced by the three peaks is alleviated by the high resolution available with germanium detectors, which minimizes the

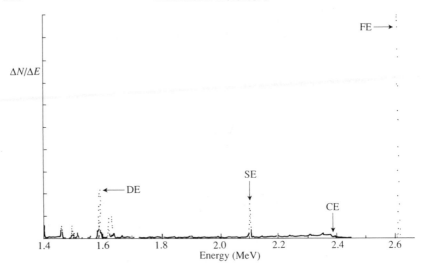

Figure 5.20 A gamma-ray spectrum from a 108 cm³ coaxial Ge(Li) detector of efficiency ratio 18.5 per cent. A 2.615 MeV full-energy (FE) peak is shown, together with the single (SE) and the double (DE) escape peak (at 2.104 and 1.593 MeV, respectively) arising from the pair production mechanism. The Compton edge (CE) about 0.25 MeV below the FE peak is also shown. The source is $^{232}_{90}$Th, and the 2.615 MeV gamma comes from the last member ($^{208}_{82}$Pb) of its decay chain. Peaks from other gammas in the decay chain are also visible in the vicinity of the DE peak. (By courtesy of I.R. McAulay, Trinity College, Dublin.)

possibility of confusion in line identification when many gamma energies are present. We note that for Si(Li) detectors the spectral situation is simpler since usually neither annihilation photon is stopped, the atomic number being lower, and so only a double-escape peak appears. On the other hand germanium has a higher pair production cross-section than silicon, although the factor is only about four since the cross-section varies roughly as Z^2 rather than Z^5 as for the photoelectric effect.

At low energies yet another escape effect occurs, which similarly decreases with increasing volume. As with silicon, the X-ray emitted from a germanium atom after a photoelectric process can then sometimes escape from the detector. This is more likely with germanium, since firstly the absorptions for a given incident X-ray energy occur nearer the surface in germanium than in silicon, and secondly a characteristic K X-ray (for example) from germanium is more penetrating and therefore more likely to escape in any case because of its much higher energy, 11 keV, instead of 1.8 keV in silicon. Hence the resulting escape peak, 11 keV below the main peak, is much more prominent for germanium detectors, and the intrinsic full-energy efficiency in this region is noticeably reduced below 100 per cent, as Figure 5.18 shows.

We have already said that germanium detectors cannot be used at room temperature because of excessive thermally generated currents, so they must be cooled during operation. In addition it improves the resolution if we cool the first stage of the preamplifier as well, as we shall see in Chapter 7. The detector and its preamplifier first stage, as with cooled Si(Li) systems, are usually surrounded by a vacuum jacket. Again a thin vacuum-tight entrance window is required for the radiation, but unless the lowest energies are to be detected the window need not be excessively thin and fragile. In the past, difficulties have in any case arisen with the use of germanium detectors much below 10 keV because of the effects of surface dead layers. We shall find, however, in Section 5.9 that newer types of germanium detectors can be used even for very-low-energy X-rays.

Germanium detectors are less satisfactory for electron detection, because with the higher atomic number more electrons than with silicon will be backscattered without depositing their full energy. At high energies around 10 MeV, however, this consideration may be overridden because of the higher stopping power offered by germanium. For the same reason germanium detectors have occasionally been used with high-energy ions, for example with alpha particles up to 200 MeV or protons up to 60 MeV.

As we have already said, many of the remarks in this section apply not only to Ge(Li) detectors but also to any detector fabricated from germanium. A major disadvantage peculiar to Ge(Li) detectors is that they must be cooled not only during operation but also in storage. Unlike in silicon, lithium diffuses rapidly in germanium at room temperature and so the compensated region in a Ge(Li) detector would be lost. Redrifting by the manufacturer is possible, but expensive, and not always entirely satisfactory. There is therefore an obligation at all times to keep the liquid-nitrogen dewar refilled, either manually once or twice a week or automatically.

5.9 HPGe detectors

The disadvantages of Ge(Li) detectors have led to the development since 1970 of detectors made from germanium of such high purity that the need to compensate for impurities by lithium drifting is eliminated. Highest-purity material tends to be p-type; concentrations of residual acceptor centres as low as 10^{16} m^{-3} have been achieved, and in some cases the centres are due primarily to crystal defects rather than any remaining acceptor impurities. This is an example of how the requirements of radiation detection have led to some remarkable advances in materials processing and technology.

This is still far too high to make a true conduction counter even when cooled to 77 K. However, equation (5.11) (modified for p-type material) shows that the thickness of the depletion layer of an n^+–p junction made with this material is (with 16 as the dielectric constant) 13 mm at a reverse bias of 1000 V. In this way high-purity, or HPGe, detectors can be made with active volumes comparable with those of Ge(Li) detectors. In operation cooling is still needed in order to reduce the thermal current, but room-temperature storage is now quite permissible. Once again a liquid-nitrogen cryostat is normally used with the detector. Broadly speaking the same excellent gamma response is shown with HPGe as with Ge(Li) detectors, although it must be remembered that HPGe detectors have tended to cost more because high-purity material is expensive.

In the construction of a planar HPGe detector an n^+ contact on one side is often formed by lithium doping, and a p^+ contact is formed on the other side by metal evaporation or ion implantation. With p-type bulk material the n^+ contact is rectifying and the p^+ contact ohmic. There is an advantage in making the p^+ contact the entrance window, since it can be made much thinner (about 0.3μm if ion implanted) than the lithium contact (perhaps 600μm). The detector should therefore be strongly overdepleted (Section 5.5.5) in order to maintain a high collecting field at the p^+ ohmic contact, as shown in Figure 5.21. This improves the charge-collection properties, and also the timing properties, which tend to be somewhat worse than in planar Ge(Li) detectors. This is because an HPGe detector is a junction detector, with therefore a spatially non-uniform field (even when overdepleted). In contrast the field in a perfectly compensated Ge(Li) detector would be uniform. Planar HPGe detectors, as with their Ge(Li) counterparts, are used for energies up to a few hundred keV.

X-ray detection at energies below about 3 keV can be more difficult with germanium than with Si(Li) detectors. In germanium the photoelectric stopping cross section is so much larger that much of the ionization can sometimes be produced in the electrode thickness and any germanium dead layer underneath, instead of being deposited in the active volume of the detector. Good charge collection has, however, been observed down to as low as 200 eV in small-area HPGe detectors with special evaporated metal contacts, as described in reference 6.

Coaxial HPGe detectors are also available. They must, however, always have a central hole or well so that an inner contact can be made on its wall. This is because there is no equivalent of the undrifted central core of germanium which, as we saw in Section 5.8, could be used as a contact in a Ge(Li) detector. The inner contact in an HPGe detector should be the ohmic one, with the rectifying contact around the outer circumference. With the contacts this way round, the $1/r$ electric field radial variation which would be shown by a perfectly compensated coaxial Ge(Li) detector can be arranged to be partly counterbalanced by the linear rise in the numerical size of the field expected on going from an ohmic to a rectifying contact, as

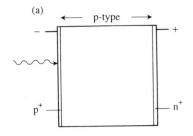

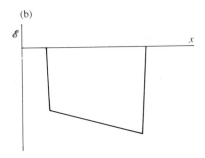

Figure 5.21 (a) Structure of a planar HPGe detector. (b) Variation of electric field $\mathscr{E}$ with distance through the detector.

Figure 5.22 shows. This optimizes charge collection, and so timing properties can actually be better with coaxial HPGe than with Ge(Li) detectors.

The disadvantage of the arrangement shown is that with p-type material the outside, rectifying, contact is the thick lithium one, and so (unless a well counter arrangement is desired) the radiation has to enter through it. It is now possible, however, to obtain ultrapure n-type germanium. A coaxial detector using this material has the thin p^+ ion-implanted contact on the outside surface, since the implantation now forms a rectifying, not an ohmic, contact. The thicker conventional n^+ lithium contact, which is now ohmic, is on the inside. With such a thin window, photon energy spectrometry from 3 keV to 10 MeV in a single detector is possible. An additional advantage will become apparent when we consider the question of radiation damage in Section 5.11.

We close this section with an extreme example of how radiation detection requirements have influenced manufacturing techniques for germanium detectors. Experiments in deep mines which look for very rare events (for example, dark matter searches) require ultralow background count rates,

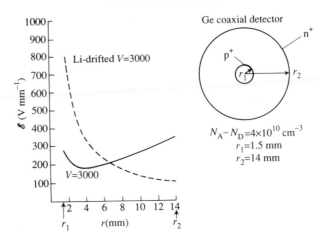

Figure 5.22 Variation of size of electric field through a coaxial HPGe detector made from p-type material (full curve) with an ohmic inner contact at r_1 and rectifying outer contact at r_2, compared with a similar plot for a Ge(Li) detector (broken curve) at the same bias. The bias was chosen to make the field in the HPGe detector as uniform as possible. Note that in both cases the field is directed towards the centre. (Adapted from Llacer, J. (1972), *Nuclear Instruments and Methods*, **98**, 259–68, Figure 3.)

and this can place extraordinarily stringent requirements on the detectors used. Radioisotopes are continuously formed in the germanium through nuclear reactions involving cosmic rays. However, their intensities, and thus the radioisotope production rate, are much higher on the surface than underground, that is, in the ore before manufacture and afterwards in the mine containing the experiment concerned. By minimizing the exposure received during manufacture, this contamination by radioisotopes was brought to very low levels. Reference 7 describes the many interesting procedures used in rushing the fabrication of two germanium detectors for such work, starting from the mining of the germanium ore. These (together with specially designed detector housing and shielding) resulted in a system with background count rates several orders of magnitude lower than in commercially available 'low-background' arrangements as used in, say, ordinary environmental measurements.

5.10 Semiconductor detectors for neutrons

Before discussing how semiconductor detectors can be used with neutrons, we will recall some of the facts about neutrons we noted in previous chapters as well as adducing some additional information. Neutrons can be detected basically by two methods. The first is by the ionization produced by recoiling charged particles elastically scattered by the neutron, and the second is by detection of charged particles from reactions initiated by it.

The first of these is, for obvious reasons, restricted to fast neutrons: the second can be used for a whole range of energies, but is especially important for low-energy and thermal neutrons. We recall that three reactions involving light nuclei are particularly useful in this second technique: $^{10}_{5}B(n,\alpha)^{7}_{3}Li$, $^{6}_{3}Li(n,\alpha)^{3}_{1}H$, and $^{3}_{2}He(n,p)^{3}_{1}H$. All produce energetic products in the MeV region, and have cross-sections which rise sharply at the lower-energy end. The fission process can provide a fourth reaction with comparable cross-sections at low energies and yielding, of course, charged particles (i.e. fission products) with energies in the 100 MeV region. In the following paragraphs we shall be discussing both the simple counting of neutrons and their spectrometry (that is, the measurement of their energies).

On their own, silicon and germanium devices do not prove to be particularly efficient detectors for neutrons. For fast neutrons this is because very little energy is transferred to the scattered atom due to the relatively large mass of this atom in the case of silicon, and even more so for germanium. A standard kinematical 'billiard ball' calculation shows that a non-relativistic neutron can transfer any fraction of its kinetic energy to the atom being struck, up to a maximum of $4x/(1 + x)^2$. Here x is the ratio of the mass of the recoiling atom to that of the neutron. (Note that x is close to the mass number A of the nucleus involved.)

For silicon, where $x \approx 28$, this fraction is 0.13, so for a typical 2 MeV fast neutron from the fission process, the maximum silicon recoil energy is only 270 keV. Furthermore the pulse-height defect in a detector due to subsequent atomic collisions (Section 5.5.8) is for this silicon ion energy about 100 keV, and so the maximum net pulse height produced is quite low. In germanium, where $x \approx 73$, the fraction is 0.05, and the pulse-height defect is larger, so the response is smaller still. For slow neutrons the situation is even worse, as there is no reaction of any significance with silicon and germanium corresponding to the three reactions with high cross-sections mentioned above.

We can produce acceptable neutron counters using semiconductor detectors by using some form of more efficient intermediate or 'converter' material to produce charged particles detectable by them. One simple method is to dope a detector either by diffusing $^{10}_{5}B$ acceptors into the surface of n-type silicon, or $^{6}_{3}Li$ donors into p-type silicon. Sufficient dopants can easily be introduced to give detection efficiencies for thermal neutrons of up to a few per cent. Alternatively, we can coat the surface of a junction detector with a fissile material like $^{235}_{92}U$, or have an adjacent coated foil, and count the fragments produced.

We now turn to neutron spectrometry, that is, the measurement of their neutron energies, and obviously this is possible only for fast neutrons. In this case we shall start with the use of reactions, which of course are also valid for fast as well as slow neutrons. In principle the energy of the incoming neutron can be obtained from that of the outgoing particles, provided that we know the Q value of the reaction concerned. (This, unfortunately, eliminates the $^{10}_{5}B(n,\alpha)$ reaction, as it does not have a unique Q value, due to the possibility of the reaction proceeding to either the ground or an excited state of the product nucleus. The fission reaction, which shows a substantial variation in the energy release from event to event, is also unsuitable.) The Q values for the $^{6}_{3}Li(n,\alpha)$ and $^{3}_{2}He(n,p)$ reactions are, respectively, 4.78 and 0.764 MeV. (These values would, of course, swamp any addition due to the neutron kinetic energy if the latter is too small, which is why the technique works for only fast neutrons.)

Because of the possibility of escape of one of the reaction products from a single detector, it is usual to use two detectors in a 'sandwich spectrometer' configuration, as shown in Figure 5.23. In this a thin converter layer of either $^{6}_{3}Li$ (in fluoride or other suitable chemical form) or $^{3}_{2}He$ gas under pressure is sandwiched between the entrance windows of two silicon junction detectors which are facing each other. The two products from each reaction fly apart in approximately opposite directions, and the two pulses, one from each detector, are summed to give the total energy (neutron (a)).

A two-detector arrangement has an additional advantage in that, as described in Section 6.8, the observation of a coincidence in time of pulses from the two detectors reduces problems from background radiation; a pulse from one detector unaccompanied by one from the other is rejected. Figure 5.23 shows how coincidence measurements can, however, occasionally miss true events especially when the neutron energy is large (neutron (b)), because

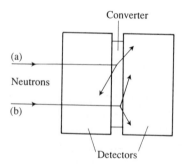

Figure 5.23 A sandwich spectrometer for neutrons. In the case of neutron (b) no reaction product enters the left-hand detector, and so this event fails to record a coincidence of pulses from the detectors (see text).

then the reaction products must have considerable net forward momentum and their paths will therefore be far from collinear.

The detection efficiency increases for larger thicknesses of the converter layer. This, though, is at the expense of energy resolution, because the reaction products lose a larger, and undetermined, fraction of their energy in the layer, depending on the position of the reaction site and the angles made by the paths of the reaction products with the plane of the converter. $^6_3$LiF thicknesses of 1.5 g m^{-2} (150 mg cm^{-2}) typically give fast neutron efficiencies of 10^{-6} and resolutions of about 200 keV. The less common $^3_2$He sandwich spectrometers have better efficiency and resolution. This occurs for a number of reasons, among them the fact that the $^3_2$He cross-section is for many neutron energies appreciably higher than that for $^6_3$Li, and because the reaction products have lower atomic numbers, which leads to lower energy losses in the converter. The lower Q value, however, makes discrimination against background considerably harder.

Sandwich spectrometer efficiencies are in general much smaller than in another possible type of neutron detector, in which $^3_2$He is used directly as the gas in a proportional counter. The main advantages of a semiconductor sandwich are that it is compact and that the linearity with energy of its response is very good.

We now discuss elastic scattering using a converter. In this a thin layer of hydrogenous material like polyethylene is mounted in front of a detector. The protons elastically scattered out of the layer have a range of energies up to, as before, a fraction $4x/(1 + x)^2$ of the incident neutron energy. In this case $x \approx 1$ and so the fraction is close to unity instead of a few per cent as with heavier materials. This is the principle of the proton recoil detector, and a semiconductor device can be used to detect the protons. With care, neutrons of energies down to only 1 keV can be detected. If the hydrogenous layer is placed immediately in front of the detector, efficiencies are very high. Any energy measurement is, however, then relatively difficult, because for a given neutron energy E_n the protons have a whole range of energies E_p depending on the angle θ at which they are scattered.

Alternatively the detector can be placed at a distance from the layer. This defines the angle θ (with respect to the incident neutron direction) for the direction of the scattered protons which are detected, as shown in Figure 5.24, but at the expense of reduced efficiency. This assumes of course that the incident neutrons are all collimated, at least approximately. It is easy to show then that E_n is given by $E_p/\cos^2 \theta$. In order to maximize E_p this angle θ is kept low. Often an $E - \Delta E$ telescope arrangement (Section 5.5.5) is used, as is shown in the figure. With the two detectors operated in coincidence (Section 6.8) this enables events due to protons arriving from the layer to be accepted and background events rejected. The sum of the coincident responses from the two detectors then gives E_p. In practice resolutions of a few hundred

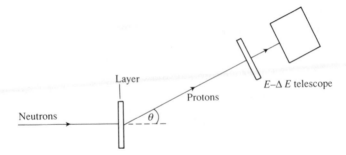

Figure 5.24 Neutron spectrometry using an E-ΔE telescope for protons elastically scattered from a hydrogeneous layer.

keV are obtainable but with efficiencies of only 10^{-5}–10^{-6}. The latter can be improved with a thicker hydrogenous layer or smaller distance between layer and detector, but naturally the resolution is then worse.

An important disadvantage of the use of semiconductor detectors for neutron detection is the radiation damage caused by neutrons (if they strike the detector) and by the reaction products themselves. The effects of the damage build up as such a detector is irradiated, and eventually it becomes unusable. It is to the topic of radiation damage in such detectors in general that we must now turn.

5.11 Radiation damage in semiconductor detectors

As mentioned previously, semiconductor detectors, unlike gas-filled ionization chambers and their derivatives, are susceptible to irreversible damage by the radiation to which they are exposed. We have seen that silicon and germanium have the outstanding features of high purity and crystal perfection, with long carrier lifetimes leading to good charge-collection properties in detectors. Heavy charged particles in particular will spoil this picture by disrupting the lattice through atoms being dislodged from their sites in the crystal. Prolonged exposure leads to an increase in the leakage current of a detector, worse energy resolution, multiple peaks in a pulse height spectrum for monoenergetic particles, and degraded timing characteristics.

For front-face irradiation of a surface barrier detector (that is, on the gold rectifying contact), a significant deterioration in detector performance first appears after the detector has received an integrated particle flux of approximately $3 \times 10^6 \, \text{mm}^{-2}$ for fission fragments, $10^9 \, \text{mm}^{-2}$ for alpha particles, $10^{11} \, \text{mm}^{-2}$ for protons, and $10^{12} \, \text{mm}^{-2}$ for fast electrons. It can be seen that the heaviest particles, as expected, cause the most damage. In contrast, electrons below about 250 keV cause little or no disruption to the crystal because of their small momentum. Thus, many secondary electrons produced by gamma rays cause little effect, with perhaps 10^{13} mm^{-2} of gammas required before noticeable damage occurs.

Even though they are not directly detectable, fast neutrons are damaging because of the heavy recoil atoms of silicon produced; the threshold before the performance deteriorates is only about 3×10^9 mm^{-2}.

In general, totally depleted charged-particle detectors are more resistant than partially depleted devices because the average electric fields through the sensitive volumes of the former tend to be higher. Any reduction in carrier lifetimes thus has less effect. Conversely, Si(Li), Ge(Li), and HPGe detectors, with their low fields and large thicknesses, are much more susceptible to radiation damage because of the long transit times for carriers across their depletion regions. Nevertheless the electrons and X- or gamma rays which they detect do not themselves normally cause a problem. However, accompanying fast neutrons in, for example, an accelerator environment significantly degrade detector performance after a flux of 10^7 mm^{-2} or less.

Coaxial n-type HPGe detectors are an exception to this, being much more resistant to fast neutrons as we now proceed to show. Because of the radial geometry in coaxial detectors, more interactions are likely to occur nearer the outer contact than the inner one. From this it turns out that the majority of charge pulses are induced primarily by the carrier which drifts towards the central electrode (as we noted in Section 3.3.2 for electrons in a cylindrical ionization chamber); for only those few events in the small volume neighbouring the inner contact is this not true. This type of asymmetry in pulse formation can be shown to have no counterpart in planar detectors, even for a linearly decreasing field in an HPGe detector, or indeed for any other field profile; it is a consequence solely of the non-planar geometry. In any case (apart from well-type counters) many of the gamma-ray detection events in a coaxial detector are likely to occur near the outer surface because of the 'exponential attenuation' law (equation (2.8)).

These asymmetries, which reinforce each other, can be exploited by making the carrier which drifts inwards to be an electron, as in an n-type HPGe (since it has, as Section 5.9 describes, the n$^+$ contact on the inside), rather than a hole, as in a p-type detector. Since fast neutrons primarily generate hole (rather than electron) trapping centres, the former detector, in which the signal is normally induced primarily by electrons, is more radiation resistant than the latter.

5.12 Detectors made from special materials

Most semiconductor radiation detectors use either silicon or germanium because of the very good charge transport properties shown by both materials. However, they possess two major problems. One is that the band gaps of 1.1 and 0.7 eV, respectively, are too small to allow room-temperature operation except in the case of silicon detectors with small volumes, as used for electrons or heavier charged particles. The second is that the atomic number (Z) of even germanium, which, at 32, is much higher than the value of 14 for silicon, is still not high enough to enable the construction of detectors with high peak efficiencies for the detection of gamma rays of higher energies.

Many alternative materials have been investigated to see if one or both of these difficulties can be overcome. Diamond is an obvious candidate, since it accompanies silicon and germanium in group IV of the periodic table and has the same crystallographic structure. Despite its high cost it has in fact been investigated more or less continuously since 1945. Its atomic number of 6 is very low, but its large band gap of 5.4 eV makes it an insulator, and this enables simple conduction counters to be made and used at or even above room temperature. They have a fast response, but are limited by cost to small sizes. They can also suffer from effects caused by charge trapping and polarization, which after a time act to reduce the collecting field and efficiency of a counter.

Other materials studied early on included cadmium sulphide, CdS, and silver chloride, AgCl. These gave conduction counters with materials of high atomic number, but showed severe charge trapping. In addition AgCl required cooling to 77 K to reduce the high leakage current shown at room temperature.

More recently many other materials have been studied, such as the compound semiconductors gallium arsenide, selenide, and phosphide, GaAs, GaSe, and GaP. Much work has gone into developing detectors made from cadmium telluride, CdTe, since the atomic numbers (49 and 52) are quite high. Only moderate resolution (1.1 keV FWHM at 5.9 keV) has been achieved, and polarization has been a persistent problem.

Mercuric iodide, HgI_2, has come closest to overcoming the difficulties we saw that silicon and germanium have. The atomic numbers of 80 and 53 are both high, as is the band gap of 2.14 eV. Its photoelectric absorption cross-section is thus substantially greater than in germanium, because of the high atomic numbers, and its room-temperature resistivity can be as high as $10^{12} \, \Omega$ m, because of the high band gap. It can hence be used to make compact 'conduction counter' detectors which operate at room temperature (or slightly below), which have high efficiencies for the detection of X- and gamma rays, and which have good resistance to radiation damage. Good resolution can be shown particularly for small X-ray detectors; reference 8 quotes a figure of 175 eV FWHM at 5.9 keV energy with the detector at room temperature and cryogenic cooling for the preamplifier first stage. However, as with some of the other compound semiconductors, it is difficult to grow large crystals, and there are problems arising from charge traps and the low mobility of holes. For low-energy X-rays, which are absorbed near the surface, hole-collection problems can be minimized if the entrance window contact is negatively biased. This ensures that it is electrons, with their better mobility, which cross the detector thickness and which thus generate the output signal.

References

1. Martoff, C.J., Kaczanowicz, E., Neuhauser, B.J., Lopez, E., Zhang, Y., and Ziemba, F.P. (1991). Operation of a high-purity silicon diode alpha particle detector at 1.4 K. *Nuclear Instruments and Methods*, **A301**, 376-9.
2. Gatti, E. and Rehak, P. (1984). Semiconductor drift chamber — an application of a novel charge transport scheme. *Nuclear Instruments and Methods*, **225**, 608-14. The original paper on drift detectors.
3. Luke P. N. (1988). Low noise germanium radial drift detector. *Nuclear Instruments and Methods*, **A271**, 567-70.
4. Delaney, C.F.G. and Finch, E.C. (1984). Some simple considerations on pulse shapes in radiation detectors. *American Journal of Physics*, **52**, 351-4.
5. von Ammon, W. and Herzer, H. (1984). The production and availability of high resistivity silicon for detector application. *Nuclear Instruments and Methods*, **226**, 94-102.
6. Cox, C.E., Lowe, B.G., and Sareen, R.A. (1988). Small area high purity germanium detectors for use in the energy range 100 eV to 11 keV. *IEEE Transactions on Nuclear Science*, **35**, 28-32. Includes comparisons with the performance of Si(Li) detectors.
7. Brodzinski, R.L., Miley, H.S., Reeves, J.H., and Avignone III, F.T. (1990). Further reduction of radioactive backgrounds in ultrasensitive germanium spectrometers. *Nuclear Instruments and Methods*, **A292**, 337-42.
8. Iwanczyk, J.S. (1989). Advances in mercuric iodide X-ray detectors and low noise preamplification systems. *Nuclear Instruments and Methods*, **A283**, 208-14. Published in an issue (number 2 of the volume) which concentrates exclusively on compound semiconductors for detectors.

Further reading

Bertolini, G. and Coche, A. (ed) (1968). *Semiconductor detectors*. North-Holland, Amsterdam. Still an important reference book.

Dearnaley, G. and Northrop, D.C. (1966). *Semiconductor counters for nuclear radiations*, (2nd edn). Spon, London. A well-known early account of the field.

Debertin, K. and Helmer, R.G. (1988). *Gamma- and X-ray spectrometry with semiconductor detectors*. North-Holland, Amsterdam. A specialist text on the silicon and germanium detectors used in the area.

Rosenberg, H.M. (1988). *The solid state*, (3rd edn). Clarendon Press, Oxford. A good introduction to solid-state physics, with an informative bibliography of the many books on the subject.

6
Electronics for radiation detection

6.1 Introduction

We have seen how it is possible to construct electrical devices which are not merely able to count the particles of radiation incident on them, but also in some cases give us a spectrum of their energies to a very high degree of resolution. From the viewpoint of simple electrical measurements this in one sense is quite remarkable. The amount of charge created by, say, an alpha particle in a gas ionization chamber is in macroscopic terms very small indeed, only a little more than 10^{-14} C. Even in a measuring system of total capacitance as low as 20 pF the resultant voltage generated is only about 1 mV. Nevertheless, single alpha particles were detected and the quantity of ionization directly measured (i.e. without charge multiplication as in a proportional counter) as long ago as 1928.

In fact in modern electronics the quantities of charge liberated in ionization should not be thought of as all that small. Digital integrated circuits sometimes handle such small amounts of charge that any ionization from background radioactivity in, for example, the encapsulation can create so-called 'soft errors' in the logic; thus the circuit has accidentally become a radiation detector.

The answer to any difficulties about the quantity of charge liberated is, of course, to use amplification. Sometimes, as we have already seen, this is an integral part of the internal operation of a detector, as in proportional counters and in semiconductor avalanche detectors. In the case of scintillation counters, amplification is provided in the photomultipliers or microchannel plates. Every type of detector, whether charge multiplication is used or not, normally requires some additional external electronic amplification before an energy spectrum can be collected or other pulse processing performed.

So important is the question of electronic amplification and pulse processing that, in addition to outlining the main principles in Sections 6.2, 6.3 and 6.4 in this chapter, we shall devote a large portion of Chapter 7 to considering how to optimize the procedure. By this we mean how the ratio of the detector signal to the background electronic 'noise' (spurious electrical disturbances of a basic nature) can be made as large as possible. Without going too far into electronic details we shall then be able to appreciate the extent to which high-resolution detector and electronic techniques combined together have now developed.

Sections 6.5 and 6.6 consider how the pulses from a detector, amplified as necessary, can be counted or sorted according to pulse height to give a spectrum of the energy deposited. In Section 6.7 we discuss the rather different techniques that are needed to exploit the fast timing capabilities that many detectors possess. After discussions of coincidence techniques and of how the housing and packaging of electronic units have been standardized, we conclude with an example of a complete electronic system which illustrates how some of the various units described in this chapter can be used in conjunction with one another.

The method of interconnecting units together with coaxial cables and the technicalities involved are considered in Sections 6.3.6 and 6.7.2.

6.2 Amplification principles for radiation detectors

We need first to make some general remarks about the process of amplification. Amplifiers for conventional signals coming from, say, an audio or video source are usually classified either as voltage-sensitive amplifiers or current-sensitive amplifiers. The former, for use naturally with voltage sources, have a high input impedance so that any impedance present in the source will not degrade the transfer of the voltage signal to the amplifier. Current-sensitive amplifiers, on the other hand, have a low input impedance to allow current from the current source to flow freely into the amplifier input. We make the same distinction between amplifiers for radiation detectors, but in a slightly different manner.

The position as regards a detector is that it produces not a continuous signal but a series of discrete pulses of current (occurring randomly in time). We use Figure 6.1 to discuss the detector/amplifier interaction. Here the dashed triangular symbol is the conventional one for an amplifier: Z_L is whatever external load it is called upon to drive, while the left-hand dashed box represents the detector. In the latter a current source i_s represents the momentary flow of current in the detector: in the usual voltage mode of operation this current quickly charges up both the detector capacitance C and the input capacitance of the amplifier C_i effectively in parallel with it (see p. 75). The resistor R represents the detector load resistor, while R_i is the input resistance of the amplifier—again both in parallel, and of combined value R_{in}, say. The action of the voltage-sensitive amplifier is to amplify the voltage v_s appearing on C_{in}, the total stray capacitance $(= C + C_i)$. Here, as throughout this chapter, lower-case letters (for example, v, i, and q) are used to denote time-varying quantities (namely,

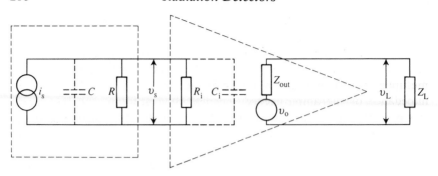

Figure 6.1 Equivalent circuit for the detector/amplifier interaction.

voltage, current, and charge, respectively, in this case), whereas capital letters (V, I, Q) will imply that there is no time dependence.

(We note in passing that there can be another contribution to C_{in} – the cable capacitance – which need not concern us here, but which we shall have to discuss in Section 6.3.1).

In the present mode of operation we wish this stray capacitance C_{in} to be charged completely, or nearly completely, before we discharge it in readiness for the next pulse. Therefore the discharge time constant, given by $R_{in}C_{in}$, must be very much greater than τ, the characteristic collection time of the charge, while small enough to ensure adequate discharge between pulses. This means that R_i must be reasonably large (as well, of course, as R), so the requirement of high input resistance for a voltage-sensitive amplifier appears also in the detector case – although for a somewhat different reason than for continuous audio or video signals. For this reason an FET (field-effect transistor) with its intrinsic high input impedance is almost invariably used as the first transistor in the amplification chain. The effective input resistance R_i is then the FET gate resistor (R_g in Figure 3.3), which can then be chosen appropriately. By comparison a bipolar transistor has an intrinsic low input impedance, which itself would therefore determine the effective input resistance.

Note also that the input voltage produced by the current i_s is larger if C_{in} is small (as the usual equation for capacitor voltage

$$v_s = \frac{q_s}{C_{in}} = \int_0^t \frac{i_s dt}{C_{in}}$$

shows, where q_s is the charge deposited on C_{in}). We thus require that C_i (and, of course, the detector capacitance C) be as small as possible. We can say alternatively that the impedance of C_i should be as large as possible – so we are indeed talking of a high-input-impedance amplifier and not merely a high-resistance one.

(The other components of the amplifier on the output side are a voltage generator of size $v_o = Gv_s$, which merely symbolizes that the amplifier multiplies the input signal v_s by an amount G, its gain, and the output impedance of the amplifier, Z_{out}, which indicates how well this amplifier performs in driving an external load. Clearly the output voltage v_L developed across the external load Z_L is highest, that is, $v_L \approx Gv_s$, when the size of Z_{out} is very much less than that of Z_L.)

We now proceed to make some further remarks about voltage-sensitive amplifiers for detectors, leaving to Section 6.3.3 a discussion of how we can use a current-sensitive amplifier – that is one with a low input impedance – in those cases where we wish to display the actual form of the detector current pulses ('current mode'). It would, firstly, seem reasonable to make the gain of a voltage-sensitive amplifier high, because the input pulse from a detector is, as we have seen, often very small. Secondly, because of the different time-scales of the pulses from various detectors, it would also seem reasonable to ensure that the bandwidth (that is, the range of speeds or frequencies over which the amplifier operates) is broad.

There are, however, several possible difficulties which we have overlooked. One very general principle is that high gain is obtained only at the expense of bandwidth and vice versa. Even if the restrictions this imposes turn out to be not serious, arbitrary increase in amplifier gain will quite possibly lead to fluctuations and instabilities in the gain and the output voltage delivered, unless there is suitable feedback of part of the output voltage back to the input, as considered later. In addition we shall have to examine critically the bandwidth which should be used, as this is central to the question of the reduction of electrical noise which always accompanies the detector pulses (see Chapter 7). The effects of gain changes and of noise must be minimized if we are to resolve detector signals of different pulse heights as clearly as possible.

It is common, although not essential, to separate physically a detector amplifier into two units: a preamplifier and an amplifier. We shall see that the preamplifier contains a 'feedback' system which helps to define the mode or character of the amplification to be used. The possibilities to be considered are (a) the voltage mode, in which the height of the output pulse is proportional to the charge deposited in the detector, (b) the charge-sensitive mode, which is a variant of the voltage mode and which is used when (as sometimes happens) the input capacitance varies with conditions, and (c) the current mode, in which, as we said, the output voltage is intended to follow the current which flows while the charge is being collected.

The preamplifier is also crucial in noise considerations, as we shall see in Chapter 7, since any noise generated at this stage is amplified by all the succeeding stages in the system. It is therefore important to place the preamplifier as close as possible to the detector in order to minimize any noise or signal losses associated with the capacitance of the input cable from the detector. A short cable also minimizes any external interference which may occur in this region, where signal amplification has yet to take place. (Indeed, where possible, it is often advisable to mount the preamplifier in the same shielded enclosure with the detector.) It is for precisely the same reasons that a preamplifier is attached right on to an aerial for a television or radio when the signal received is weak, even if the aerial is outside on the roof of the house.

We would, however, find it highly inconvenient to have to tune to different radio or television channels by making adjustments to a circuit attached to an aerial on the roof. Similarly a detector preamplifier may have to be in an awkward or inaccessible position, possibly in an environment with high radiation levels from an accelerator, reactor, or large radioactive source. Just as the television main amplifier and tuner are situated in a more convenient situation indoors, so the detector main amplifier is separated from the preamplifier and placed perhaps in a control room. In addition we shall find that this main amplifier unit often also performs the function of pulse processing in order (among other reasons) to maximize the ratio of the signal pulse height to the noise. In a sense this is analogous to a radio or television tuner; in each case the bandwidth, or frequency range, is limited just to the region of interest. (Sometimes with detector electronics, however, as in some fast timing applications, this processing may be omitted.) Even if, as can sometimes happen, the preamplifier and main amplifier are not physically separated, we shall still find it convenient to divide the discussion in this way.

6.3 Preamplifiers

6.3.1 *Voltage-sensitive preamplifiers*

We have just seen that it is desirable for the gain of any amplifier to be stable. In the technique of negative feedback a fraction of the output is fed back to the input so as to oppose the signal there. We shall see that this makes the value of the gain of an amplifier, and indeed the nature of the amplification itself, dependent not on the active components, that is, the transistors, but on much more stable passive components such as resistors and capacitors. Several types of feedback are possible; here we are interested in the commonest type, technically known as 'voltage-derived voltage feedback'. We shall see that this can be used to produce a stable voltage-sensitive preamplifier for a radiation detector, for which the output voltage is accurately proportional to the input voltage. Firstly, however, we need to consider two examples of how in general this type of feedback is realized.

The first example is shown in Figure 6.2. We start with some amplifier whose voltage gain, A, is high and positive. A fraction of the output voltage is being fed back so as to oppose the input voltage, which as we saw is always the situation with negative feedback. The amplifier output appears across sone external load (not shown). The output impedance can usually, as we shall see shortly, be assumed to be much lower than the impedance of this load. The output voltage is then just the open-circuit value $v_o = Gv_s$, as shown, where v_s is the input voltage to the fed-back amplifier circuit and G the gain after feedback. Because of this feedback, if the output voltage v_o for some reason drifts upwards, the voltage v_i at the amplifier input falls. Hence in turn v_o falls, provided, as we assume, that the gain $A = v_o/v_i$ of the amplifier before feedback is positive. This tends to stabilize the output voltage and the gain of the fed-back preamplifier circuit, and is one of the primary virtues of any form of negative feedback.

The actual value of G can be easily derived if we assume that (a) the input impedance of the preamplifier before feedback is high enough for the current fed back to the amplifier input from the join of R_f and R_f^* to be negligible compared with the current flowing through R_f and R_f^* between the output and ground and (b) the gain A before feedback is so high that v_i

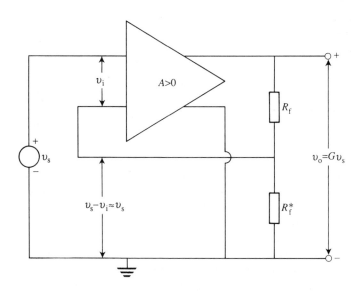

Figure 6.2 A general 'non-inverting' amplifier configuration using negative feedback.

can be quite small compared with the source voltage v_s (i.e. $G \ll A$), and yet produce a reasonable output voltage v_o. If these two assumptions hold then the voltage across R_f^* is v_s closely, and we can immediately derive the fed-back circuit gain $G = v_o/v_s$ from the potential divider formula applied to R_f and R_f^*:

$$G = \frac{v_o}{v_s} = \frac{R_f + R_f^*}{R_f^*} = 1 + \frac{R_f}{R_f^*}. \tag{6.1}$$

We have achieved our objective of making the circuit gain depend only on passive component values rather than on the gain of the initial amplifier before feedback. Since $v_i \ll v_s$ we have considerably reduced the gain available, but the advantages of negative feedback more than compensate for this. Indeed the stability of the gain as obtained through negative feedback is one of the primary factors which make radiation spectrometry possible.

As we noted a moment ago in our derivation of equation (6.1), the value G of the circuit gain must be considerably less than the gain A of the amplifier before feedback. If more circuit gain is required then we can simply feed the output into the input of a second similar fed-back amplifier. This can be done with this circuit arrangement, because it can be shown that its input impedance is even higher and its output impedance even lower than before feedback is applied. The arrangement therefore satisfies the criterion given in Section 6.2 for the input impedance in a voltage-sensitive amplifier, and furthermore is likely to satisfy the condition that the load impedance greatly exceeds that of the output, which we assumed a moment ago.

The arrangement we have been considering is known as a non-inverting configuration, since the input and output voltages have the same sign. Figure 6.3 shows our second example of how negative feedback can be applied, this time to an amplifier of high negative gain to give an inverting voltage-sensitive amplifier. We assume as before that (a) the input impedance of the preamplifier before feedback is high enough for the current fed back to the amplifier input from the join of R_f and R_f^* to be negligible compared with the current flowing through R_f and R_f^* between the output and ground and (b) the gain A before feedback is so high that v_i can be quite small compared with the source voltage v_s (i.e. $G \ll A$), and yet produce a reasonable output voltage v_o. The input voltage v_i will under these circumstances be so small that the input terminal to which feedback is being applied is almost at ground potential (in fact we often speak of this point being a 'virtual earth'). The voltage across R_f^* is therefore v_s and across R_f is v_o (which is negative). If the amplifier input impedance is also high, then, as before, the same current flows almost entirely through R_f and R_f^*. The gain G of the circuit arrangement after feedback (if we assume that R_f^* is

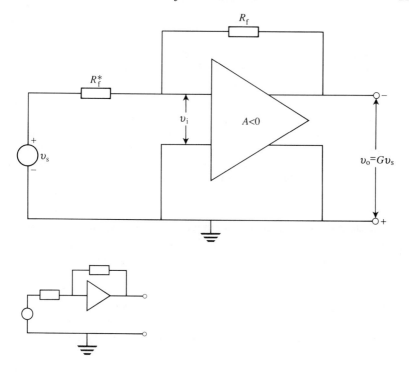

Figure 6.3 A general 'inverting' amplifier configuration using negative feedback, and (inset) a simplified representation often used.

large compared with the source impedance) thus follows from the potential division ratio between R_f and R_f^*:

$$G = \frac{v_o}{v_s} = -\frac{R_f}{R_f^*}.$$

(6.2)

The circuit gain G, now negative, is once again defined by passive component values, and, as noted, is much lower than A, the amplifier gain before feedback.

(Notice that here, unlike the previous situation in Figure 6.2, both the input and the output of the initial amplifier before feedback each have one of their leads earthed. The inset to Figure 6.3 shows how in cases such as these the amplifier triangular symbol in a circuit diagram may be alternatively placed between the signal leads only, with the earthed leads shown in a simplified manner — or indeed in other more schematic diagrams, omitted altogether. For the moment we shall use the full representation as in the main part of Figure 6.3, although later the other conventions will also be used.)

The output impedance of the inverting configuration in Figure 6.3, like that of the previous non-inverting arrangement, can also again be shown to be low. However, this time the input resistance of the circuit is given by R_f^*, since the right-hand side of this resistor is almost at earth potential. There is a limit to how high we can make the value of R_f^*, since, from equation (6.2), R_f^* must be a small fraction of R_f if our circuit is not to possess too little gain. Furthermore we cannot resolve this problem by allowing R_f to be very high, otherwise the current flowing through R_f^* and R_f would become so small that their potential divider action would be appreciably loaded even by the small input current to the amplifier itself. In practice R_f^* may be of order $10^3 \, \Omega$, which makes the input impedance low for a voltage-sensitive amplifier. In contrast we have seen that the non-inverting configuration of Figure 6.2 does not suffer from this disadvantage. We note, however, that there is an important analogue of the inverting configuration which is treated in Section 6.3.2.

Feedback of one form or another is used in both the preamplifier and the main amplifier for signals from a radiation detector. The non-inverting configuration described above would for example be very suitable for a voltage-sensitive preamplifier. This is because its input impedance can be extremely high and thus (provided the detector load resistance is similarly high) the value of the discharge time constant $R_{in} C_{in}$ of Figure 6.1 (where we assume that the dashed symbol now represents just a preamplifier) will be more than adequate to ensure that C_{in} is fully charged by the signal. Indeed it is usual to keep R_g, the gate resistor of the input FET, outside the feedback loop and directly connected across the source (as in Figure 3.3). This conveniently allows us to choose a suitable value for it so as to fulfil our other condition that $R_{in} C_{in}$ (which now equals $R_g C_{in}$ if R_g is reduced appropriately) should be small enough to allow C_{in} to discharge between pulses. (We recall in this connection from Section 3.2.3 and Figure 3.6 that R_g and R, the detector load, may be, in some important cases to be discussed in the present chapter, one and the same resistor.)

In contrast the inverting configuration has, as we have just seen, a much lower input resistance ($\approx R_f^*$). On its own it would thus be unsuitable for defining the voltage mode in a voltage-sensitive preamplifier since now $R_i \approx R_f^*$, and $R_{in} C_{in}$ would be too short to allow C_{in} to charge up fully. Some form of impedance 'buffer' must precede it. This buffer normally consists of a 'source follower', that is, a unity gain amplifier with a single fed-back FET, whose input resistance (which is therefore that for the whole preamplifier) is once again that of its gate resistor R_g.

Now the full amplitude of the voltage developed by a detector is just Ne/C_{in}, as we saw in Chapter 3, where N is the number of charge pairs collected and C_{in} is the total capacitance on which the voltage appears. The voltage-sensitive preamplifiers that we have been considering simply produce an output voltage accurately proportional to this input, which then decays with a time constant $R_{in}C_{in}$, normally much shorter than the intervals between pulses. Such preamplifiers can be used with scintillator/ photomultiplier combinations or with gaseous counters, but there is an important problem which precludes their general use in high-resolution spectrometry, particularly with semiconductor detectors. This is that there are several contributions to the capacitance C_{in} which depend on the particular operating conditions. For example, the connection between the detector and preamplifier (which, as described in Section 6.3.6, is usually a coaxial cable) itself has capacitance, the value of which depends on the type and length of connection. Adding 10 or 20 cm of cable to whatever length was previously used or specified would increase the capacitance already present by perhaps 10–20 pF. Since detector capacitances are often also of this magnitude it would be quite easy, for example, to double the magnitude of C_{in} by using a longer connection while still keeping the length reasonably short, as required from our earlier discussion. The result of this would be to halve the measured voltage.

Even if the cable length is kept the same, the detector capacitance for a partially depleted semiconductor detector can itself depend on the applied bias according to (for example) equation (5.21). Changing the bias would cause the preamplifier output voltage also to change. Even if the bias is always set to the same nominal value the voltage may still drift. For many years voltage-sensitive preamplifiers were the only type available, but their use with semiconductor detectors used in high-resolution spectrometry would, we now conclude, be in general unsatisfactory.

6.3.2 *Charge-sensitive preamplifiers*

A variation on a voltage-sensitive preamplifier to remedy the problem just described was first devised in the 1950s, and consists of using a different system of feedback so as to make the output voltage not only directly proportional to the input charge but also independent of the input capacitance. This results in what is known as a charge-to-voltage converter, or simply a charge-sensitive preamplifier. To achieve this the configuration in Figure 6.4 is used. Here a feedback capacitance C_f takes the place of the resistance R_f in Figure 6.3, while the other capacitance C_{in}, as before, represents the sum of the detector, cable and preamplifier input capacitances.

The input to the preamplifier is the current i_s from the detector, which would normally be expected to quickly charge C_{in}. However, the action of C_{in} and C_f is analogous to that of R_f^* and R_f in Figure 6.3, thus keeping v_i

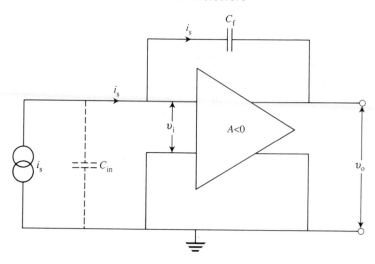

Figure 6.4 The basic charge-sensitive preamplifier configuration for a radiation detector.

small (our virtual earth again), and a more detailed analysis confirms this. The capacitance C_{in} is therefore hardly charged at all, and instead essentially all the charge is drawn on to the feedback capacitance C_f. Thus the final output voltage step V_{cs} induced by the collection of N charge pairs is given by

$$V_{cs} = -\frac{Ne}{C_f} \tag{6.3}$$

which is independent not only of the preamplifier gain before feedback, as before, but also now of C_{in}.

The conditions for correct functioning of the charge-sensitive preamplifier may be determined by analogy with the resistor feedback case, where the magnitude of the gain A of the preamplifier before feedback must be much greater than that after feedback $(= R_f/R_f^*)$. The impedance of a capacitor is inversely proportional to its capacitance, so we may reasonably conjecture that the corresponding condition here would be $A \gg (1/C_f)/(1/C_{in})$, or $A \gg C_{in}/C_f$, which is confirmed by closer analysis. For $A = 1000$ and $C_{in} = 10\,\text{pF}$ a suitable value of C_f would be about 1 pF.

The great advantage is thus (when $C_{in} \ll AC_f$) that the output always gives a signal proportional to the energy deposited in a detector, with the constant of proportionality always the same regardless of the capacitance at the input. Charge-sensitive preamplifiers have proved their worth in so many situations that they are often used even when the detector capacitance is not expected to change.

The gain after feedback (in mV MeV^{-1}) is actually quite modest. This is, as before, the price we have to pay for achieving the benefits of negative feedback. From equation (6.3) it is easy to calculate that with $C_f = 1$ pF the output voltage for each MeV of energy deposited in, say, a silicon detector is about 44 mV, and in a germanium detector about 54 mV. Here the energy per charge pair is, from Section 5.2, taken as 3.62 and 2.96 eV, respectively. Some further amplification is often provided in the preamplifier and either polarity of output may be found.

A charge-sensitive preamplifier thus produces, for each charge pulse in, an output voltage step, the height of which is proportional to the charge collected. The rise-time of the step is determined by the time it takes to induce this pulse in the input circuit of the preamplifier. The pulses will of course appear randomly in time and the fast rising steps will in general be of different heights. In Figure 6.5 we have made them fairly closely of the same height, as would occur for a well-resolved gamma-ray line.

It can be shown that, if the gain A of the preamplifier before feedback is very large, the charge drawn on to the feedback capacitor will remain there almost indefinitely. Consequently the output voltage step decays away only very slowly, with the result that successive input pulses produce a staircase of steps until the voltage saturates at the largest value, V_m say, that the charge preamplifier can produce (Figure 6.5). After this the preamplifier can no longer respond to incoming charge pulses. It is thus necessary to place

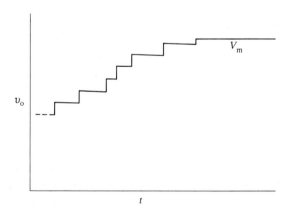

Figure 6.5 Saturation of the output voltage of the basic charge-sensitive preamplifier of Figure 6.4.

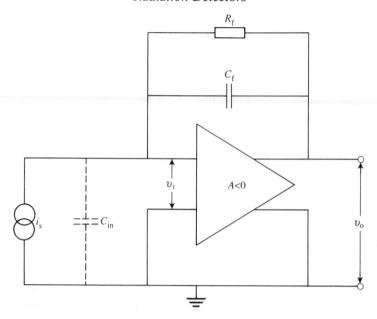

Figure 6.6 A charge-sensitive preamplifier modified so as to guard against saturation of the output.

directly across C_f a large resistor, R_f, as shown in Figure 6.6, to at least partially discharge C_f after each voltage step (as we shall see later). In fact R_f now plays the role for C_f that R_{in} did for C_{in} in the voltage-sensitive preamplifier case in Section 6.3.1. (This resistor also takes care of the small leakage current of the detector, which charges C_f and thus produces a voltage drift even in the absence of input pulses. Without R_f, if the leakage current is, say, 1 pA, then with a 1 pF capacitor the voltage changes at $1\,\mathrm{V\,s^{-1}}$ until, as before, the preamplifier is driven into saturation, in this case in the order of seconds.)

6.3.3 *Current-sensitive preamplifiers*

Another type of preamplifier, the current-sensitive preamplifier, results when the input resistance R_i in Figure 6.1 is deliberately reduced to a low value. Under these circumstances the signal current flows directly through R_i without appreciably charging up C_{in} ($= C + C_i$). Thus the shape of the pulse of current can be monitored, rather than the total charge carried by this current. This type of preamplifier is particularly suitable for fast-timing measurements, as we shall see in a moment.

Just as in the case of the voltage-sensitive preamplifier, we can use negative feedback with an initial amplifier (here current sensitive) to produce a superior stable amplifier with, in this case, an input impedance even lower than before feedback, which of course is exactly what we want. The circuit arrangement (technically known as a 'voltage-derived current feedback amplifier') is shown in Figure 6.7. The output voltage v_0 is here simply related to the input current i_s by

$$v_0 = -i_s R_f.$$

The preamplifier therefore, as before, monitors the rate of flow of charge from a detector rather than the charge itself.

Figure 6.8 shows an arbitrary current waveform i_s and the corresponding charge waveform q_s. Although the rise-time of the charge pulse may in many case be quite fast, the charge waveform, since it is given by $\int_0^t i_s \, dt$, may nevertheless be expected to rise more slowly in general than the current waveform. This is why a current-sensitive preamplifier can be expected to facilitate fast-timing measurements.

6.3.4 *Mathematics of current and voltage modes*

What are the relationships in practice between the shape of the current pulse from a detector and the output pulse shape from a real (rather than idealized) current- or voltage-sensitive preamplifier to whose input the detector pulse

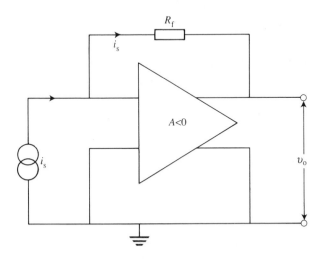

Figure 6.7 A current-sensitive preamplifier configuration for a radiation detector. The gain of the initial amplifier before feedback is negative.

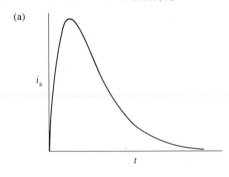

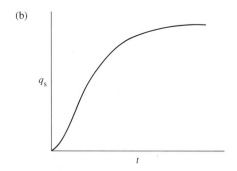

Figure 6.8 (a) An arbitrary current waveform i_s. (b) The corresponding charge waveform $q_s = \int_0^t i_s \, dt$.

is applied? We shall not attempt to examine every possible situation which may occur, but shall here answer this question primarily for a particular shape of the input current pulse, which in fact occurs in several actual cases. As a bonus the analysis will also help us when we come, in Section 6.4.2, to investigate pulse processing in main amplifiers, since there are formal similarities in the mathematics presented in that sub-section and in the present one.

We start by suggesting an appropriate form for the time variation of the detector current i_s. The expression

$$i_s = \frac{Q}{\tau} \exp\left(\frac{-t}{\tau}\right) \tag{6.4}$$

as well as occurring in several detector situations leads to an analysis which is mathematically relatively tractable. Here τ is the time constant of the decay. We can easily prove that the definite integral $\int_0^\infty i_s \, dt$ is equal to Q, which shows that Q is the total charge which flows from the detector.

One example of a system for which equation (6.4) can be valid is a scintillator–photomultiplier combination. As considered in Section 4.4.5, the photomultiplier views what in many cases is an exponentially decaying light output from the scintillator following a gamma-ray or particle interaction. Consequently the photomultiplier anode current has the form of equation (6.4), where Q is the total charge collected by the anode. The magnitude of the decay constant τ can be anything from about 1 ns for a plastic organic scintillator through to about 1 μs for inorganic materials like CsI(Tl).

Another system for which equation (6.4) can be valid is a semiconductor junction detector which is partially or just totally depleted. We have seen from Section 5.5.9 that if an amount of charge Q is liberated by short-range radiation near the front (high-field) face of the detector, then the charge induces a current given by equation (6.4) (which is the same as equation (5.26)) as it drifts away from the front face in the linearly decreasing electric field of the detector. The value of τ in these cases is, as we also saw, given as $\tau_d w/d$, where τ_d is the dielectric relaxation time of the detector material, w is the detector slice thickness, and d ($\leqslant w$) is the thickness of the depletion layer. Values for τ_d are, as we said previously, often in the general range of 0.5–5 ns. The cases of an overdepleted detector and a detector with a uniform field (like a Ge(Li) or Si(Li) detector), for which equation (6.4) does not hold, are discussed at the end of this sub-section.

It can immediately be seen that the idealized current pulse shape as given by equation (6.4) has a rise-time of zero, which makes it obviously attractive for timing measurements. By comparison the pulse shape obtained by a perfect integration of equation (6.4)

$$q_s = Q \left[1 - \exp\left(\frac{-t}{\tau} \right) \right] \tag{6.5}$$

as would be obtained approximately from a voltage- or charge-sensitive preamplifier, has a non-zero rise-time governed by the time constant τ.

In practice, of course, it is often impossible to monitor exactly the current pulse shape from a detector because the rise-time of the current-sensitive preamplifier may not be negligible compared with τ. Similarly the voltage pulse shape may not be exactly monitored because the fall time of the voltage-sensitive preamplifier may not be large compared with τ. Therefore, following the objectives stated earlier for this sub-section, let us analyse the effect of a general preamplifier on the current pulse given by equation (6.4). Our results will be applicable to current-sensitive and voltage-sensitive preamplifiers, and we shall afterwards indicate how to extend them to charge-sensitive preamplifiers.

Figure 6.9 shows a preamplifier equivalent input circuit which is appropriate for our needs. This circuit is a simplified version of the input circuit

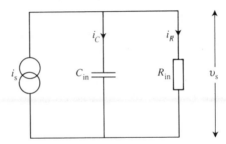

Figure 6.9 The equivalent input circuit for a general preamplifier.

for the amplifier in Figure 6.1. The flow of charge from the scintillation or semiconductor detector, represented by the current generator i_s (equation (6.4)) is shared between R_{in} (the detector load resistance R and amplifier input resistance R_i in parallel) and C_{in} (the capacitances of the detector, connections, and preamplifier input all summed together). Obviously when R_i is low compared with R, as for example in the current-sensitive case, the effect of R is negligible, and R_{in} is approximately given by just the amplifier input resistance R_i. Note that in the case of a semiconductor junction detector we must assume that it is just totally depleted so that $\tau = \tau_d$ (where τ_d is the dielectric relaxation time, as above). The presence of any undepleted material would necessitate a more complicated equivalent circuit, the analysis of which, however, can in many cases lead to quite similar results to what we shall derive, as reference 1 shows.

Let us therefore examine the currents flowing in the circuit of Figure 6.9. The current from the detector generates a voltage v_s across R_{in} and C_{in}, such that the currents i_R and i_C through R_{in} and C_{in}, respectively, are

$$i_R = \frac{v_s}{R_{in}} \tag{6.6}$$

and

$$i_C = C_{in}\frac{dv_s}{dt}. \tag{6.7}$$

Equation (6.7) follows because

$$v_s = \frac{q_s}{C_{in}} = \frac{1}{C_{in}}\int i_C\,dt$$

where q_s, as before, is the charge at any instant on C_{in}, so

$$\frac{dv_s}{dt} = \frac{i_C}{C_{in}}$$

which leads directly to equation (6.7).

Since $i_s = i_R + i_C$, equations (6.4), (6.6), and (6.7) give us

$$\frac{Q}{\tau} \exp \left(\frac{-t}{\tau} \right) = \frac{v_s}{R_{in}} + C_{in} \frac{dv_s}{dt}. \tag{6.8}$$

If this equation is multiplied on both sides by $[\exp(t/R_{in}C_{in})]/C_{in}$, it can be integrated directly to give, after rearrangement, v_s as a function of t:

$$v_s = \frac{QR_{in}}{R_{in}C_{in} - \tau} \left[\exp \left(\frac{-t}{R_{in}C_{in}} \right) - \exp \left(\frac{-t}{\tau} \right) \right] \tag{6.9}$$

where the boundary condition assumed is that $v_s = 0$ at $t = 0$. The open-circuit output voltage from the preamplifier is then given simply by $v_o = Gv_s$, where G is, as before, the preamplifier gain. The type of expression given by equation (6.9) is, incidentally, quite common, and we shall refer to it again in Section 6.4.2.

The form of equation (6.9) is shown in Figure 6.10. Here at short times the voltage time dependence has the form $(1 - \exp(-t/T_R))$, where the rise-time constant T_R is the smaller of τ and $R_{in}C_{in}$. At long times the voltage takes the form $(\exp(-t/T_F))$, where the fall time constant T_F is the larger of τ and $R_{in}C_{in}$. We show the form of v_s as if it were a positive-going pulse. The sign of Q may of course be negative, in which case v_s would be negative-going. This would be the case, for example, for electrons being collected at a photomultiplier anode, or for electrons being collected at the ohmic contact of an n-type surface barrier detector.

Let us first consider the situation where $R_{in}C_{in} \ll \tau$. In this case equation (6.9) reduces approximately to

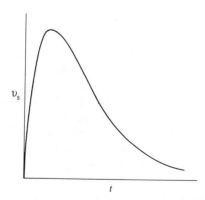

Figure 6.10 Form of v_s in equation (6.9).

$$v_s = \frac{QR_{in}}{\tau} \exp\left(\frac{-t}{\tau}\right). \tag{6.10}$$

Since $i_s = v_s/R_{in}$, we thus return to the true current-sensitive condition described by equation (6.4), except that the exact equation shows that the rise-time is degraded slightly from zero up to the small value $R_{in}C_{in}$. (We mentioned this earlier in Chapter 4). This is the optimum situation for timing measurements, as we said a moment ago.

In order to achieve this situation we must choose a preamplifier with a low input resistance and a detector with a low input capacitance. As we discussed in Section 4.4.5, the current-sensitive situation is not difficult to achieve for scintillation counters with $\tau \approx 1\,\mu s$. With $C_{in} = 20\,pF$, say, an amplifier resistance R_i of as high as $5\,k\Omega$ (which determines R_{in} if we assume the detector load resistor, R, to be large) makes $R_{in}C_{in}$ ten times lower at $0.1\,\mu s$. If, however, $\tau \approx 1\,ns$, as for a fast scintillator, or for a silicon detector of resistivity $10\,\Omega\,m$ (equation (5.27)), and $C_{in} = 20\,pF$ still, R_i must be only $5\,\Omega$ for $R_{in}C_{in}$ to be ten times lower, at $0.1\,ns$. For these very fast signals a matched system in the sense of Section 6.7.2 must be used, and in this context the lowest value of R_i usually is, for technical reasons, $50\,\Omega$. Of course it may not be felt necessary to strive to make $R_{in}C_{in} \ll \tau$ if fast timing is the objective and τ is already short enough. Indeed it is even possible to use the rising edges of charge pulses for semiconductor detectors for fairly fast timing, as we shall mention again in Section 6.3.6.

When $R_{in}C_{in} \gg \tau$, equation (6.9) shows us that we return to the pulse shape for the voltage sensitive case as indicated by equation (6.5), except that the fall time is degraded slightly from infinity down to a large value $R_{in}C_{in}$ (just as we indicated earlier in this and the previous chapter). This follows because (apart from the large but finite fall time) the voltage signal v_s is given by the value of q_s in equation (6.5) divided by C_{in}, that is,

$$v_s = \frac{Q}{C_{in}}\left[1 - \exp\left(\frac{-t}{\tau}\right)\right]. \tag{6.11}$$

Returning to the current-sensitive preamplifier, we note that, although they are very suitable for fast-timing measurements, they also have some important limitations. In the general case (equation (6.9)) the maximum voltage to which the preamplifier input signal rises is reduced below that of the voltage signal input in the ideal voltage-sensitive case (equation (6.11)). (The fractional loss in signal height is called the 'ballistic deficit'). In the true current-sensitive case, when $R_{in}C_{in} \ll \tau$, the input signal maximum as given by equation (6.10) can be seen to be the ideal one (Q/C_{in}) multiplied by a factor $R_{in}C_{in}/\tau$, that is, by the ratio of the two time constants. If the speed of a current-sensitive preamplifier is required, then its smaller voltage signal input must be accepted. The voltage-sensitive output, with its larger size and slower fall ($R_{in}C_{in}$), is more suited for pulse-height measurements.

The analysis of the effect of a charge-sensitive preamplifier on an input current can be shown to produce the same types of results as above, except that the circuit time constant $R_{in}C_{in}$ should be replaced by the time constant $R_f C_f$, where, as described in Section 6.3.2, C_f is the feedback capacitance and R_f is the resistance across it. Of course only the case $R_f C_f \gg \tau$ interests us here, since we want the full charge signal to be developed. (The condition $R_f C_f \ll \tau$ in fact relates to the complementary case of Figure 6.7, where C_f is now the small stray capacitance across R_f.)

Finally we shall refer to the case of detectors which do not produce exponentially decaying current pulses of the form given by equation (6.4). We have noted in Chapter 5, for example, that lithium-drifted and hyper-pure detectors of large volumes, as used with gamma and X-rays, would produce a whole variety of pulse shapes depending on the detector geometry and the field configuration, and on where the ionization is deposited in the detector. If it were possible to observe these in the current mode, the rise-time, as before, would be limited only by the value of $R_{in}C_{in}$, and thus fast. In practice this situation is a little more complicated, as these detectors normally have (because of noise and cooling considerations) a preamplifier mounted as an integral part of the overall assembly. Since the normal application is energy spectrometry this preamplifier is charge sensitive, and so the output is effectively the integral of the current pulse. It is nevertheless also possible here to use this pulse for timing purposes, as the rise-time, although longer than for an integrated pulse from a small-volume junction detector, is still quite short. However, these rise-times show variations corresponding to the different current pulse shapes which would have occurred, and we shall have to discuss in Section 6.7.4 how to deal with these variations.

6.3.5 Detector biasing

Frequently the detector bias, that is, the high voltage necessary for detector operation, is applied through a terminal on the preamplifier housing, inside which is an associated circuit described below. The bias is then transmitted over the preamplifier input lead on to the detector. It is important to realize, however, that the use of the preamplifier in this way is purely a matter of convenience. The arrangement is not an essential part of the operation of a preamplifier; the bias supply can (and sometimes must) be handled separately.

The standard circuit, for say, a proportional counter or Geiger–Müller

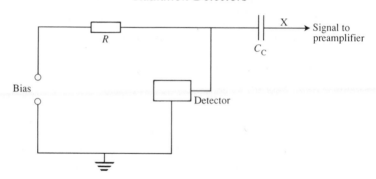

Figure 6.11 Standard biasing circuit for a proportional counter, Geiger–Müller counter, or charged particle semiconductor detector.

counter is shown in Figure 6.11 (as we saw in Section 3.2.3). The signal is taken through a coupling capacitor C_C from the anode. The cathode, which is the outer cylinder, is earthed, as is best from both safety and shielding considerations. A similar circuit can be used with semiconductor detectors for charged particles, either with positive or negative bias. Because the detector is capacitively coupled to the preamplifier circuit input, only the signal (plus any noise or fluctuations) is passed, whereas the steady leakage current of the detector is not. However, this current passes through the load resistor, R, the value of which is normally chosen to give a resultant voltage drop across it which is a small fraction of the overall bias voltage. Since one terminal of the detector is in this arrangement earthed, the bias resistor R and coupling capacitor C_C can be easily mounted inside the preamplifier earthed housing, with the detector signal and also the bias voltage being applied through appropriate connectors, as shown in Figure 6.12.

For all semiconductor detector systems of low leakage and high resolution, such as those for gammas or X-rays, the detector and load resistor reverse places to give the arrangement shown in Figure 6.13. Since the d.c. voltage above ground at the junction between these two components is now small, the coupling capacitor C_C previously used can be omitted, which gives direct d.c. coupling between the detector and the preamplifier input.

The full advantages of the arrangement in Figure 6.13 are best appreciated only after a study of Chapter 7 on noise. For convenience we shall here briefly mention the most important points. In the standard bias circuit shown in Figure 6.11 a resistor is connected at X before the first stage of the preamplifier. This resistor is the R_i of Figure 6.1 or the R_f of Figure 6.6. Because the coupling capacitor C_C can be omitted in Figure 6.13 the load resistor R can be combined with this resistor, as will be shown in

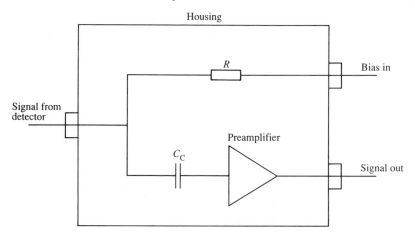

Figure 6.12 Incorporation of the circuit components of Figure 6.11 into the housing of a preamplifier.

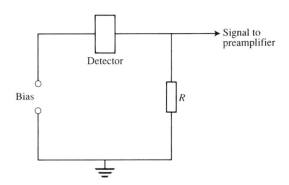

Figure 6.13 Biasing circuit for a high-resolution semiconductor detector.

Figure 7.5, where $R_i = R_g$, and in Figure 7.6, the corresponding charge-sensitive case. As there are fewer components the stray capacitance to ground is reduced, and (as we shall see in Chapter 7) this improves the signal-to-noise ratio. Furthermore with a special charge-sensitive arrangement we shall see how to remove entirely the remaining resistor, and with it all resistor noise.

Care must be taken with either configuration (unless special protection circuits are included) to increase or remove the bias slowly with, for example, a multi-turn potentiometer. Otherwise switching transients (which, like the detector signals themselves, are transmitted through any coupling capacitors present) can damage the transistor in the first stage of the preamplifier.

6.3.6 *Other considerations*

As well as performing the crucial tasks of initial amplification of small signals and of defining the feedback character, a preamplifier also has to transmit the amplified signal to the main amplifier. As we saw in Section 6.2, if the full amplitude of the amplified voltage signal is to be transferred to the load (i.e. the main amplifier), the preamplifier should have a much lower output impedance than the input impedance of the main amplifier. The preamplifier therefore acts as an impedance buffer or interface between the detector and the main amplifier.

Normally signals are transmitted between electronic units over coaxial cables. As shown in Figure 6.14, such a cable consists of a central wire carrying the signal, which is separated by a dielectric insulator from an outer braided cylindrical sheath of wires coaxial with the central wire and forming the electrical return path. The sheath also shields the signal wire from unwanted electromagnetic interference and pick-up voltages. Such shielding is particularly important with the small pulses between the detector and preamplifier when these are not in the same shielded enclosure, and in any case between the preamplifier and main amplifier. The coaxial construction also provides a well-defined geometry for the connections so that, for

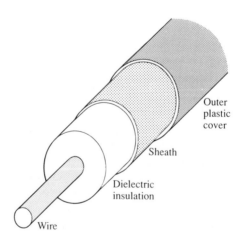

Outer plastic cover

Sheath

Dielectric insulation

Wire

Figure 6.14 Structure of a coaxial cable.

example, we can know and control their electrical characteristics such as capacitance and inductance. We shall find that this is an especially important feature when we come to consider fast-timing measurements in Section 6.7.2.

We have seen that current-sensitive preamplifiers are particularly suitable for fast counting or timing measurements. Charge-sensitive preamplifiers usually also provide some facility, albeit often inferior, to enable such measurements to be made. Sometimes this is simply a second output identical to the main one, since in the case of semiconductor detectors at least, the rise of even the charge pulse is fast enough for measurements to an accuracy of a few nanoseconds.

Most charge-sensitive preamplifiers also provide an input labelled 'test pulse'. This input is designed to receive pulses of fixed height from a pulse generator. The charge thus deposited on a small internal standard capacitor is transferred to the input of the actual preamplifier configuration. In this way a lower limit to the resolution of the system, namely that part due to the preamplifier itself (see Chapter 7), can be set. Such resolution measurements may be performed with the detector present, or with a capacitor, of value equal to that of the detector, connected instead. This is because the detector capacitance is usually a substantial part of the overall capacitance, and thus affects the resolution. (With the detector present, the resolution loss due to its leakage current is included in our measurement.)

6.4 Main amplifiers

6.4.1 *The functions of a main amplifier*

We commence this section with a little recapitulation. For detectors which, like scintillation counters, produce a reasonably large output signal, a very simple preamplifier–amplifier system will often suffice. We could, for example, employ a voltage-sensitive preamplifier and choose the value of $R_{in}C_{in}$ appropriately so that it was both sufficiently greater than τ, the detector signal rise-time, to integrate the signal, and yet sufficiently short that each pulse would have time to return to the baseline before the next one arrived. In other words, $\tau \ll R_{in}C_{in} \ll 1/\bar{n}$, where $\bar{n}$ is the average rate of arrival of pulses. In these circumstances the pulses are in a form to be counted or otherwise processed, after whatever additional amplification is necessary to bring them to the required level (usually in the 1 to 10 V range). So here the shaping of the pulses is done at the detector/preamplifier interface, and the main amplifier has no role except that of simple amplification.

The situation is very different for a semiconductor detector where no internal amplification exists, and where, particularly for the spectrometry of low-energy gamma and X-rays, considerable external amplification is needed. First of all, for the reasons stated earlier, we shall choose a charge-sensitive rather than a voltage-sensitive preamplifier. In addition the small-ness of the signals now being dealt with means that the reduction of electronic 'noise' from the preamplifier is a priority. One particular source of noise, as we shall see in Chapter 7, is any resistor located at the input of the preamplifier. We shall also see that, to reduce this contribution as much as possible, it is necessary to make such resistors as large as is practical. In the charge-sensitive case this will apply to R_f, the resistor which discharges the feedback capacitor C_f after each pulse (Section 6.3.2). Consequently it will usually be impossible at normal count rates to ensure that the decay con-stant $R_f C_f$ (the analogue of $R_{in} C_{in}$ in the voltage-sensitive case) will be short enough to allow each pulse to return to the baseline before the next arrives, and they will in fact 'stand on each other's tails'. The main amplifier is now given the task not only of increasing the size of the pulses, but also of unscrambling and shaping them into a form suitable for counting or other measurements. We discuss how this is done in Section 6.4.2. It should, however, be remarked that the combination of a charge-sensitive pre-amplifier with pulse shaping in the main amplifier is now a standard arrange-ment and is widely used with all detector types (not just semiconductor detectors) even when not strictly required. (Main amplifiers designed for fast-timing measurements constitute a special case and are mentioned in Section 6.7.3.)

We return now to discuss the problems arising when we choose a large value of R_f. Suppose, for example, with a semiconductor detector we have $R_f = 10^9 \, \Omega$ and $C_f = 1 \, \text{pF}$, giving $R_f C_f = 10^{-3} \, \text{s}$—a value typically found in charge-sensitive preamplifiers of good resolution. A train of pulses arriving at a mean rate of even only a thousand per second will therefore produce considerable pulse overlap. The first point to note is that a mean d.c. current of $\bar{n} N e$ passes through R_f, where $\bar{n}$, as before, is the average rate of pulse arrival and N, as usual, is the number of charge carriers collected per pulse. This current in turn produces a mean d.c. output voltage of $\bar{n} N e R_f$ (on which the pulses sit), which can be much larger than the output step size $N e / C_f$ generated by each pulse. This is shown in Figure 6.15. It is not difficult to show that we are thus limited by this effect to values of $\bar{n}$ lower than about $10^4 \, \text{s}^{-1}$ for radiation of a few MeV energy. Otherwise the preamplifier output saturates at the maximum d.c. level that it will supply, which is typically a few volts. To deal with high count rates, then, we must reduce R_f, which means increased noise and thus lower resolu-tion in our energy spectra. We shall discuss in Chapter 7 an optical feed-back method which will help resolve this conflict between count rate and resolution.

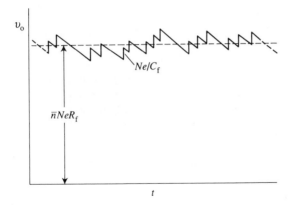

Figure 6.15 Output waveform from a charge-sensitive preamplifier at high input count rates.

In the meantime we shall continue our present discussion for moderate count rates. First, we must remove the d.c. level developed, before it is further amplified. Even when this is done (by passing it through a coupling capacitor at some stage after the charge-sensitive loop) we are still left with the variations in heights above the baseline to which pulses of identical heights rise, due to their random times of arrival and consequent random position on the tail of the previous pulse. Thus, as we have seen, the pulses, as well as probably requiring additional amplification, need to be shaped into a more suitable form before they are transmitted to a subsequent unit. Several different shaping techniques exist, but broadly speaking they are all designed to achieve two aims. One is to eliminate the effects just described of the long tails through some sort of low-frequency cut-off process. The other aim is to smooth out rapid noise fluctuations through application of a high frequency cut-off.

We shall postpone to Chapter 7 a full treatment of how the effects of noise are best reduced. The other aspects of pulse processing can, however, be analysed now. We shall concentrate here on one simple form of pulse processing by resistance-capacitance circuits.

6.4.2 *Principles of resistance–capacitance shaping*

If we put an ideal voltage step from a preamplifier (which is attached to, say, a semiconductor detector) across the capacitance–resistance circuit shown in Figure 6.16, the step itself is reproduced at the output, since it takes time to build up a voltage difference across the capacitor C_1. (Alternatively we

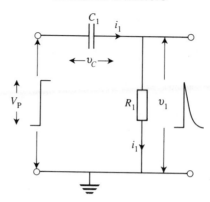

Figure 6.16 Removal of long tails from a preamplifier output step by a capacitance–resistance shaping circuit.

might say that a capacitor offers zero impedance to a rapidly changing voltage.) As the capacitor charges up and the voltage v_C across it increases, the output voltage v_1 across the resistor R_1 decreases as the charging current flowing through R_1 falls towards zero. To determine the mathematical form of v_1 we first write down that v_1 and v_C are related to the step height of the input pulse V_P by

$$v_1 + v_C = V_P \tag{6.12}$$

where the voltage v_C across the capacitor is (like what we said in Section 6.3.4) given by

$$v_C = \frac{q_1}{C_1} = \frac{1}{C_1}\int i_1 \, \mathrm{d}t = \frac{1}{R_1 C_1}\int v_1 \, \mathrm{d}t . \tag{6.13}$$

Here q_1 is the instantaneous value of the charge on the capacitor and $i_1 = v_1/R_1$ the current flowing. Substituting for v_C from equation (6.13) into equation (6.12) and differentiating with respect to t we get

$$\frac{\mathrm{d}v_1}{\mathrm{d}t} + \frac{v_1}{R_1 C_1} = \frac{\mathrm{d}V_P}{\mathrm{d}t} = 0 . \tag{6.14}$$

The solution of this is the well-known exponential fall-off with time of v_1, such that

$$v_1 = V_P \exp\left(\frac{-t}{T_1}\right) \tag{6.15}$$

where T_1, the time constant of the fall, is given by $R_1 C_1$, and the boundary condition is that $v_1 = V_P$ when $t = 0$. We assume in all the above that the input step is ideal, i.e. that its rise-time is much shorter than T_1, and its fall much longer than T_1.

If therefore we make T_1 short enough we can effectively return the pulse quickly to the baseline (zero volts) before a following pulse arrives, as Figure 6.17 shows. The step height has, however, been faithfully transmitted. However, we shall find in Chapter 7 that the detector and preamplifier noise which is transmitted by this simple shaping technique is rather high. The addition of a resistance–capacitance circuit as shown in Figure 6.18 enables a better noise performance overall to be achieved. From the order of the components the complete system is known as '*CR–RC*' shaping.

The unity gain amplifier in Figure 6.18 acts as an impedance buffer or interface between the *CR* and *RC* circuits. Its function is to enable the full voltage v_1 from the first circuit to be applied to the input of the second, regardless of the output and input impedances concerned. In other words, the input impedance of the amplifier is much greater than the output impedance of the *CR* circuit, and the output impedance of the amplifier is

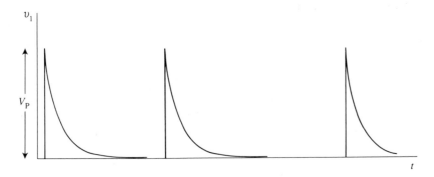

Figure 6.17 Sequence of pulses from the output of the circuit of Figure 6.16.

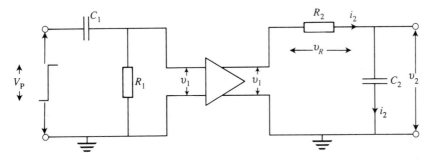

Figure 6.18 A complete *CR–RC* main amplifier pulse-shaping system.

much less than the input impedance of the RC circuit, as explained in Section 6.2. The gain of the amplifier can, if wished, be in practice greater than one. There will also (not shown) normally be further amplification stages following the CR–RC shaping circuit, with, of course, voltage feedback loops as appropriate to ensure stability, as was noted in Section 6.3.1.

For unity gain of the amplifier in Figure 6.18, therefore, the voltage v_1 at the RC circuit input is the sum of the voltages v_2 and v_R developed across C_2 and R_2, respectively, and so

$$V_P \exp\left(\frac{-t}{T_1}\right) = v_2 + i_2 R_2 = v_2 + R_2 C_2 \frac{dv_2}{dt}$$

where the current i_2 through the capacitor (and therefore either component) is given by $C_2 dv_2/dt$. The product $R_2 C_2$ gives the time constant, say T_2, of the second, resistance–capacitance circuit. It governs the circuit time response, just as T_1 governs that of the first, capacitance–resistance circuit. (Normally $T_1 \geqslant T_2$ for reasons that will become clear shortly.)

We can therefore write

$$V_P \exp\left(\frac{-t}{T_1}\right) = v_2 + T_2 \frac{dv_2}{dt}. \tag{6.16}$$

The overall form of the response of two combined circuits of time constants T_1 and T_2 is mathematically like that described in Section 6.3.4 in connection with preamplifiers, as can be seen from the formal similarity of equations (6.8) and (6.16). The solution of equation (6.16) is therefore analogous to equation (6.9), as the appropriate boundary condition in the present case is that $v_2 = 0$ at $t = 0$, which is like the one used previously. Thus

$$v_2 = \frac{V_P T_1}{T_1 - T_2}\left[\exp\left(\frac{-t}{T_1}\right) - \exp\left(\frac{-t}{T_2}\right)\right]. \tag{6.17}$$

This expression holds for $T_1 \neq T_2$. When $T_1 = T_2$ it becomes indeterminate, and the solution can then be shown in this particular case to be given by

$$v_2 = \frac{V_P}{T_1} t \exp\left(\frac{-t}{T_1}\right). \tag{6.18}$$

The general form of the output pulse v_2 for $T_1 \neq T_2$, as given by equation (6.17), is similar to that of Figure 6.10. The rise-time T_R of this pulse is governed by the smaller of T_1 and T_2, which we said earlier was in this case T_2, and the fall time T_F by the larger of T_1 and T_2, here T_1. The pulse shape given by equation (6.18) is also generally similar in appearance, but with rise and fall governed now, of course, by the same time constant.

When $T_1 \gg T_2$, for example, equation (6.17) becomes, when $t \ll T_1$,

$$v_2 = V_P \left[1 - \exp\left(\frac{-t}{T_2}\right) \right]$$

and, when $t \gg T_2$, becomes

$$v_2 = V_P \exp\left(\frac{-t}{T_1}\right).$$

We can also show that when $T_1 \gg T_2$ the signal rises to almost the full height V_P of the input from the preamplifier. Essentially the output from the first (CR) circuit (equation (6.15)) is then hardly altered by the second (RC) circuit.

In contrast, when T_1 is not very large compared with T_2, a ballistic deficit is incurred similar to that described in Section 6.3.4. The ratio of the maximum voltage attained to that reached when $T_1 \gg T_2$ (namely, V_P) can be shown, from equations (6.17) and (6.18), to have values of 0.83, 0.74, 0.50, and $\exp(-1) = 0.37$ for values of T_1 of $16T_2$, $8T_2$, $2T_2$, and T_2, respectively. In each case, however, the maximum voltage reached still remains a valid measure of the step V_P from the preamplifier (that is, the two voltages are proportional to each other) provided that the time constants T_1 and T_2 are stable and well defined in value.

What are the best values for T_1 and T_2? Clearly, for example, the fall time should not be too long, otherwise at high count rates successive pulses might overlap. However, as we have already indicated, one of the most important factors influencing the choice of time constants is that of electrical noise. If high-resolution pulse-height analysis is our aim then a proper choice of constants will minimize the amount of preamplifier noise (as a fraction of the signal) passed by the main amplifier. It can be shown that for this to be achieved the value of T_1 should be reduced down to that of T_2 and both set to a particular value determined by detector and preamplifier parameters. This common time is, for gamma-ray spectrometry with germanium detectors, normally a few microseconds. A main amplifier will almost always have provision for changing these time constants in order to deal with different situations. (On the other hand for scintillation and proportional counters with large internal multiplication the choice of time constants in a main amplifier does not significantly affect the resolution achieved.)

We can now see why in practice T_1 is not normally set to be less than T_2. If this were done then both the ballistic deficit and the noise performance would become worse.

We frequently refer to the combination of C_1 and R_1 in Figure 6.18 as a differentiating circuit. This is because its output across R_1 varies with time in proportion to the time derivative of the input voltage after a time long compared with $R_1 C_1$, as elementary circuit theory shows. For example, the voltage input in Figure 6.16, being constant, then gives an approximately zero output. Likewise R_2 and C_2 in Figure 6.18 are called an integrating circuit. This is because its output across C_2 varies with time in proportion to $\int_0^t v_1 \, dt$ at times short compared with $R_2 C_2$. If a step voltage were applied, for instance, with $v_1 = 0$ for $t < 0$ and $v_1 = V$ for $t > 0$, a linear 'ramp' voltage proportional to t would then be initially produced at the output. The times $R_1 C_1 \, (=T_1)$ and $R_2 C_2 \, (=T_2)$ are therefore frequently referred to as the differentiation and integration time constants, respectively. It must be remembered that the circuits, despite their names, do not accurately differentiate and integrate at other than long and short times, respectively. Yet other names are 'high-pass filter' and 'low-pass filter', which describe their operation on sinusoidal signals of different frequencies.

A conclusion reached in Chapter 7 is that shaping with one differentiating plus one integrating circuit, as just described, still does not provide the optimum means of noise reduction. We shall find that other circuits, for example ones which use further integration, can offer even better noise performance.

In the remainder of Section 6.4 we shall discuss two techniques which are of particular importance at high count rates. We shall find that under these conditions the resolution with which we can measure pulse heights deteriorates because of effects in the main amplifier. There is in fact an inevitable tradeoff between resolution and count rate which is already evident even at as low as a few hundred counts per second. The two topics to be discussed represent attempts to mitigate the effects of this tradeoff.

6.4.3 *Pole-zero cancellation*

Except in the special optical feedback method to be discussed in Chapter 7, the top of the step output from a charge preamplifier is not flat but instead decays away slowly with a time constant of perhaps up to a millisecond (given by $R_f C_f$, as described in Section 6.4.1). A flat-top input to the differentiating circuit of the main amplifier (shown in Figure 6.16) gives at its output, as we have seen, the exponential decay given by equation (6.15) and shown in Figure 6.17. With a slowly decaying input, however, the output is changed to the form shown in Figure 6.19. At first the output falls as before with the same time constant $T_1 = R_1 C_1$. It then undershoots to negative values and finally decays up to zero with the long time constant of the decay-

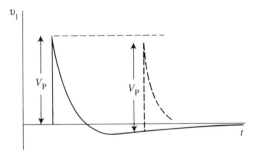

Figure 6.19 The output from the differentiating circuit of a main amplifier for a slowly decaying step input. The decay constant of the output undershoot is shown rather small for clarity; in practice it is usually even longer than shown compared with the initial fall time constant T_1.

ing input from the preamplifier. The next pulse may therefore well occur before the undershoot has decayed away. In this case this pulse would still step up through the correct voltage difference, V_P, but, having started from a negative value, it would fail to reach a high enough voltage level, as Figure 6.19 also shows. Reducing the preamplifier decay time constant in order to shorten the duration of the undershoot is no solution, because this will increase the noise, as we shall find in Chapter 7; it will also increase the magnitude of the voltage undershoot. We can understand this last point because the areas enclosed above and below the time axis can be shown to be equal, which is equivalent to saying that the capacitor C_1 in Figure 6.16 will not transmit a d.c. voltage. The full analysis leading to the mathematical form of the output shown in Figure 6.19 is similar to that which will be given in another context in Section 6.7.6.

The technique known as 'pole-zero cancellation' provides a solution to the problem. In it we connect a resistor R_{PZ} across the capacitor C_1, as shown in Figure 6.20. We can regard the action of R_{PZ} and R_1 in series somewhat as if they fed a fraction of the input signal to the output signal v_1. If the value of $R_{PZ}C_1$ is made equal to the time constant of the preamplifier fall, then a full analysis shows that this positive fraction of the input cancels out the undershoot (hence 'cancellation' in the name of the technique). The output in these circumstances can then be shown to be given by

$$v_1 = V_P \exp\left(-t/fR_1C_1\right) = V_P \exp\left(-t/fT_1\right) \tag{6.19}$$

where $f = R_1/(R_1 + R_{PZ})$. In other words, a true exponential fall without undershoot is obtained. In practice R_{PZ} is adjusted empirically until the undershoot on v_1 is observed on an oscilloscope just to disappear. We also notice that the fall of the output has been speeded up by this technique, so

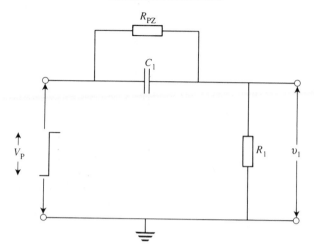

Figure 6.20 Pole-zero cancellation circuit.

we have certainly succeeded in removing the deleterious effects of the long decay following the step voltage V_P.

6.4.4 *Baseline restoration*

Even with pole-zero cancellation the output pulses at high count rates from an amplifier may fail to start from the zero-level baseline. This is because most main amplifiers have at some stage a coupling capacitor, if only in the CR part of the shaping circuit. A capacitor cannot transmit a d.c. signal, which means that the amplifier output cannot ever be exactly what is called unipolar, that is, it cannot remain always on one side of the time axis. This effect is quite negligible until the time between pulses becomes comparable with their width. As shown in Figure 6.21 (which is drawn for idealized rectangular pulses passing through a long time constant CR circuit), the pulses in this case shift downwards in voltage from their position in the widely spaced condition. Because no d.c. signal can be transmitted by the capacitor, the area enclosed above the baseline or time axis by the pulse train is equal to that enclosed below.

If the pulses were regularly spaced and all of the same shapes, as in Figure 6.21, no problem would arise, because we would merely have to determine the baseline shift and add it on to the observed pulse heights. However, we know from Section 1.4.3 that pulses from radioactive disintegrations and other nuclear counting situations occur not regularly but at random. Thus, as the spacing between pulses increases during a lull, as it were, and decreases

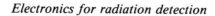

(a)

Voltage

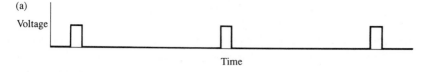

Time

(b)

Voltage

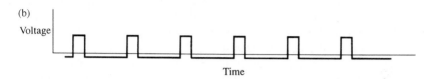

Time

Figure 6.21 (a) Widely spaced pulses showing negligible baseline shift. (b) Closely spaced pulses showing baseline shift.

during a burst, the shift decreases and increases, respectively. The net effect is to introduce a jitter in pulse height and therefore to broaden the resolution observed, since exact compensation for the baseline shift is no longer feasible in the way just mentioned.

Several solutions are possible. Clearly all must be applied near the output of the amplifier, where no further shifts will be introduced. One method uses diode clamping, the simplest variety of which is shown in Figure 6.22. The diode, if perfect, prevents the output voltage from ever going negative, since it conducts strongly in this circumstance. The heights of the voltage steps at the input, which are transmitted faithfully by the capacitor, appear therefore at the output 'clamped' to zero. Of course no diode is perfect, since conduction effectively takes place only when a finite threshold forward voltage

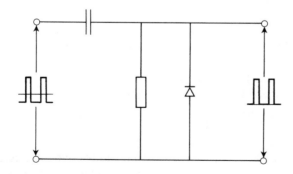

Figure 6.22 Baseline restoration using diode clamping.

of a fraction of a volt has been exceeded, so the baseline restoration is correspondingly somewhat imperfect.

Another solution which avoids this problem involves the introduction of undershoots to the pulses with two *CR* differentiating circuits in succession, each with a *short* time constant (Figure 6.23). The buffer amplifier between the two circuits is required for impedance matching purposes. As we said in Section 6.4.3, the areas enclosed above and below the time axis must be equal. Consequently the average value of the output is zero, and so the baseline is *quickly* 'restored', or no longer shifted. A pulse which undershoots is called 'bipolar', as opposed to 'unipolar'. The mathematics of bipolar pulse production are given in another context in Section 6.7.6.

Yet another method is to shape pulses by using a long length of cable. We anticipate the discussion in Section 6.7.2 by pointing out that the velocity of propagation in a cable of the step output from the preamplifier (as with any other signal) is not infinite. As a result the step can be reflected at the end of a cable to produce, as it were, a delayed 'echo', in analogy with an acoustic echo when, say, one's hands are clapped together in a long corridor. We shall not discuss the detailed principles of 'delay line shaping', but it is possible to shape pulse steps into not only unipolar but also bipolar pulses of finite duration by the use of deliberately created reflections of this nature. The same advantages for high count rates apply to these bipolar pulses as to those described in the preceding paragraph.

Normally we use unipolar pulses if the count rate is not too high, since the use of bipolar pulses tends to introduce somewhat more noise (in the sense described in Chapter 7). An amplifier will often have provision for switching between unipolar and bipolar shaping.

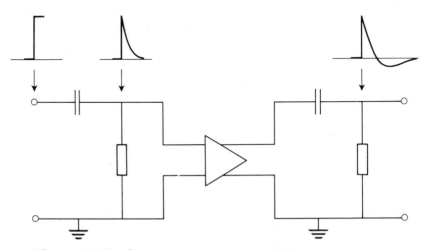

Figure 6.23 Baseline restoration using two differentiating circuits.

6.5 Simple counting and analysis systems

6.5.1 *Overall concepts*

Broadly speaking, pulses are derived from a radiation detector and then amplified for one of four main purposes. One is to examine the shape of the pulses on, say, an oscilloscope, about which we shall say more in Section 6.7.1. The other three are to count the number of pulses in a given time interval, to determine accurately the heights of the pulses for energy analysis, and to measure the times of occurrence of the pulses with respect to each other or to reference pulses from another detector.

Pulse counting, which forms the subject of this section, is in some ways an undemanding aim compared with the other three. Nevertheless, before we feed the pulses into a counting (or 'scaling') circuit we must first select which pulses we want to count. It might be all of them, but we must still ensure that any small noise pulses also present do not get counted as well. We perform this selection with some form of 'discriminator'. As a result our complete electronic system looks as shown in Figure 6.24. Provided that the signals are clearly distinguishable from the noise, the requirements placed on the amplifier are not as stringent as with pulse-height analysis systems.

As with our previous descriptions of amplifiers, we shall not in general make detailed references to the circuit functioning of the units to be described. Exceptions to this will be made only where we need to know as a user some particular internal feature so as to enable us to employ the unit properly.

6.5.2 *Discriminators*

These are electronic units based on Schmitt trigger or similar circuits. A discriminator in its simplest form give an output pulse of standard shape for every input pulse which exceeds a voltage threshold of our choice. Thus in

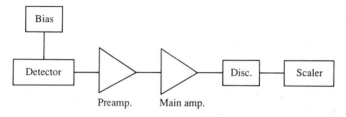

Figure 6.24 Block diagram of a complete pulse counting system.

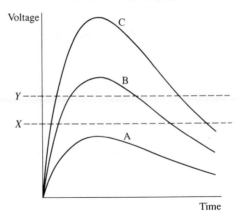

Voltage

C

B

Y

X

A

Time

Figure 6.25 Selection of pulses according to their height with a single threshold level on a discriminator (two possible levels shown), and with time walk effects shown (see text).

Figure 6.25 pulses B and C produce an output with either threshold, X or Y, but pulse A does not with either of them. The threshold level is normally set by using a calibrated precision potentiometer on the front panel of the unit. The value of the threshold ranges from close to zero up to typically 10 V.

The output pulse is rectangular in shape and of fixed height and width, perhaps a few volts and about 1 μs, respectively. This type of standardized pulse is called a 'logic' or 'standard' pulse; its presence indicates that the desired feature (in this case, the crossing of the threshold) has occurred, and no information is otherwise carried by its height or width. This is in distinction to an 'analogue' or 'linear' pulse, for which the pulse dimensions such as height do carry information. The output pulse from an amplifier is an example of such a pulse.

The output pulse from a discriminator appears at or just after the time instant when the input pulse crosses the threshold. Thus the time of appearance varies, or 'walks', with respect to the beginning of the pulse when either the pulse height or the threshold level changes. In Figure 6.25, for example, pulse B crosses threshold X later than the larger pulse C, showing walk due to a change in pulse height, and pulse B crosses threshold X earlier than threshold Y, showing walk due to a change in threshold level. These effects are of little, if any, consequence in counting applications, but it means that when we come to consider timing measurements a different type of discriminator will be needed.

The full name for the type of unit described is an integral discriminator, since output pulses are produced for all events above the threshold. This is in distinction to differential discriminators, which we shall now consider.

6.5.3 *Single-channel analysers*

These are sometimes called differential discriminators, because, as indicated in Figure 6.26, an output pulse is produced when the input lies between a lower and an upper discriminator level (shown as broken lines in the figure). Single-channel analysers, or SCAs for short, have two independent potentiometers controlling these levels. They can be labelled 'E_1' and 'E_2' as indicated to denote lower and upper energy settings, to which the pulse heights are normally proportional. Often, however, the two potentiometers are linked so as to control instead the lower level and the difference between the levels. The lower level can then be labelled 'E', and the difference or window labelled 'ΔE', as also shown in Figure 6.26. Thus, in either method, pulses in a chosen height range, possibly around a particular peak in a spectrum, can be selected and the others rejected. We return to these devices in Section 6.5.5.

6.5.4 *Scalers and ratemeters*

Scalers are units which count the number of input pulses and which often display continuously the number recorded on some form of electronic

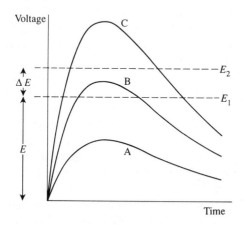

Figure 6.26 Selection of pulses according to their height with two threshold levels on a discriminator (single-channel analysis); the middle pulse (B) is here accepted and pulses A and C rejected.

display. The term 'counter' is sometimes found, although as we have seen, this term can also be used for a radiation detector itself. Scalers accept logic pulses from discriminators and SCAs. They will often work directly with amplifier pulses, but this is not good practice unless, as is sometimes the case, the scaler unit itself has a built-in discriminator at the input which is adjustable so that noise can be rejected. The number of counts can often be read out electronically into some other data recording equipment, in which case the display is then sometimes omitted.

Frequently a timing unit is built in or can be connected. In this case two facilities usually offered are timing up to a preset count and counting up to a preset time. In both cases the count rate can then be determined. The preset count mode is valuable since we can specify in advance the statistical accuracy we require (for X counts the statistical error is $\sqrt{X}$, as we saw in Section 1.4.3), and then count until this accuracy has been reached.

It is also possible to obtain 'ratemeters' which display directly the count rate from the deflection of a needle on a moving-coil meter. Because of the statistical fluctuations inherent in the counting of radioactive disintegrations, the deflection must be smoothed out if it is not to fluctuate too strongly, particularly at low count rates. The smoothing is performed electronically with a resistance–capacitance combination whose time constant is adjustable; the larger the constant the greater the smoothing, but this will also mean that the time to respond to genuine changes in the rate will be longer.

We must not forget when making counting measurements to make any necessary corrections such as the subtraction of background counts which do not come from the source of interest. In addition at high count rates the dead time of the system will affect the rate observed. This dead time may be limited by the detector (a Geiger–Müller counter, for example, has an especially long dead time, as we saw in Chapter 3), or by the resolving time of the electronics, or by a combination of both.

Let us examine dead-time effects a little further by generalizing the introductory analysis already given in Section 3.4.4. These effects refer to the inability of a counting system at high count rates to keep up, as it were, with the counts coming in. In many cases we can quantify this by saying that a system, having received a count, is unable to record a second count until the dead time has passed. If this time is T_D, say, and the observed count rate is m, then we can calculate the true rate n by noting that the system is dead for a fraction mT_D of the time. The rate at which events are lost is therefore mnT_D. The loss rate is also, however, given by the expression $n - m$, so we can equate the two expressions and solve for the true rate n to give

$$n = \frac{m}{1 - mT_D}. \tag{6.20}$$

We note that at low rates ($m \ll 1/T_D$) the observed rate m is approximately equal to n. At high rates, when $m \approx 1/T_D$, n can no longer be determined. This is because m (which from equation (6.20) is given by $n/(1 + nT_D)$) can be seen eventually to saturate at the value $1/T_D$ regardless of the value of n.

We see that in general we must known T_D in order to determine n. The value of T_D for a particular system can be experimentally determined by the use of the two-source method mentioned in Section 3.4.4 from the equation

$$n_3 = n_1 + n_2$$

where n_1 and n_2 are the true count rates with source 1 and source 2 respectively, and n_3 is the true rate with the two sources together. From equation (6.20) it thus follows that in terms of the corresponding observed rates m_1, m_2, and m_3

$$\frac{m_3}{1 - m_3 T_D} = \frac{m_1}{1 - m_1 T_D} + \frac{m_2}{1 - m_2 T_D}$$

from which T_D can be calculated.

In some systems the dead time T_D is not constant but depends on n or perhaps on the pulse height or shape from the detector and amplifier. In such cases corrections can still be made but clearly their evaluation will be more complicated.

6.5.5 *Measurements of energy spectra*

Let us now consider one largely obsolete but historically important example of the use of the complete counting system shown in Figure 6.24, in which the discriminator ('Disc') is here chosen to be of the SCA (single-channel analyser) type. Suppose that radiation of a range of energies is incident on the detector. We shall feed the amplifier output into the SCA, and set it for the 'E, ΔE' or 'window' mode, which is the second way of using an SCA that we described in Section 6.5.3. We then record with a scaler the number of counts in a given time interval, with the discriminator level $E = 0$ V and the window $\Delta E = 0.25$ V, say. This will represent our first 'channel' of data. We than repeat this for 39 more channels, say, with the channel level $E = 0.25$, 0.50, 0.75, 1.00, 1.25, ... up to 9.75 V, keeping the time interval for each channel the same and the channel width $\Delta E = 0.25$ V. We can then plot a histogram of the number of counts in each channel against the channel number. Figure 6.27 shows what might be obtained if we used a scintillation counter and a radioisotope emitting gamma rays of a single energy in the region of 1 MeV.

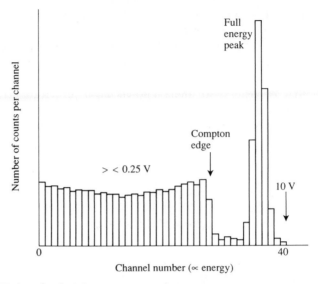

Figure 6.27 A pulse-height spectrum from a scintillation counter exposed to gamma rays of a single energy of value about 1 MeV, with the heights sorted into 40 channels.

In this way we have determined the pulse-height or energy spectrum from the detector. This type of method was, up until the 1950s, the only way to measure such a spectrum. It assumes that the spectrum does not vary (due to some electronic drift, perhaps) during the period it is being taken, and in addition is time consuming because it involves during any one count the rejection of all the rest of the counts coming in. In the example given 39/40ths or 97.5 per cent are rejected at any one time. Section 6.6 describes how with multichannel analysers no counts need be rejected in this way.

6.6 Multichannel analysers

6.6.1 *Principles*

We have just seen in the previous section that a single-channel analyser can be used to build up a spectrum of pulse heights from a detector, but that such a measurement is very tedious, since data is counted only one channel at a time. A much quicker method of collecting the data in, for example, Figure 6.27 would be to have 40 single-channel analysers suitably arranged to cover the whole energy range in 40 adjacent steps of ΔE, together with 40 scalers, one for each analyser. In this case every arriving pulse would be processed and routed to the appropriate scaler, with a large reduction in the time required to produce a spectrum. Such a device is an example of a 'multichannel analyser', or MCA for short. However, this method is not very satisfactory because of the complexity of such devices especially for larger numbers of channels, and because of the difficulty in eliminating voltage drifts in the channel positions. We shall discuss shortly one or two of the methods which are instead used in MCAs.

Historically, larger numbers of channels became necessary as soon as the resolution of detectors improved. In the case of the scintillation counter spectrum of Figure 6.27 we can see that a relatively modest number of channels can fairly adequately define the full-energy peak and Compton edge position, together with the shape of the remaining part of the spectrum. Smaller channel widths and larger numbers of channels for the same full-scale voltage lead to little or no significant new information since the detector resolution is already worse than the channel width.

Let us suppose, however, that we use a 40-channel analyser with a germanium detector, for which the resolution for gamma rays of around 1 MeV might be 2 keV. We can see from Figure 6.28 that the channel

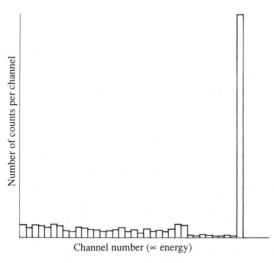

Figure 6.28 A pulse-height spectrum from a germanium detector exposed to gamma rays of a single energy of value about 1 MeV, with the heights sorted into 40 channels.

width ($\approx 1\,\text{MeV}/40$) is now many times larger than the detector FWHM, and consequently more channels are required if we are to resolve all the details of such a spectrum. As a general rule of thumb we can say that four or five channels are about right for covering the width of a peak (as in Figure 6.27). This will enable us to estimate properly its position and FWHM, while at the same time the channel width will not be excessively small. For a signal resolution of 2 keV this implies that at least two or three thousand channels are required over the full scale. This contrasts with the 50 keV resolution that a typical scintillation counter might have at the same energy, for which fewer than a hundred channels full scale could suffice. Most MCAs, whatever their principle of operation, offer the facility for changing the number of channels full scale, which range typically through factors of 2 from about 128 to 8192.

(In passing we also notice the much greater extent to which the full-energy peak in Figure 6.28 stands out above the lower-energy Compton continuum compared with the situation in Figure 6.27. This would remain true even for a larger and thus more appropriate number of channels in Figure 6.28. We have already referred in Section 5.8 to this advantage possessed by germanium detectors over their scintillation counterparts.)

If the total number of channels is increased and the channel width reduced (keeping the full-scale voltage into the analyser the same), the number of counts X per channel collected in a given time from a given spectrum of interest will decrease. There will thus be relatively larger statistical fluctuations in these reduced numbers of counts. This is because, as we mentioned in Section 1.4.3, the statistical error in an observation of X counts is $\sqrt{X}$, giving a relative fluctuation of $\sqrt{X}/X$ or $1/\sqrt{X}$, which increases as X decreases. It is therefore important to have no more channels full scale than is necessary, since the additional complexity leads to no new information. Instead it is preferable to smooth these fluctuations by maintaining the total number of channels at a value just large enough to display the peaks properly, as already described.

Of course we may nevertheless want to cover a peak with more than four or five channels, if, for example, there is a problem with the shape of a peak. From what we said in Section 1.4.3, we would normally expect this to approximate to a Gaussian distribution. There may, however, perhaps be charge collection problems, giving a long low-energy tail to the peak which we wish to examine in detail. It would be wrong, however, to suggest that this examination could be achieved only by increasing the overall number of channels in the spectrum. We could instead arrange that channel zero corresponds not to zero energy but to some threshold energy just below the region of interest. This displacement or offset of the calibration could produce a result as in Figure 6.29, where here a peak (shown in this case as a symmetrical Gaussian) is arranged to stretch over all the channels available.

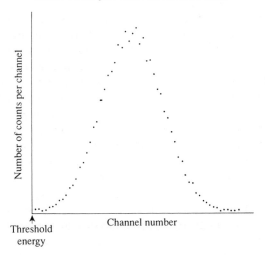

Figure 6.29 A pulse-height spectrum which has been offset and expanded, showing a single peak, and enabling any details of interest to be examined.

(Of course we fail to collect all energies below the threshold.) In this figure the dots represent the tops of the histogram columns such as we saw in the earlier spectra of Figures 6.27 and 6.28. Other examples occur in figures in the earlier chapters. Most MCAs offer this offset option; alternatively the so-called 'biased amplifier' which produces this effect may be included in the main amplifier.

The basic technique of a modern MCA is that the height of each input pulse from the main amplifier is first converted into digital form in an analogue-to-digital converter, or ADC for short. The digitized pulse is next stored in a memory similar to a computer memory. The contents of the memory are then displayed on a screen, using for example dots as in the manner shown in Figure 6.29. We shall now use this framework to examine in more detail how an MCA works.

6.6.2 *Analogue-to-digital converters*

The 'Wilkinson ADC' is probably the oldest type still commonly used. In it a capacitor is quickly charged up to the voltage of the input pulse, and then discharged linearly, that is, at a constant rate. During the discharge time of the capacitor a scaler counts the number of cycles produced by an oscillator whose frequency is stabilized. When the capacitor voltage reaches zero the number reached in the scaler is proportional to the discharge time and hence is a measure of the input pulse height. An analogue-to-digital conversion has therefore, as desired, been performed. The dead time of ADCs (that is the time in which it is processing a pulse and thus unable to count a further one) depends on the numbers of oscillator cycles counted and thus on the pulse height; the higher the pulse the larger the dead time. The number of cycles counted eventually defines a particular channel number in a pulse-height spectrum, so with a 4096-channel Wilkinson ADC with a 100 MHz oscillator (i.e. 10^{-8} s period), for example, we would expect the dead time to vary from a low value with small input pulses up to a time of 4096×10^{-8} s or about 40 μs for the largest pulses analysable.

The 'successive approximation ADC' is the other type which is in common use. Here the channel number is generated not by counting pulses but by making successively closer approximations to the height and thereby generating a binary number. To illustrate this suppose we wish to digitize a 10 V high input pulse on a scale running in 1 V steps up to a maximum of 16 V. We first compare the input pulse height with an internally generated voltage of 8 V, which is half the full-scale value. Since the input is larger, a '1' is entered into what is going to be the most significant bit of a binary scaler. The circuit then subtracts 8 V from the 10 V input, and compares the 2 V remainder with an internally generated voltage of half the 8 V half-scale value, namely, 4 V. Since the remainder is smaller, a '0' is entered as the next most significant bit. Because this bit is '0', the circuit then decides not to subtract the quarter-scale value of 4 V from the 2 V remainder, which it would do if the bit were '1'. The 2 V remainder is then compared with half the quarter-scale value, that is, 2 V. The remainder is this time not smaller, so a '1' is entered as the next bit. Finally the circuit decides to subtract the eighth-scale value of 2 V from the 2 V remainder, and compares the new remainder of 0 V with half the eighth-scale value, which is 1 V. The new remainder is smaller, so a '0' is entered as the least significant bit. The number read into the binary scaler is thus 1010, or 10 in decimal notation. Thus with four stages of successive approximation channel numbers up to 16 are generated. Twelve stages, for example, would cover a range of 2^{12} or 4096 channels.

It can be seen that the dead time of this type of ADC remains constant with channel number for a given number of channels, although the dead time increases as the number of bits and thus channels increases. The successive approximation type is inherently faster, but its linearity between pulse height and channel number tends to be not quite as good because of cumulative errors in the subtraction process.

6.6.3 *Spectrum storage and analysis*

We can see that both types of ADCs have yielded from an input pulse a count number which is registered in a scaler. In order to build up a pulse-height spectrum the number is transferred to the memory of the MCA. More precisely, the memory basically has a number of addresses or locations. The content of each address is itself a number. The address corresponding to the number in the scaler is selected and its contents incremented by one. In this way a pulse-height spectrum can be built up and shown on a display monitor.

In practice even a modest MCA will have more facilities than this bare minimum. Usually there will be channel markers and address locations and contents of the marked channels available on the display. The pulse-height calibration in terms of energy can be incorporated, and quantities such as the total number of counts between two particular marked channels, or the area, centroid, and width of a peak can be determined. (We remember from Chapter 1 that the area under a peak is a measure of the total number of counts associated with that peak.)

Background subtraction is another facility which is usually provided. In this the source, accelerator beam, or other origin of radiation introduced for the experiment is removed, a spectrum of the background radiation is taken for the same length of time as with the source present, and the background spectrum subtracted from the main observed spectrum to give a spectrum without background. A timer is usually provided which can be set to measure either elapsed time (true time as measured by any clock) or live time. This latter is the time which has passed for which the analyser has been available to record a spectrum. It is lower than the elapsed time by an amount equal to the accumulated dead time of the analyser. In background subtraction, for example, live times must be used when two spectra are accumulated, otherwise more counts will be lost during the high-count-rate source measurement than during the low-count-rate background measurement. A final spectrum can usually be also read out to an external printer, graph plotter, magnetic tape cassette recorder, or magnetic disc recorder.

A disadvantage of a multichannel analyser as just described is that it lacks flexibility. The tasks are performed by external controls such as buttons and potentiometers attached to internal circuits, but no modifications or additions to the procedures can be easily made. It is usually possible to transfer a spectrum into a computer for further analysis, if desired. Nowadays the output of an ADC (or indeed several ADCs) is usually transferred directly to an ancillary or even personal computer which has been adapted and programmed for the complete task of multichannel storage and spectrum analysis. The programming can either be bought as a package or be written by the user. The data is of course stored in the memory of the computer. In this way the analysis of the data either as it comes in or afterwards can be tailored for the specific needs of a particular experiment. Instructions to the system are entered through the computer keyboard.

A disadvantage which has occurred with analysers based on some simple computers is that the system is not powerful enough simultaneously to collect data and to provide a display updated sufficiently frequently. However, many systems of all degrees of sophistication are now available from manufacturers, and their own brochures provide some of the best sources of information available.

MCAs, whether computer based or not, can enable spectra to be 'stabilized'. The instabilities referred to are the drifts with time that inevitably occur in the pulse-processing electronics during the course of a measurement. These drifts can be due to temperature or other changes in the detector or its electronics, producing changes in gains or voltage levels in the system. As a result peaks in a pulse height spectrum can be broadened or distorted, thus causing a loss of resolution. In a scintillation counter system, for example, small changes in the photomultiplier bias voltage can cause relatively large changes in the photomultiplier gain. Even high-resolution measurements with germanium detectors can be susceptible to shifts especially if, as in environmental measurements, the source activities are low and the counting times long. A 'spectrum stabilizer' in some way senses the position of a peak and compares it with an accurately known reference voltage level. From this comparison a signal is generated and fed back to the electronics so as to change the gain and restore the peak to its correct position.

One way of sensing the drift of a peak as displayed in an MCA spectrum would be to set the window of a single-channel analyser over part of the peak. If the peak drifted with respect to the SCA window, then the count rate from the SCA would change. From this the signal to be fed back could be derived. Of course it would have to be known or assumed that the true count rate from the source stayed constant with time. More usually two SCAs are used, each set over different parts of the peak, one on one side of the peak centre and the other on the other so that each SCA count rate is equal and their difference zero. Any peak drift destroys the equality of the count rates by tending to raise one rate and lower the other. The count rate difference, now no longer zero, can therefore be used to provide the signal to be fed back. The principle is shown in Figure 6.30, where now the curves represents smooth fits to the data points on the MCA screen. This method does not require the count rate to be constant with time. A variation on the last method described is periodically to illuminate the detector with a pulsed source of light so as to introduce each time a fixed number of charge carriers. Hence an extra peak in the spectrum is created, which the spectrum stabilizer

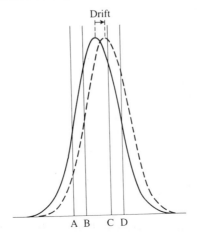

Figure 6.30 Detection of spectrum drift. Two SCAs, one with window set between A and B below a peak centroid and the other above it between C and D, are adjusted to give equal count rates. If the peak drifts upwards (broken curve) the count rate from the first SCA falls below the rate from the second.

can use. (This assumes that the detector is photosensitive, which is usually the case with semiconductor detectors and of course with photomultipliers.) This is especially useful if no clearly defined peak is present in the spectrum of interest.

Obviously, whatever method is used, it is important for the SCA levels themselves to be extremely stable. The SCAs and associated circuitry could be incorporated into the MCA itself. Section 6.10 will show a system in which a spectrum stabilizer is used.

We conclude by mentioning one rather different type of facility usually offered on MCAs, which is known as multichannel scaling or simply multi-scaling. In this the incoming pulses are sorted not according to their size but according to when they occur. Channel one records the total number of counts occurring between times $t = 0$ and $t = T$, say. Channel two records the count between $t = T$ and $t = 2T$, channel three between $t = 2T$ and $t = 3T$, and so on. The time T, known as the 'dwell time', can be selected as required. This mode of operation is very useful for studying time-dependent phenomena such as the decay of the activity, this is, the count rate, from a short-lived radioactive source, or the decay of fluorescence as monitored by, say, a photomultiplier.

6.7 Timing systems

6.7.1 *Introduction*

We have seen that the pulses from radiation detectors, especially fast scintillators and small-volume semiconductor detectors, can be as short as a few nanoseconds or even less. As well as, or instead of, measuring the height of a pulse, we can therefore determine to a high degree of accuracy the time of occurrence of the pulse with respect to that of some other event in the experiment. We may, for example, wish to determine the flight time of a particle between two detectors, both of which are triggered by it. Alternatively the time elapsed between an 'initiating' event, perhaps a light flash, and the subsequent appearance of a related pulse of radiation may be required. A variation on these themes might be that we have only one pulse and wish to determine its width or the time taken, say, to fall to half its maximum value. These are all examples of the last purpose described in Section 6.5.1 for which pulses are derived from a detector. In all these cases special timing electronics are needed. This forms the subject of this section; more information than will be given here appears in reference 2, which gives a specialized account of this topic.

A very direct way that can be used is to display the pulse or pulses on a fast oscilloscope. Rise-times of a few nanoseconds or better are readily available from oscilloscopes with bandwidths in the region of 1 GHz. With digitizing or analogue storage versions even a single pulse can be recorded and displayed on a screen for subsequent examination, or alternatively the trace from a conventional oscilloscope can be photographed. In this way the time difference between two pulses from different detectors, displayed on separate traces, can be directly determined, or the details of a single trace measured.

This method can find only limited application, since it would be extremely tedious to build up a time-of-flight spectrum, say, between two detectors for particles which had a range of velocities. Even if the particles had supposedly the same velocity we might still want to see the width of the peak in a time spectrum. In this section we shall describe the types of electronic units used for such measurements. Before we do this, however, we must first discuss some new considerations which should be observed whenever any fast-timing units (including the oscilloscopes just described) are interconnected.

6.7.2 *Cables and impedance matching*

The velocity of propagation of a pulse down a typical coaxial cable is, because of its associated capacitance and inductance, only about two-thirds of that of light in a vacuum. Since this latter velocity is about $3 \times 10^8\,\mathrm{m\,s^{-1}}$, we can see that the pulse takes 5 ns to travel along 1 m of this cable. This time is not negligible compared with time-scales that we wish to measure: a 5 MeV alpha particle, for example, takes 65 ns to travel 1 m *in vacuo*. Knowledge or measurements of cable delays will therefore be necessary in any absolute measurement of the flight time. Indeed we shall see that it is useful sometimes deliberately to introduce lengths of delay cable into a timing circuit.

Now that propagation times are no longer negligible compared with the times of interest, we have to consider in a timing system the effects of pulse reflections and their relatively gradual decay. We shall not give a full treatment of the principles involved, but a summary should help us to understand the essentials.

When a step of voltage V travels down a long coaxial cable, it is accompanied by a step of current I, as shown in Figure 6.31. The ratio of the voltage to the current depends on the geometry of the conductors and on the dielectric insulating medium between them, and is thus characteristic of the cable used. It is therefore called the 'characteristic impedance', Z_0, of the cable, and can be regarded as the load placed on a source by this long cable. Values of 50 Ω and 93 Ω are commonly used. It is important to realize that this impedance leads to no energy loss as heat; the product of V and I gives instead the power being propagated down the line by the step. Significant heat generation in the cable can, for example, occur if the conductivities of the conductors are not large enough, but we shall assume that such resistive losses are negligible.

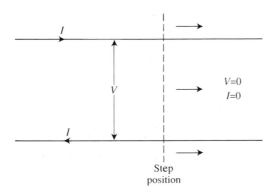

Figure 6.31 Representation of a step of voltage travelling down a coaxial cable. To the right of the step there is as yet no voltage difference between the signal line (inner wire) and the return line (outer sheath), and the current flowing in the lines is still zero.

When then happens when a voltage source launches a step into the line? The first effect in general is that the size of the source step, V_s, is reduced to the step size V in the cable by the potential divider effect between the source impedance Z_s and the characteristic impedance Z_o. We treat Z_o for this purpose as an ordinary impedance and thus get that at first $V < V_s$ unless $Z_s = 0$ (or $Z_o = \infty$, which cannot happen in practice). Subsequent developments can modify the value of V, as we shall see.

After some time, perhaps several nanoseconds, depending on the cable, the step reaches the end of the cable and is transferred to the input of a fast amplifier or one of the other units to be described later in this section. If the input impedance of the unit, Z_i, is equal to Z_o, the voltage and current in the cable are in the correct ratio to pass their associated energy completely into the load, and the unit is said to be matched to the cable. If on the other hand $Z_i \neq Z_o$, the voltage and current in the cable are not in the correct ratio for the load; only a fraction of the energy enters it and the rest of the energy is reflected as a smaller voltage and current step back down the cable to the output of the unit which transmitted the step in the first place. This new step will similarly be partially reflected at the source back into the cable unless the source impedance Z_s equals Z_o, in which case the energy of the returning step will be fully transferred back into the source.

We should note that Z_s, Z_o, and Z_i are for our purposes considered to be purely resistive. As a consequence, for example, a perfectly sharp voltage step from the source remains perfectly sharp in the cable. It is not difficult to observe on an oscilloscope significant rounding of a step when it is transmitted through even a few metres of some types of cable. This is because Z_o cannot, as in our simple picture, be taken as a pure resistance. Furthermore an amplifier, as we have already seen, like other units, often has significant capacitance at its input. We therefore refer to impedances rather than resistances for the various Zs, even in our simplified treatment.

We can now distinguish a number of situations. If $Z_s \neq Z_o \neq Z_i$, the step is reflected back and forth along the cable, rather like an echo, transferring energy at each reflection into the load or source. Consequently the waveform received by the load looks like a series of decreasing steps, reaching the final voltage V_f in one of two different ways depending on the circuit parameters, as Figures 6.32(a) and 6.32(b) show. This is an undesirable situation in any fast-timing measurement. Eventually the 'echoes' die out, leaving the final voltage V_f across the input of the receiving unit, which is given by

$$V_f = \frac{V_s Z_i}{Z_s + Z_i}. \tag{6.21}$$

This is exactly what would be expected in the absence of the cable from the

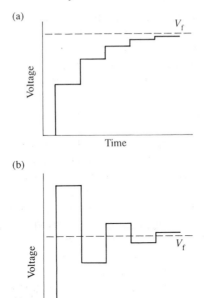

Figure 6.32 Possible waveforms received by a load which is not matched to a cable. If $Z_i < Z_0$, the resulting waveform can be shown to look like (a); if $Z_i > Z_0$, then (b) results.

usual d.c.-type potential divider formula. The characteristic impedance Z_0 does not appear, because it is only of significance while the reflections are propagating. Thus, if the cable delays are small compared with the time-scale of the pulses, our main concern is only that $Z_s \ll Z_i$, so as to give us $V_f \approx V_s$. This criterion is not necessarily the best, however, for fast pulses viewed on short time-scales, or, for that matter, for microsecond pulses like those from shaping amplifiers if they are propagated over hundreds of metres of cable. (For very similar reasons, reflections and characteristic impedances are important in audio cable telephony, where signal time-scales are in the region of 1 ms and cable lengths are many kilometres.)

However, there are two or three situations in which reflections or their effects are eliminated. One is the case, already mentioned, of when the load is matched to the cable, that is, if $Z_i = Z_0$. In this case equation (6.21) is valid as soon as the step reaches the load, not merely when a long time has elapsed. In particular, when $Z_s \ll Z_0 (= Z_i)$ almost the full voltage step as delivered by the source is launched into the cable and received by the load after the propagation delay. There are no reflections to cause any signal distortion, and so this is a desirable situation to work in. If $Z_i > Z_o$, as is normally the case, we can easily place a resistor in parallel with the load Z_i to bring the impedance down to Z_o. If Z_s is not small then a smaller voltage signal is launched into the cable, but there are still no reflections when this signal is absorbed by the load. We speak of the cable being correctly terminated at the load.

Another satisfactory situation is when the source is matched to the cable, that is, if $Z_s = Z_o$. This time there is a reflection at the load if $Z_i \neq Z_o$, but the reflected step returns to and is absorbed by the source. No further change occurs in the voltage across the load, nor is there any signal distortion. We would therefore expect from equation (6.21), that when, in particular, $Z_i \gg Z_o \, (=Z_s)$, the full voltage step as delivered by the source should be received by the load after the propagation delay, and a full analysis of the situation confirms this. If $Z_s < Z_o$, as is usually the case, a resistor in series with Z_s can bring the total source impedance up to Z_o. If Z_i is not very large then a smaller voltage signal is delivered to the load, but the reflected signal is still fully absorbed when it returns to the source. In all these cases the cable is said to be correctly terminated at the source.

In practice residual reflections may occur even though an attempt is made to terminate the cable correctly at one of its ends. For example, we have already seen that the impedances concerned may not be purely resistive. In such cases it is worth trying to match both the source and the load to the cable, that is, putting $Z_s = Z_o = Z_i$. Here, however, we see from equation (6.21) that $V_f = V_s/2$, that is, only half the voltage step height from the source is delivered to the load.

Reflections can occur even when a fast pulse is split and divided between two cables. If the splitting is performed as in Figure 6.33(a), with an input cable and two output cables all of characteristic impedance Z_o and with the centre conductors all directly connected together, then the characteristic impedances of the two output cables appear in parallel to each other. Thus an input signal sees at the splitting their combined impedance of $Z_o/2$, and hence part of the signal will be reflected back into the input cable even if the output cables are correctly terminated. To overcome this the three centre conductors can be connected together through a 'star' network of three resistors each of value $Z_o/3$, as shown in Figure 6.33(b). As viewed from the input cable at A, the equivalent circuit of the splitter with the two output cables connected looks as shown in Figure 6.33(c). From this it is easy to show that the impedance viewed from A is just Z_o, and so matching has been achieved and reflections at the junction eliminated. Reflections at the far end of the output cables will of course be removed by correct terminations there.

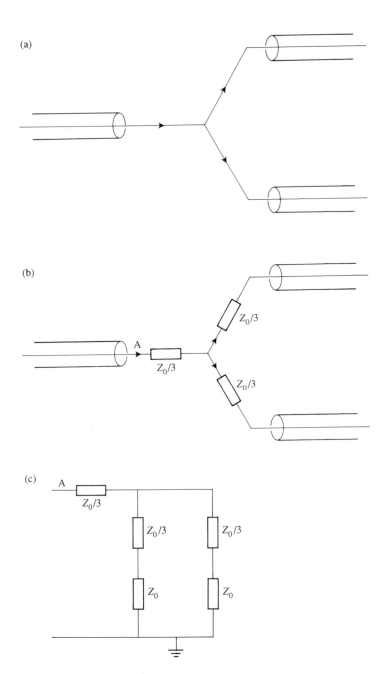

Figure 6.33 Splitting a fast pulse into two: (a) simple arrangement with short direct connections between the centre conductors of the three cables; (b) short connections through three resistors each of value $Z_0/3$; (c) equivalent circuit looking from A in figure (b). (The characteristic impedance of all three cables is Z_0 and the sheaths are all earthed.)

So, as we foreshadowed in Section 6.3.6, coaxial cables are important not merely because of their shielding properties but because with them we can obtain a well-defined characteristic impedance and thus ensure the matching which is so important for fast signals.

Now that we have considered how to connect or interface fast-timing units together, we can proceed to discuss some of the units themselves.

6.7.3 *Fast amplifiers*

A main amplifier output, with shaping designed for optimum noise reduction as described in Sections 6.4.1 and 6.4.2, is not normally used for fast-timing measurements, since the time constants provided, being usually in the microsecond region, are too slow. Instead amplifiers with faster integration and differentiation shaping circuits are used with the pulses from a suitable preamplifier. In this way pulses with fast rise-times (and sometimes fall times) are obtained, with, of course, less than optimum noise reduction. Alternatively, very fast amplifiers without any shaping and operating directly on current pulses from the detector are available. D.c. coupling between stages removes problems due to baseline shift. In such fast applications input and output impedances of typically $50\,\Omega$ enable matching to cable characteristic impedances to be made.

6.7.4 *Timing discriminators*

Just as with counting techniques, as described in Section 6.5.1, we need to pass our amplified fast pulses into a discriminator. In this way we select only those pulses whose heights exceed the discriminator threshold set by ourselves. The output pulses, which are of standardized shape, also enable us to make timing measurements from them more easily. However, the question then arises, at which instant of time should an output pulse appear? We saw from Figure 6.25 in Section 6.5.2 that if we use the instant at which the leading edge of the input pulse crosses the discriminator threshold as our timing reference point, then this time varies or 'walks' with respect to the beginning of the pulse when either the pulse height or the threshold level changes. If we do not want this walk then we must find an alternative to this 'leading edge' method of time determination, as it is called.

One possibility, which is called the 'constant fraction timing' technique, is shown in Figure 6.34. It assumes that the leading edge is, at least approximately, linear, as can sometimes be the case for a detector with a uniform collecting field (such as a lithium-drifted detector), and that the pulse fall time is relatively long. The input pulse (Figure 6.34(a)) is split into two to give an inverted and attenuated version (Figure 6.34(b)) together with a full-amplitude replica delayed by a time T_D which is longer than the rise-time T_R

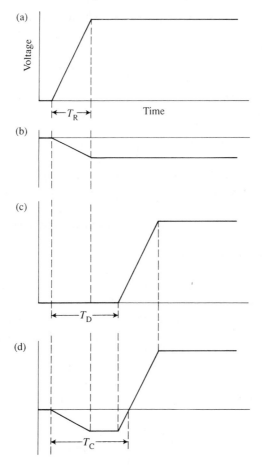

Figure 6.34 Pulses in a constant fraction discriminator: (a) input pulse; (b) inverted and attenuated version of input pulse; (c) delayed input pulse; (d) waveform obtained by addition of pulses in figures (b) and (c).

(Figure 6.34(c)). The time delay is introduced by a suitable cable as described in Section 6.7.2. The pulses in (b) and (c) are added to give the pulse shown in Figure 6.34(d). If the pulse rise is assumed to be linear and the top flat then the pulse in Figure 6.34(d) crosses the time axis at an interval $T_C = T_D + f T_R$ after the beginning of the pulse, regardless of its amplitude. Here f is the numerical factor or fraction by which the original pulse in Figure 6.34(a) is multiplied in order to give the attenuated version (Figure 6.34(b)). If T_R does not vary, then the zero-crossing point can be used as a reference time which is invariant to changes in pulse height. This point can be recognized by a suitable trigger circuit which generates an output pulse. We see that this reference point occurs at a time T_D after the pulse rises to a fraction f of the full pulse height.

If T_R varies, the zero-crossing point defines a time which is not invariant with respect to the pulse beginning. This feature of the zero-crossing point being dependent on the rise-time can be turned to advantage in other applications, as Section 6.7.6 will illustrate. Note that the fraction of the full pulse height reached at a time T_D before the zero-crossing point is still constant at the value f—hence the name 'constant fraction'.

Germanium detectors with large volumes can present particular timing problems, since the pulses from them can show relatively large variations in both rise-time and shape, due to the features we discussed in Sections 5.6 and 5.8. One approach is to use a technique related to constant-fraction timing which is called 'amplitude and rise-time compensated' (ARC) timing. Here the input pulse is split, as in constant-fraction timing, into inverted and delayed versions (Figure 6.35 (a)–(c)), but here the time delay T_D is much shorter than the rise-time T_R. The addition of the pulses in Figures 6.35(b) and 6.35(c) gives a pulse as in Figure 6.35(d), which crosses the time axis at an interval $T_C = T_D / (1 - f)$ after the beginning of the pulse. Here f is as before and the rise is again assumed to be linear. We have also assumed that the zero-crossing time occurs before the delayed pulse reaches its maximum. This implies that T_D must not be too large and f must not be too near to unity. We thus look only near the beginning of the pulse, and so, as we see from the expression just given, the zero-crossing point is now independent not only of the pulse height but also of the rise-time and its shape.

Real pulses, of course, as we saw earlier, only approximate to the idealized shapes used in the previous discussions. In practice, therefore, empirical adjustment of the fraction f can be made in order to optimize the timing resolution.

Most timing discriminators provide two or more outputs. At least one output will give fast standard pulses indicating the timing reference point. Another output usually gives a rather slower pulse which is designed to drive, for example, a scaler rather than a timing unit.

6.7.5 *Time-to-amplitude converters*

A 'time-to-amplitude converter', or TAC for short, has two inputs, each usually fed from a timing discriminator. The height of the output pulse is proportional to the time difference between the input pulses. The output pulse can be fed into a multichannel analyser and a spectrum of the time differences obtained. The way a TAC usually works is that the pulse at one of the inputs, which is labelled 'start', starts the charging of a capacitor with a

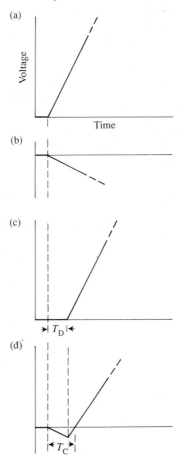

Figure 6.35 Pulses in an amplitude and rise-time compensated discriminator: (a) input pulse; (b) inverted and attenuated version of input pulse; (c) delayed input pulse; (d) waveform obtained by addition of pulses in figures (b) and (c).

constant current. The charging is stopped when a pulse is received at the input labelled 'stop'. The height to which the capacitor voltage rises is proportional to the time difference between the 'start' and 'stop' pulses, and this voltage directly defines the output pulse height. In this way time intervals from about 10 ns to 1 ms or more can be measured. The spectrum of the time differences under investigation can be obtained by feeding the output pulses from the TAC into an MCA.

Sometimes the time spectrum observed has a large, flat background of events, or alternatively the spectrum itself is entirely featureless and flat.

This usually means that there are uncorrelated pulses which happen to arrive within the interval chosen on the TAC for the full time-scale. Uncorrelated pulses occur, for example, when the background count rate from either or both detectors is high, and in such cases it is important to cut out (or reduce as far as possible) the background with the discriminators.

It is possible to eliminate the intermediate step of converting the time interval into an amplitude (at the cost of some flexibility) by using a 'time-to-digital converter' (TDC). In a TDC the start and stop pulses are used to start and stop a scaler counting pulses from an oscillator whose frequency is stabilized. It is best suited to time intervals longer than 1 μs. With much shorter time intervals the percentage accuracy becomes seriously limited by the speed of the oscillator. Even if its frequency is as fast as 500 MHz the period is still as long as 2 ns, and so the timing accuracy cannot be any better than this. This problem does not arise with a TAC and separate ADC. In the case of a Wilkinson ADC, for example, the number of oscillations counted is dependent on a capacitor discharge time, which in principle can be as long as we want (at the expense of a longer dead time, of course).

Figure 6.36 shows a typical system using a TAC. The delay unit, which can be placed in either line, consists of various lengths of coaxial cable which can be inserted so as to move the spectrum along the time axis.

An interesting situation arises if the 'start' pulses are accompanied by a large background count rate against which discrimination is not possible on grounds of size. In such a situation the dead time would be unnecessarily high since every start pulse would operate the TAC. However, if this line with the high count rate has sufficient delay inserted so that its pulses occur later than those in the other line, and the connections to the 'start' and 'stop'

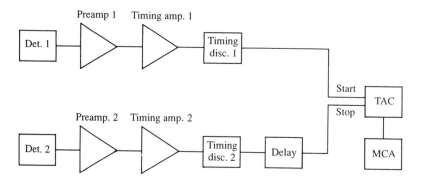

Figure 6.36 A typical system using a time-to-amplitude converter, in which the spectrum of the time differences between pulses from two detectors is measured and recorded.

lines interchanged, then all the high-rate pulses are ignored except those of interest which are immediately preceded by a 'start' pulse. Thus the time sequence of the input pulses has been reversed, and the actual time difference between the detector pulses is the inserted time delay minus the time difference recorded by the TAC.

6.7.6 *Pulse-shape discrimination*

We referred in Section 6.7.1 to the possibility of examining the shape of a pulse. We there discussed the use of a suitable oscilloscope, but discriminators can also be used. For example, we can take a pulse from a voltage- or charge-sensitive preamplifier, split it (after fast amplification, if necessary) as described near the end of Section 6.7.2, and pass each resulting pulse into a timing discriminator, as shown in Figure 6.37. In our first application we use constant-fraction discriminators for this purpose. We saw in Section 6.7.4 that this type of discriminator triggers at a time determined by when the input pulse reaches a constant fraction of its full pulse height. One discriminator can be set to trigger on a low fraction near the beginning of the pulse rise, and the other can be set for a high fraction near the top. If the outputs are fed into the 'start' and 'stop' inputs, respectively, of a TAC, the resulting output therefore shows the spectrum of the rise-times of the different pulses. This can be used, for example, to discriminate between different charged particles which, however, deposit the same energy in the detector. This is possible because the charge-collection process, on which the rise-times depend, may be affected by the particle range and the distribution of ionization along it. Alternatively the density of ionization deposited may affect the rise-time through the plasma effects described in Section 5.5.9. In these ways groups of different types of particles, such as alphas and heavier ions, may be separated.

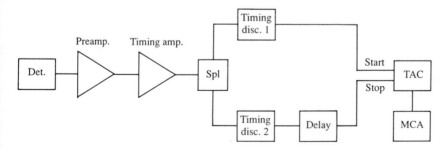

Figure 6.37 A system for discrimination of pulse shapes from a single detector: two applications are discussed in the text. Spl = splitter (Section 6.7.2.).

Another example occurs in scintillation counting. The fall times τ of fast pulses from some scintillators can be affected by the density of deposited ionization, as described in Chapter 4, so particle identification is again possible. A technique that can be used here is to pass the fast pulse through a differentiating circuit so as to produce a bipolar pulse like that at the output in Figure 6.23. Such a pulse, as we shall show, crosses the time axis at a moment independent of the pulse amplitude but dependent on τ—or, more generally, on the pulse shape. A discriminator which gives an output pulse when the zero crossing occurs is then used in the 'stop' line of a configuration like that of Figure 6.37, while a leading edge discriminator set as low as possible provides the 'start' pulse to the TAC. The time interval recorded by the TAC is now a measure of τ, and therefore particles with different fall times may be separated. For example, gammas produce lightly ionizing electrons and hence may give pulses from a scintillator with small fall times, whereas neutrons produce heavily ionizing protons and larger fall times. Thus fast neutrons can be counted while accompanying gammas can be vetoed, even when their respective pulses cannot be separated by conventional pulse-height analysis.

We now proceed to give the mathematical background to the zero-crossing property of bipolar pulses. The anode current pulse from the photomultiplier which views the scintillator is given by

$$i_s = I_o \exp\left(\frac{-t}{\tau}\right) = \frac{Q}{\tau}\exp\left(\frac{-t}{\tau}\right)$$

as we can see from equation (6.4) and the analogous equation in Section 4.4.5. This current produces a voltage at the input of an attached amplifier of size $I_o R_{in} \exp(-t/\tau)$. (Here R_{in} is as defined in Section 6.3.4: it is called simply r in our introductory treatment in Section 4.4.5). After amplification we can thus write the voltage produced as

$$v_o = V_P \exp\left(\frac{-t}{\tau}\right) \tag{6.22}$$

where V_P is the amplifier output at $t = 0$. We have assumed that the input time constant $R_{in} C_{in}$ (called rC_s in Chapter 4), with C_{in} as defined in Section 6.3.4, is much smaller than τ, as required for current-mode operation. As shown in Figure 6.38, the voltage given by equation (6.22) is applied to a differentiating circuit of C_1 and R_1 in series (similar to the circuit in Figure 6.16 in Section 6.4.2), such that we look at the output voltage v_1 across R_1. Just as for the case in Section 6.4.2, the form of v_1 is given by the solution of an equation analogous to equation (6.14), which this time becomes

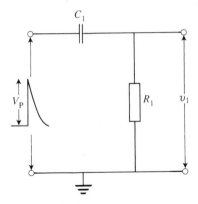

Figure 6.38 A CR differentiating circuit with an exponentially decaying pulse at its input.

$$\frac{dv_1}{dt} + \frac{v_1}{R_1 C_1} = \frac{dv_o}{dt}. \qquad (6.23)$$

Last time, however, the right-hand side was dV_P/dt, which, since V_P was there a steady step height, could be equated to zero. This time the right-hand side is non-zero, since v_o varies with time, so when we substitute from equation (6.22) into equation (6.23) we get

$$\frac{dv_1}{dt} + \frac{v_1}{R_1 C_1} = -\frac{V_P}{\tau} \exp\left(\frac{-t}{\tau}\right). \qquad (6.24)$$

This equation shows formal similarities with equation (6.8) and equation (6.16), and it can be solved in a similar way after multiplying both sides by the integrating factor $\exp(t/R_1 C_1)$. The capacitor C_1 transmits instantly the initial step of voltage, V_P, which this time gives the boundary condition that $v_1 = V_P$ at $t = 0$. The solution of equation (6.24) now takes the new form

$$v_1 = \frac{V_P}{T_1 - \tau}\left[T_1 \exp\left(\frac{-t}{\tau}\right) - \tau \exp\left(\frac{-t}{T_1}\right)\right] \qquad (6.25)$$

where, as in equation (6.15), $T_1 = R_1 C_1$. Note that equation (6.25) is unaltered if we interchange τ and T_1; either time constant can be larger than the other. The form of equation (6.25) is shown in Figure 6.39. (We do not treat the case when $\tau = T_1$, but the zero-crossing properties which we discuss below still then hold.)

In this zero-crossing method we look at the instant when the bipolar waveform of v_1 crosses the time axis, that is, when $v_1 = 0$. This

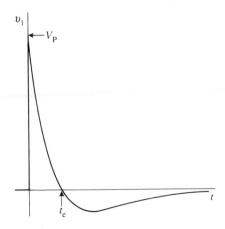

Figure 6.39 Waveform at output of circuit in Figure 6.38.

zero-crossing time, t_c, is given by equating the expression inside the large brackets in equation (6.25) to zero, which gives

$$t_c = \frac{T_1 \tau \ln\left(T_1/\tau\right)}{T_1 - \tau}.\tag{6.26}$$

We can see that this time always exists (in other words, there is a finite positive value for this time), and furthermore it is independent of the size of V_P. In contrast, the time when the bipolar waveform v_1 takes any value other than zero is not independent of V_P. We can further easily prove from equation (6.26) that as τ increases (regardless of whether $\tau <$, $=$, or $> T_1$) t_c increases monotonically. If we detect t_c with an appropriate discriminator in the 'stop' line, that is, with a circuit which gives an output when the voltage returns to zero, a TAC output dependent on τ (but not, as we said, on V_P) can be generated, and hence with the pulse-shape discrimination circuit we have been discussing particle discrimination can be achieved. In the example of gamma/neutron discrimination given earlier, we would expect two peaks to appear in the output time spectrum, the gamma peak coming first.

Of course in practice the input pulse to the circuit of Figure 6.38 will have a finite rather than zero rise-time, due to the effect of R_{in} and C_{in}, as described in Section 6.3.4. Equation (6.22) should in this case be replaced by one with two exponentials and two time constants, τ, the fall time (as before), and the rise time $R_{in}C_{in}$, and equation (6.25) by one with three

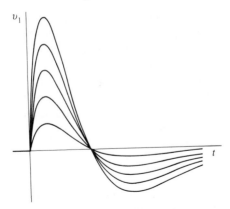

Figure 6.40 Waveforms at output of circuit in Figure 6.38 for four input pulses of different amplitudes, all of whose shapes are determined by the same three time constants.

exponentials and three constants, τ, $R_{in} C_{in}$, and T_1. However, the general conclusions about the invariance of the crossing point with respect to amplitude remain unaltered. Figure 6.40 shows how such pulses can produce a zero crossing time which is well defined, and any actual bipolar pulses displayed on an oscilloscope show this effect clearly.

This is not, as we noted earlier, the first time that we have seen a waveform given by equation (6.25). Figure 6.23 showed one of our baseline restoration techiques, in which a voltage step was passed through two differentiating circuits. Since the first circuit in this figure produces an exponentially falling voltage as in equation (6.22), equation (6.25) gives the waveform produced by the second circuit. However, in this earlier application we were interested not in the zero-crossing invariance properties but in the equality of the areas enclosed by the waveform above and below the time axis. This type of waveform also appeared in Section 6.4.3 in our discussion of pole-zero cancellation.

6.8 Coincidence units

A TAC (Section 6.7.5) allows us to measure the time separation between two pulses. If, however, all we need to know is whether a coincidence in time has occurred or not then a simpler device, a 'coincidence unit', is used instead. In such a unit an output pulse of standard shape appears whenever two standard pulses are applied to the two inputs simultaneously, or in practice within a 'resolving time' chosen by the user. There is now no distinction between the inputs as there is with a TAC. In electronic terms we can in fact think of a coincidence unit as a logic 'AND' gate, in which an output appears only when both inputs are present, whereas when neither or only one input is present, no output is produced.

The resolving time Δt of a coincidence unit should be set no longer than is required to capture the pair of pulses, so as to minimize the number of chance or accidental coincidences from background pulses. The coincidence rate m_C for uncorrelated random pulses of rates m_1 and m_2 appearing at each input is given by

$$m_C = 2m_1 m_2 \Delta t$$

since approximately $2m_2\Delta t$ pulses at the second input occur in the times Δt before and after each of the m_1 pulses per unit time at the first input. We therefore minimize m_C by making Δt (and if we can, m_1 and m_2) as small as possible, while still of course keeping it large enough to capture each genuine pair of pulses. Δt can not, of course, be reduced all the way to zero because of the intrinsic resolving time of the coincidence circuitry itself.

Coincidence units often have more than two inputs so that, if desired, pulses have to be present at three, four, etc., inputs within the resolving time before an output pulse appears. Usually any of the input circuits can be modified to become an anticoincidence or 'veto' input. In this mode an output appears only when pulses are present at the 'coincidence' inputs and absent at the 'anticoincidence' inputs.

Coincidence units have many applications. For example, Figure 6.41 shows two detectors triggered by the same radiation. The coincidence unit ensures that only this type of event is counted in the scaler, and that radiation

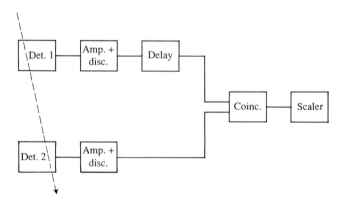

Figure 6.41 The counting of coincident pulses from two detectors.

striking only one detector is rejected. If the flight time of the radiation between the detectors is not small compared with Δt, then the delay unit shown can be set to the flight time to allow for this effect. A variation on these arrangements is when we wish to count in coincidence not the same radiation but different radiations from the same nuclear event. An example would be the detection of pairs of annihilation photons recoiling in opposite directions from positrons decaying in matter placed between the two detectors. In all these cases the aim is to accept the desired events and reject unwanted background. Because of their usefulness, coincidence circuits are frequently built into scalers and multichannel analysers, rather than being used as physically separate units.

Figure 6.42 shows an example of an anticoincidence arrangement, in which a detector viewing a source, the counts from which are to be recorded is surrounded by a second detector, which acts as an 'active shield'. This shield (with the anticoincidence circuit) inhibits data collection every time it is triggered by external radiation. Such radiation may otherwise have been also improperly recorded by the main detector inside. We have already mentioned this in Section 4.5.2, and have seen an example in Section 3.4.2 for counting feebly active beta sources, in which both detector and shield are Geiger counters.

A slightly different application for the same arrangement occurs in gamma-ray spectrometry, as also mentioned earlier in Section 4.5.2. In this case the detector is a scintillator or germanium detector, while the shield is usually a scintillator. The shield (as well as vetoing external radiation) suppresses or minimizes the recording of Compton scattering and other events which do not contribute to the full-energy peaks in the spectrum. In this case the simultaneous escaping radiation has a definite probability of being recorded by the active shield, which inhibits data collection as before. The full-energy peaks, which carry the spectrometric information are thus rendered more prominent.

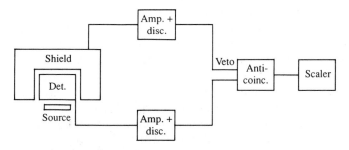

Figure 6.42 An anticoincidence counting arrangement.

We shall consider lastly a special type of coincidence unit, although this would be an inadequate name to describe its function. A 'linear gate' has an 'analogue' input and a 'gate' input. The pulse at the analogue input appears unchanged in shape or size at the output provided that a pulse is present at the gate input for the duration of the other pulse. The gate pulse is of standardized height but variable width, depending on the width of the main pulse to be 'gated' through. An application will be shown in Section 6.10.

6.9 Housing of electronic units

Originally most of the various electronic units we have described — amplifiers, discriminators, scalers, and so on — were housed in separate 19 inch wide units that were mounted into relay racks of this width. The relay racks remain to this day, and full-width units are sometimes still used. Most units were, however, considerably reduced in size as soon as transistors and integrated circuits replaced thermionic valves. There are two main standards now in use to which units are built. In both of these the units are fixed in and normally electrically plugged into a 'bin' or 'crate' which itself is mounted in a relay rack. The bin provides through connectors the d.c. voltages which supply power to the units; it houses a mains transformer and rectifiers from which these d.c. voltages are derived. The standards ensure that bins and units from different manufacturers are compatible both mechanically and electrically.

The 'nuclear instrument module', or NIM, standard is widely used for small-scale electronic systems handling pulses from one or a few radiation detectors. In the NIM system most pulses are transmitted between units through coaxial cables connected to the front or back of the units. Up to 12 single-width units can fit into a standard NIM bin. Sometimes units are of double or treble width and thus occupy two or three spaces in a bin. Nearly all the units we have described are available in the NIM format.

The other standard is called CAMAC (computer automated measurement and control). It is designed primarily for the construction of large-scale systems with many detectors which require to be interfaced to computers. A CAMAC crate has 25 spaces or stations for separate units or modules, and again double- or treble-width modules are found. The crate provides not only power for each module but also a 'dataway' over which most of the signals pass from one module to another. The signals on the dataway are mostly standardized logic and control pulses; the standard seeks in this way to avoid a multiplicity of external cables. The signals all have to pass through a crate controller which usually occupies two of the crate stations and which also functions as an interface to any external equipment. The CAMAC

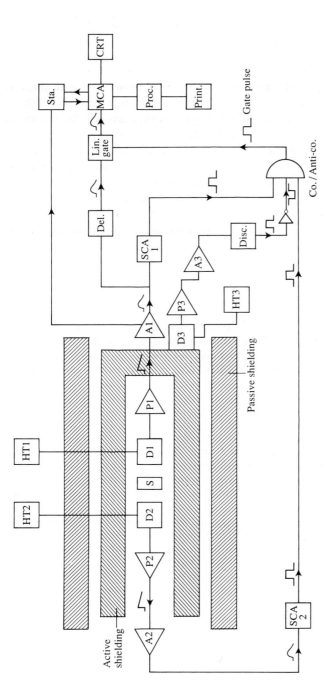

Figure 6.43 A complete low-level counting system: S, sample; D1, D2, detectors viewing sample; D3, detector (typically one of a number of photomultipliers viewing a large scintillator used as the active shielding); P1, P2, P3, preamplifiers; A1, A2, A3, amplifiers; HT1, HT2, HT3, bias voltage supplies; SCA1, SCA2, single-channel analysers; Disc., discriminator; Co./Anti-co., logic circuit producing an output pulse when two signal pulses present in coincidence (top and bottom inputs) and anticoincidence veto pulse from active shield absent (middle input); Del., delay; Lin. gate, linear gate; MCA, multichannel analyser; CRT, cathode-ray tube display (or similar); Proc., data processor; Print., printer, plotter, tape, disc, etc.; Sta., spectrum stabilizer.

system is more costly than the NIM system, and in simple applications only the latter is likely to be found.

6.10 A complete system

Finally we present in Figure 6.43 an example of a complete system which could be used for the coincidence detection and analysis of low levels of radiation, such as might be associated with environmental samples. The detectors are surrounded not only by passive shielding, such as lead blocks, but also an active shield like that described in Section 6.8.

A study of this figure will illustrate ways in which many of the units described in this chapter can be used. The system depicted is meant to serve only as an example, and in a real system not all the units shown may be present.

References

1. Delaney, C.F.G. and Finch, E.C. (1984). Some simple considerations on pulse shapes in radiation detectors. *American Journal of Physics* **52**, 351–4.
2. Spieler, H. (1982). Fast timing methods for semiconductor detectors. *IEEE Transactions on Nuclear Science*, **NS-29**, 1142–57. A valuable tutorial article.

Further reading

Delaney, C.F.G. (1980). *Electronics for the physicist*. Ellis Horwood, Chichester. Includes applications to radiation detectors.

Millman, J. and Grabel, A. (1987). *Microelectronics*, (2nd edn). McGraw-Hill, New York. An authoritative general book on modern electronics: one of many such texts available.

Nicholson, P.W. (1974). *Nuclear electronics*. Wiley, London. A thorough account of the field.

7

Energy resolution in radiation detectors

7.1 The necessity for good energy resolution

Nearly all the detectors we discussed in the previous chapters have the ability not only to indicate by an output pulse the passage of an ionizing particle, but also to allow determination of its energy from the amount of charge produced in the pulse. Among other things this ability opens up the important field of gamma and X-ray spectrometry whereby we are able to identify the various radioisotopes present in a sample from the position of their characteristic full-energy peaks in a pulse height, that is energy, spectrum. If, however, a great number of isotopes are present, as in say an environmental sample, then since the peaks are always of a finite width they may overlap and run together, thus making positive identification difficult or impossible. The widths of the peaks will depend, among other things, on the detector type being used. The narrower the peaks the less, obviously, is the confusion. This was dramatically illustrated in Figure 5.17, where the same spectrum is shown taken by a scintillation counter (which as we shall see in detail later in this chapter has only moderate energy resolution) and by a germanium detector (which we shall see has excellent energy resolution). It is obvious that there is a vast amount of extra information available in the latter spectrum. Our purpose therefore in this chapter is to investigate the factors that control the energy resolution in radiation detector systems, and the steps that we can take to improve it.

7.2 Sources of instrumental linewidth

Although considerations on resolution are by no means restricted just to gamma and X-radiation, we shall continue to discuss the matter, at least for the present, in these terms, both to enable us to focus our attention on one particular problem and also because gamma-ray spectrometry is probably the major area in which energy resolution is of critical importance. When we look then at the full-energy peak in a typical gamma-ray spectrum from a radioactive isotope (such as Figure 5.17) it is pertinent to ask whether the apparent spread in energy, as indicated by the finite width of the line, is a real effect coming from the gamma rays themselves, or whether it is 'instrumental' in origin, that is coming from processes in our detector and related electronics. In fact we shall be able to show that essentially all of this linewidth can be accounted for by instrumental processes; that is such gamma rays are monoenergetic to a very high degree – as indeed other types of measurement confirm.

We shall invoke two phenomena to explain this instrumental linewidth. The first is the statistical fluctuations which occur in the processes whereby a detector converts the initial energy of a gamma ray into an output pulse. For example, only a part of the original gamma-ray energy goes to produce photons in a scintillator crystal, and the actual fraction converted suffers a slight statistical variation from one gamma photon to the next, thus leading to a spread in the size of the final output pulse from the detector. Other statistical effects producing further spreads occur in the photomultiplier itself.

The second effect producing linewidth is electronic 'noise'. This noise occurs with the high-gain amplifiers needed in conjunction with some radiation detectors, and manifests itself as irremoveable small random fluctuations on the amplifier output. This means that even if the statistical fluctuations in the detector mentioned in the previous paragraph turned out to be negligible, the output pulses, being superimposed on the noise, would not be of uniform size, and thus the linewidth would still not be zero. Because these two effects both produce the same result – a broadening of the linewidth – the statistical fluctuations present in the detector signal are often referred to as 'noise in signal'. This is a useful enough term as long as we remember that the origin of the effect is quite different from electronic noise. We move on then to deal in more detail first with noise in signal and then with electronic noise.

7.3 Noise in signal

7.3.1 *General considerations*

We will consider in this section the contribution of noise in signal to the linewidth for scintillation counters, proportional counters, and semiconductor detectors, in that order. We start with scintillation counters, although they are the most complex from the point of view of calculations on resolution, because once we have made the necessary calculations we can quickly adapt them to the other two devices.

The scintillation counter is, as we already know, a three-stage device. The energy of the incoming radiation is first converted to photons in the crystal, then these photons are converted to electrons at the tube photocathode, and finally the electrons are increased in number by a large factor in the multiplier section before being collected at the anode. Statistically speaking we have a three-stage cascade process in which the numbers produced at each stage depend on the numbers coming from the previous stage. Indeed the actual multiplication of the electrons within the multiplier is itself a cascade of n stages, where n is the number of dynodes present. So let us treat this last problem first, and it will introduce us in a simple fashion to the general principles involved.

It would take us too far afield to rigorously deduce the relevant formulae for the statistics of cascade processes, but those interested can consult reference 1, for example. We will find the result can be simply expressed and is quite understandable physically. Before considering it, however, we remind the reader about what was said in Chapter 1 about the mean $\bar{s}$ and the standard deviation σ of a set of counts of readings. The mean is of course the ordinary average of the readings, while the standard deviation is a measure of the spread of the readings about the average. This spread can not be determined merely by averaging the deviations of the individual readings from the mean, as this will just give zero (there being as many readings above the mean as below). However, by averaging the square of the deviations (to give us what is called the variance V) and then taking the square root we get σ, a measure of the spread. Formally $V = \overline{(s - \bar{s})^2}$ where s is a typical reading and a bar represents an average. The standard deviation σ then equals $\sqrt{V}$ (that is $V = \sigma^2$). We will be more interested in the ratio of σ to $\bar{s}$ and we introduce the *relative* variance $\mathscr{V} = (\sigma/\bar{s})^2$. The theoretical relationship between σ and $\bar{s}$ depends very much on the type of distribution we postulate to describe the events in question. We shall be making use of the Poisson distribution which applies to many problems of the type we are considering. All we need to know for our purposes about this particular distribution is, as we saw in Section 1.4.3, that if the mean is as previously $\bar{s}$ then the standard deviation can be shown to be $\sqrt{\bar{s}}$. Thus the variance is $\bar{s}$ and the relative variance $1/\bar{s}$.

7.3.2 *Statistical processes in a photomultiplier*

We can best understand the equation defining the statistics for a cascade process by looking at the actual problem we are trying to solve – that of multiplication in a photomultiplier. We will use (as in Chapter 4) the symbol m for the multiplication factor at a dynode and we remember that m is not a constant but varies statistically with a mean value $\bar{m}$. Let us consider then a single electron leaving the photocathode and striking the first dynode. There it produces (on average) $\bar{m}$ secondary electrons, and these proceed to the next dynode from which (if all the dynodes are identical) $\bar{m}^2$ electrons leave, and so on with the output signal containing $\bar{m}^n$ electrons (on average) where n is the number of multiplying dynodes. We stress that the numbers at each stage are average numbers and the spread in these averages depends on the type of statistical distribution which governs the production of the secondary electrons, a matter to which we will return shortly.

It should be intuitively obvious, in any case, that the fluctuations occurring at the first dynode, where the number of electrons is the smallest, make the greatest contribution to the overall fluctuation, and for a similar reason the statistical contribution of each successive dynode decreases steadily as we go down the chain towards the anode. These remarks should help us to appreciate the formal statements about the overall mean and overall relative variance for any general n-stage cascade process. These are

$$\bar{s} = \bar{s}_1 \times \bar{s}_2 \times \bar{s}_3 \times \ldots \times \bar{s}_n \tag{7.1}$$

$$\mathcal{V} = \mathcal{V}_1 + \frac{\mathcal{V}_2}{\bar{s}_1} + \frac{\mathcal{V}_3}{\bar{s}_1 \times \bar{s}_2} + \ldots + \frac{\mathcal{V}_n}{\bar{s}_1 \times \bar{s}_2 \times \ldots \times \bar{s}_{n-1}}. \tag{7.1a}$$

Equation (7.1) merely tells us that the overall mean number $\bar{s}$ is just the product of the means of the individual stages $\bar{s}_1, \bar{s}_2, \ldots \bar{s}_n$. For our case of a photomultiplier with n identical stages, with the same mean multiplication factor $\bar{m}$, we have

$$\bar{s} = \bar{m} \times \bar{m} \times \bar{m} \times \ldots \times \bar{m} = \bar{m}^n \tag{7.2}$$

($\bar{s}$ is of course the quantity we called M in Chapter 4). Equation (7.1a) tells us, as we already expected intuitively, that the contributions to the overall fluctuation (here indicated by the overall relative variance $\mathcal{V}$) from the relative variances of the individual stages $\mathcal{V}_1, \mathcal{V}_2, \mathcal{V}_3, \ldots$ decrease as we proceed down the stages, but also that the reducing factor at each stage is the mean from the previous stage. Applying this to our photomultiplier with identical stages we have $\mathcal{V}_1 = \mathcal{V}_2 = \mathcal{V}_3 = \ldots = \mathcal{V}_0$, say, and of course the various means are all equal to $\bar{m}$ so

$$\mathcal{V} = \mathcal{V}_0 + \frac{\mathcal{V}_0}{\bar{m}} + \frac{\mathcal{V}_0}{\bar{m}^2} + \frac{\mathcal{V}_0}{\bar{m}^3} + \ldots + \frac{\mathcal{V}_0}{\bar{m}^{n-1}}.$$

Summing the geometrical series for $\mathcal{V}$ we have

$$\mathcal{V} = \mathcal{V}_0 \frac{1 - 1/\bar{m}^n}{1 - 1/\bar{m}}.$$

Now $\bar{m}^n$ is the overall gain of the photomultiplier, which is normally very large compared with one, so the term $1/\bar{m}^n$ is entirely negligible and thus very closely

$$\mathcal{V} = \mathcal{V}_0 \frac{\bar{m}}{(\bar{m} - 1)}. \tag{7.3}$$

Sometimes the gain of the first dynode may be different from the others, either because it is run at a higher voltage to optimize electron collection, or because it alone may make use of negative-affinity material. A very small modification of our previous calculation yields in this case

$$\mathcal{V} = \mathcal{V}_1 + \frac{\mathcal{V}_0}{\bar{m}_1} \frac{\bar{m}}{\bar{m} - 1} \tag{7.4}$$

where $\mathcal{V}_1$ and $\bar{m}_1$ refer to the first dynode and $\mathcal{V}_0$ and $\bar{m}$ to all the others.

We can not proceed further without making some assumption about the values for the relative variances, and this in turn depends on what assumption we make about the basic statistical distribution involved. The simplest one to make (and one that seems to produce results which compare quite well with experiment) is that the numbers of secondaries produced at a dynode follow a Poissonian distribution with mean $\bar{m}$. We have already noted that such a distribution would have standard deviation $\sqrt{\bar{m}}$, variance $\bar{m}$, and relative variance $1/\bar{m}$. Putting then $\mathcal{V}_0 = 1/\bar{m}$ in equation (7.3) and in addition $\mathcal{V}_1 = 1/\bar{m}_1$ in equation (7.4) yields, for the case of identical dynodes

$$\mathcal{V} = \frac{1}{\bar{m} - 1} \tag{7.5}$$

and for the case of a different first dynode

$$\mathcal{V} = \frac{1}{\bar{m}_1} \left(\frac{\bar{m}}{\bar{m} - 1} \right). \tag{7.6}$$

For identical dynodes and $\bar{m} = 3$ we have from equation (7.5) $\mathcal{V} = 1/2$ while for higher gain but still identical dynodes with, say, $\bar{m} = 30$ we have $\mathcal{V} = 1/29$. With the first dynode only having $\bar{m} = 30$ and the rest $\bar{m} = 3$ equation (7.6) gives us a value for $\mathcal{V}$ of $1/20$. Note also in passing that equation (7.6) can be written $\mathcal{V} = \mathcal{V}_1 [\bar{m}/(\bar{m} - 1)]$ or $\mathcal{V}_1/\mathcal{V} = (\bar{m} - 1)/\bar{m}$, so even if $\bar{m}$, the mean for the remaining dynodes, is only 3, the first dynode is responsible for 2/3 of the overall relative variance of the photomultiplier, whatever the type of first dynode used. This dominance of the first dynode was something we intuitively expected, and is of course the reason why a surprisingly low value of $\mathcal{V}$ can be obtained with only the first dynode being of the high-gain type.

Figure 7.1 shows a computation of the distribution in size of the output pulses from a photomultiplier (with identical dynodes) for four values of $\bar{m}$, when a succession of single electrons leaves the photocathode. This would occur in practice if the photocathode (with no scintillator of course present) were illuminated by a light source so feeble that photons were arriving sufficiently infrequently to be detected as single events. For comparison the means of the three distributions (which do not in fact coincide with their

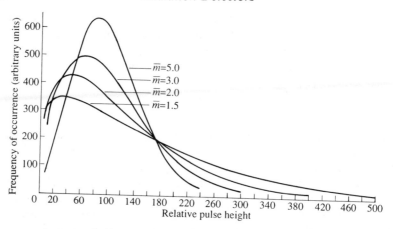

Figure 7.1 Output pulse-height distributions from a photomultiplier with single-electron input, for various values of $\bar{m}$. The scale of the horizontal axis has been renormalized for each distribution so that its mean, $\bar{m}^n$, falls at a pulse height of 100. The area under all four curves is the same. (From Lombard, F.J. and Martin, F. (1961), *Review of Scientific Instruments*, **32**, 200–1, Figure 1.)

peaks) have been adjusted to be the same. The reduction in spread in these 'single-electron spectra' as we go to lower values of $\mathscr{V}$ can be clearly seen, and this reduction will be reflected in smaller linewidths when we tackle the problem of not one, but many photons arriving simultaneously at the photocathode from a scintillator. Although we are not yet equipped to deal with this multiphoton problem in general, we will add one further illustration, Figure 7.2, showing the pulse-height distribution from a photomultiplier when electrons are liberated from the cathode mostly singly, but sometimes two at a time, and occasionally three or more at a time. (This could occur in practice if, for example, radiation of very low energy produced small numbers of photons in a scintillator.) The photomultiplier used for this experiment was one with a high-gain first dynode giving a relatively low value for $\mathscr{V}$, and thus quite a narrow width for the single-electron (and other) pulse-height distributions. Consequently we are able to see clearly the peaks corresponding to the release of one, two, and up to five electrons from the photocathode, although beyond this the peaks run together. If a photomultiplier with a larger value of $\mathscr{V}$ had been used, the greater spread in the pulse-height distribution would have meant that even the peaks corresponding to one, two, and three electrons would have run together, and no details at all would have been visible.

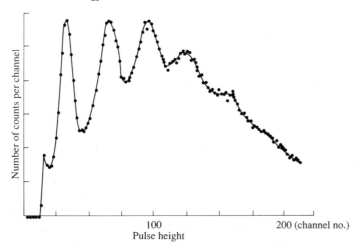

100 (channel no.)
Pulse height
200 (channel no.)

Figure 7.2 Pulse-height spectrum showing peaks corresponding to one, two, and up to five electrons. (From Morton, G. A., Smith, H. M. and Krall, H. R. (1968), *Applied Physics Letters* **13**, 356–7, Figure 1.)

7.3.3 *Statistical processes and linewidth in a complete scintillation counter*

We are now in a position to discuss the statistical processes in a complete scintillation counter. As we stated previously, this is a three-stage process with some of the energy of the gamma ray going to produce photons in the scintillator, a fraction of these photons ejecting electrons from the photocathode, and finally these electrons being multiplied at successive dynodes before being collected at the anode. All these processes involve statistical fluctuations and so we must talk in terms of mean numbers and the corresponding relative variance. Let $\bar{x}$ then be the average number of photons produced by the photoelectric interaction in the scintillator of a gamma ray of a particular energy, p the quantum efficiency of the photocathode (that is the probability of a photon ejecting an electron from it), and $\bar{M}$ the average overall multiplication by the dynodes. The average number, $\bar{N}$, of electrons produced at the anode is then clearly given by

$$\bar{N} = \bar{x}p\bar{M} \tag{7.7}$$

and the relative variance $\mathcal{V}$ in the electron number N for the three-stage cascade process is given (from equation (7.1a)) by

$$\mathcal{V} = \mathcal{V}_1 + \frac{\mathcal{V}_2}{\bar{s}_1} + \frac{\mathcal{V}_3}{\bar{s}_1 \times \bar{s}_2} . \tag{7.8}$$

Here $\mathcal{V}_1$, $\mathcal{V}_2$, and $\mathcal{V}_3$ refer, respectively, to the relative variances of the processes in the scintillator, at the photocathode, and in the multiplier, while $\bar{s}_1$ refers to the mean of the first process and is therefore equal to $\bar{x}$. Similarly $\bar{s}_2$ is equal to p.

We already know from the previous sub-section that $\mathcal{V}_3$ is equal to $1/(\bar{m} - 1)$ if we take the simple case of identical dynodes, while $\mathcal{V}_1$ can be taken as $1/\bar{x}$ if we assume, plausibly, that the production of photons in the scintillator follows a Poissonian distribution. There remains only $\mathcal{V}_2$ to be evaluated, that is the relative variance of the production of electrons from photons at the photocathode. The controlling parameter here is p, the quantum efficiency, that is the probability of a photon ejecting an electron from the photocathode — or what is equivalent, the fraction of a group of photons converted, on average, into electrons. Because there are competing processes for the absorption of photons which do not produce electrons, p is less than unity: this also means that the fraction of photons converted to electrons will not be always exactly p, but will fluctuate statistically around p with a relative variance which can be shown to be $(1 - p)/p$. Hence we put $\mathcal{V}_2 = (1 - p)/p$. Proof of this result can be found in any elementary statistics text and will not be given here, but at least we can note that if p were unity, that is all photons were always converted into electrons, then $\mathcal{V}_2 = 0$ and therefore the fluctuation should also be zero, which of course is what we expect in this case! In fact p normally varies from 0.1 to 0.3 depending on the type of photocathode in use, and also on the energy, that is the wavelength, of the photons from the scintillator.

Returning now to equation (7.8) we can write

$$\mathcal{V} = \frac{1}{\bar{x}} + \frac{1}{\bar{x}}\left(\frac{1-p}{p}\right) + \frac{1}{\bar{x}p}\left(\frac{1}{\bar{m}-1}\right) \tag{7.9}$$

or taking out the $1/\bar{x}$ term we have for $\mathcal{V}$, the relative variance

$$\mathcal{V} = \frac{1}{\bar{x}}\left(1 + \frac{1-p}{p} + \frac{1}{p(\bar{m}-1)}\right). \tag{7.10}$$

Before we proceed to insert numerical values into this last equation, let us consider how we can interpret $\mathcal{V}$, which we remember is equal to $(\sigma/\bar{N})^2$, in terms of practical gamma-ray spectrometry. For the case of a gamma ray interacting photoelectrically in a scintillator, a large number of photons, and therefore many electrons, will normally be generated at the start of the multiplying process. Compared with the single-electron case then, not only will the average pulse size obviously be greater, but for statistical reasons the distributions will appear narrower and more symmetrical — as indeed we saw in our diagrams showing full-energy peaks in Chapter 4 (e.g. Figure 4.17) Such a symmetrical 'Gaussian'-shaped peak is shown again in Figure 7.3(a).

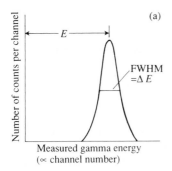
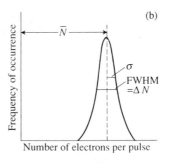

Figure 7.3 (a) Experimental Gaussian peak. (b) Graphical display of calculations.

On the y axis we have plotted the number of pulses arriving per channel, and on the x axis the corresponding measured energy of the gamma photons, derived from the pulse heights by suitable calibration. E represents the measured mean energy of the gammas, and the experimental spread in the energy distribution ΔE, traditionally taken as its width half way down from the peak is shown labelled FWHM, for 'full width at half maximum'. Figure 7.3(b) illustrates graphically the results of the theoretical calculations we have been making on the statistics of the number of electrons per pulse arriving at the photomultiplier anode: it is in fact just an alternative presentation of the data in figure 7.3(a). Once again we have shown the mean number $\bar{N}$, and the full width at half maximum ΔN. In addition we have shown σ, the standard deviation, as it is related to the quantity $\mathcal{V}$, the relative variance, which we have been investigating. For a Gaussian-shaped peak the standard deviation (which in fact occurs at the point of inflection of the curve) is as shown a little less than half the full width. In fact the full width at half maximum is 2.355 σ, as we noted in Section 1.4.3. We are now in a position to convert our calculated statistical results into a practical form. We do this as follows:

$$\Delta E/E = \Delta N/\bar{N} = 2.355\sigma/\bar{N} = 2.355\mathcal{V}^{1/2}$$

(remembering the relationship between σ and $\mathcal{V}$). The quantity $\Delta E/E$ which we will call the 'fractional full width at half maximum' (or more loosely the 'resolution') can thus be written, using equation (7.10), as

$$\Delta E/E = \text{fractional FWHM} = 2.355 \left[\frac{1}{\bar{x}} \left(1 + \frac{1-p}{p} + \frac{1}{p(\bar{m}-1)} \right) \right]^{1/2}.$$
$$(7.11)$$

We shall want to refer also to the case where the first dynode in the photo-multiplier has a different value of gain from all the others. In that case using equation (7.6) we modify the last term in equation (7.11) to give

$$\Delta E/E = \text{fractional FWHM} = 2.355 \left\{ \frac{1}{\bar{x}} \left[1 + \frac{1-p}{p} + \frac{1}{p} \frac{\bar{m}}{\bar{m}_1} \left(\frac{1}{\bar{m}-1} \right) \right] \right\}^{1/2}.$$
(7.12)

This fractional FWHM (which can be expressed either as a number less than one or as a percentage) is the usual parameter for indicating resolution in the scintillation counter. Sometimes, however (and more particularly in the case of semiconductor detectors), we use simply ΔE, the full width at half maximum itself, which we also call loosely the linewidth. The FWHM is obviously obtained by multiplying the fractional FWHM by the appropriate value of E for the gamma ray we are considering, and will be expressed in energy units (usually eV or keV).

The first thing to note about equation (7.11) is that the resolution depends on the energy of the gamma ray. If E is this gamma energy and ε the energy which it is necessary to expend on average in the scintillator to produce a photon, then the number of photons $\bar{x}$ is equal to E/ε. Consequently the fractional FWHM, which from equation (7.11) is proportional to $(1/\bar{x})^{1/2}$, is thus proportional to $1/E^{1/2}$. This means, for example, that if we have two gamma rays with energies in the ratio of 16:1 then the fractional FWHM for the less energetic photon will be four times that of the other. We must therefore specify what energy we are referring to when we quote a value for the fractional FWHM. We will, for illustration purposes, choose three energies from among the many the are used for calibration in gamma-ray spectrometry — a 1.836 MeV gamma ray which is emitted from a [88]Y source, a 0.662 MeV gamma ray from a ^{137}Cs source, and for low energies a 5.9 keV X-ray from a [55]Fe source.

Let us now put some numbers into equation (7.11) or (7.12), see what the most critical factors are, and then compare our computations with the FWHMs which can be obtained experimentally. We start with GSO (rather than NaI(Tl), for a reason that will be explained in a moment) and for this material the value of ε, the energy to produce a photon has been determined to be 97 eV. Let us imagine this scintillator coupled to the sort of photo-multiplier that would normally be used for scintillation counting with a bialkali photocathode having a value of $p = 25$ per cent, all dynodes except the first with $\bar{m} = 4$, and the first dynode running at a higher voltage to give $\bar{m}_1 = 6$. Then from equation (7.12)

$$\text{fractional FWHM} = 2.355 \left\{ \frac{1}{\bar{x}} \left[1 + \frac{1-p}{p} + \frac{1}{p} \frac{\bar{m}}{\bar{m}_1} \left(\frac{1}{\bar{m}-1} \right) \right] \right\}^{1/2}$$

$$= 2.355 \left\{ \left(\frac{E}{\varepsilon}\right)^{-1} \left[1 + \frac{3/4}{1/4} + 4 \left(\frac{4}{6}\right) \left(\frac{1}{3}\right) \right] \right\}^{1/2}$$

$$= 2.355 \left[\left(\frac{E}{\varepsilon}\right)^{-1} (1 + 3 + 8/9) \right]^{1/2}.$$

Before we proceed further it is instructive to consider the relative magnitudes of the contributions from the three sources—crystal, photocathode, and subsequent multiplication—to the resolution. These are, as we can see, in the ration of 1 : 3 : 0.9 and what is surprising, and what we probably would not have predicted in advance, is the dominance of the photocathode term in producing linewidth. This means, for example, that even if we optimized our multiplication process by using a photomultiplier with high-gain dynodes throughout, the effect on the FWHM would be far less striking than one might expect. Now returning to our equation and putting $\varepsilon = 97$ eV for a GSO scintillator and taking the standard ^{137}Cs line at $E = 662 \times 10^3$ eV, our predicted fractional FWHM works out at 0.063 or 6.3 per cent. Experimental values of 7.8 per cent have been obtained, and the agreement must be considered satisfactory in view of the idealized nature of our computations, and in particular because of the fact that no account has been taken of possible collection losses for photons in the crystal, or electrons in the photomultiplier. GSO obeys the $1/E^{1/2}$ relation rather well over a range from 20 keV to 2 MeV. We can make a spot check on this by comparing the experimental resolutions at 0.662 MeV (7.8 per cent) and at 1.836 MeV (where it is found to be 4.8 per cent). The ratio of these two FWHMs is 7.8/4.8 or 1.63, which compares with the inverse ratio of the square roots of the energies, $(1.836/0.662)^{1/2}$ or 1.665.

Putting in the appropriate values for ε and p for the case of a BGO scintillation counter yields a computed value for the fractional FWHM of 7.1 per cent at 0.662 MeV, which is not too far from the experimental value of 9.05 per cent. Again the $1/E^{1/2}$ law is found to be well obeyed over the same range as GSO. With NaI(Tl), and indeed with caesium iodide scintillators, the picture is not as simple. Although these materials are superior in resolution to those previously mentioned at all energies, they only obey the $1/E^{1/2}$ law over a range from about 5 keV to 100 keV—or to be more precise, above 100 keV the value of ε which each possesses at lower energies moves to a larger value. An explanation of this phenomenon has been attempted on the grounds that at higher energies the full-energy peak derives increasingly from one or more Compton scatterings followed by a photoelectric absorption, rather than from a single photoelectric event. If the relationship between the number of photons produced and the energy dissipated in the scintillator is an exactly linear one, then this multi-event process produces a total number of photons no different from the single-event case. If, however, the photon number/energy dissipated relationship is not exactly linear, then small (and variable) differences will appear between successive gamma rays. The FWHM will thus be increased beyond its expected value. Because of this effect our first calculation on NaI(Tl) will be at the 5.9 keV standard value. Here $\varepsilon = 22$ eV and with $p = 0.25$, $\bar{m} = 4$, and $\bar{m}_1 = 6$ we obtain

$$\text{fractional FWHM} = 2.355 \left[\left(\frac{5900}{22} \right)^{-1} (1 + 3 + 8/9) \right]^{1/2}$$

$$= 0.32 \text{ or } 32 \text{ per cent.}$$

Experimental values of 35 per cent have been recorded, so the agreement here is good. If, however, we determine the expected value at 0.662 MeV we obtain 0.030 or 3 per cent. (Indeed we could have nearly written this down from the $1/E^{1/2}$ law, as 0.662 MeV is just a little more than two orders of magnitude greater than 5.9 keV !) Experimentally the best resolution obtainable is about 6 per cent, and while this is still better than obtainable with GSO or BGO at this energy, it falls very short of what is predicted theoretically — for the reason we have already alluded to. However, it is probably fair to say in conclusion that apart from the sodium iodide and caesium iodide anomalies, the remaining gap between the best experimental resolution obtainable and that predicted by our fairly simple theoretical treatment is small enough to allow us to believe in the general correctness of our description of the mechanisms involved.

7.3.4 *Statistical processes and linewidth in the proportional counter*

We are now in a position to deal with statistical processes and energy resolution in a proportional counter, making use of the results which we took some time to develop in connection with the scintillation counter. However, unlike the scintillation counter with its three-stage operation — photon production, conversion of photons to electrons, and electron multiplication — the proportional counter is just a two-stage device with direct electron (and of course positive ion) production being followed by electron multiplication. We noted earlier that it was the conversion of photons to electrons by the photocathode of the photomultiplier which was the largest contributor to the overall variance in the scintillation counter, so we may reasonably expect that the absence of this stage in the proportional counter will lead to better energy resolution.

Corresponding to equation (7.7) with its three terms for the scintillation counter we now have for the proportional counter only two terms in the equation for $\bar{N}$, the average number of electrons reaching the centre wire, as follows

$$\bar{N} = \bar{x}\bar{M}. \tag{7.13}$$

Here $\bar{x}$ is the average number of primary electrons produced by the incoming photon (or particle), and $\bar{M}$ is the average multiplication factor for an electron on its way to collection. (Clearly the number of positive ions, whose movement we saw in Chapter 3 actually produces the output pulse, is also given by equation (7.13).) The corresponding overall relative variance $\mathcal{V}$ is given, in terms of the relative variances $\mathcal{V}_1$ and $\mathcal{V}_2$ of the two stages, simply by

$$\mathcal{V} = \mathcal{V}_1 + \frac{\mathcal{V}_2}{\bar{x}}. \tag{7.14}$$

Now $\mathcal{V}_1$ is, as before, just $1/\bar{x}$, if we assume the electron production process to be Poissonian, but the expression for $\mathcal{V}_2$ is not immediately obvious. Indeed a full treatment of this problem would be inappropriate here, but some of the relevant papers can be found cited in reference 2. Instead we will use a plausible model, which, though crude, will give us a good idea of what is happening. The avalanche in a proportional counter has some obvious analogies with the corresponding multiplication process in a photo-multiplier. Indeed we might look on the mechanism in a proportional counter as consisting of a series of 'stages' in each of which the electrons, having acquired the appropriate energy, have their number multiplied by two. Of course in reality, because of the random nature of the electron collisions, the picture is not as simple as we are suggesting here (or indeed as we developed in Chapter 3 for estimating the average overall multiplication factor). We do not expect that every original electron produced in the gas will be multiplied by exactly the same factor, but as we have implied in equation (7.13) a mean value $\bar{M}$ will exist, with a statistical spread around it. We can build this into our simple model by assuming a 'stage gain' not of $m = 2$ exactly, but one with an average value of $\bar{m} = 2$. By analogy with equation (7.5) for the photomultiplier we can then hopefully write the result for the relative variance for the complete process of avalanche multiplication as $1/(\bar{m} - 1)$ with $\bar{m} = 2$, noting, however, that implicit in the use of equation (7.5) is the assumption that the variance of each 'stage' is equal, at least approximately to that given by Poissonian statistics. In fact our result for the relative variance of $1/(\bar{m} - 1) = 1/(2 - 1) = 1$ is, rather fortuitously, that given by a more precise but still fairly simple treatment. Returning then to equation (7.14) we can write

$$\mathcal{V} = \mathcal{V}_1 + \frac{\mathcal{V}_2}{\bar{x}}$$

$$= \frac{1}{\bar{x}} + \frac{1/(\bar{m} - 1)}{\bar{x}}$$

$$= \frac{1}{\bar{x}} \left(1 + \frac{1}{\bar{m} - 1} \right)$$

and putting $\bar{m} = 2$

$$= \frac{1}{\bar{x}} (1 + 1) \tag{7.15}$$

or

$$\mathscr{V} = \frac{2}{\bar{x}}. \tag{7.16}$$

Let us now see how this result translates into actual experimental resolution in the proportional counter, and as this type of counter is, as we saw in Chapter 3, most useful for low-energy photons we will look at its resolution for our standard low-energy quanta of 5.9 keV only. The value of $\bar{x}$ is given by E/w where E as usual is the photon energy, and w here is the energy required to produce an ion pair in the gas. We saw that this latter quantity varies from 20 to 35 eV for the various gases used in proportional counters, so we will take it to be typically 25 eV. The fractional full width at half maximum is given just as before by $2.355 \, \mathscr{V}^{1/2}$, and thus using equation (7.16)

$$\text{fractional FWHM} = 2.355 \, [2(E/w)^{-1}]^{1/2}. \tag{7.17}$$

Putting in the appropriate values yields a calculated resolution for a 5.9 keV photon of 21.7 per cent (and values for other energies would follow the usual $E^{-1/2}$ relation).

There are two points that one can immediately make about this numerical result. The first, quite expected, is that the resolution is considerably better than the very best resolution we calculated for a scintillation counter at this energy, 32 per cent, which in turn was somewhat better than experimental values. This superiority of the proportional counter is, as we have already noted, due to the absence of a photocathode contribution to the relative variance. The second, and slightly embarrassing point, is the discovery that the experimental resolution obtainable for a 5.9 keV photon using a proportional counter, while the right order of magnitude, is appreciably better than we have predicted theoretically, being in the 12 to 13 per cent region as against our 21.7 per cent. Our first suspicion naturally falls on the value of unity that we have used for the relative variance $\mathscr{V}_2$ of the multiplication process, and it is certainly true that a more elaborate theory implies a reduction of this quantity down to about 0.5. However, even with this lower value, the calculated result for the resolution only reaches about 19 per cent, implying that we must look elsewhere if we hope to improve the agreement between theory and experiment. In fact it turns out that we have overestimated the value of $\mathscr{V}_1$, the relative variance concerned with the primary electron formation, but we will leave a discussion on this matter until we have dealt with the resolution of the semiconductor detector, to which this problem also applies.

7.3.5 *Statistical processes and linewidth in the semiconductor detector*

Right from the outset we would expect the resolution of the semiconductor detector to be superior to both the previous detectors discussed, as it involves only a single-stage process, the production of electron–hole pairs by the incoming photon, without even the multiplication present in the proportional counter. We write then

$$\bar{N} = \bar{x} = E/w$$

where, as usual, $\bar{N}$ is the average number of carriers (of a particular sign) collected, $\bar{x}$ the average number produced, E the photon energy, and w in this case the energy required to produce an electron–hole pair. The overall relative variance $\mathscr{V}$ is given by

$$\mathscr{V} = \mathscr{V}_1 = \frac{1}{\bar{x}}$$

(assuming Poissonian statistics), that is

$$\text{fractional FWHM} = \Delta E/E = 2.355\,\mathscr{V}^{1/2} = 2.355\,(1/\bar{x})^{1/2} \qquad (7.18)$$

or

$$\text{fractional FWHM} = 2.355\,(E/w)^{-1/2}. \qquad (7.19)$$

The FWHM itself which is the parameter usually employed with semiconductor detectors is, of course, just E times the fractional FWHM, and thus is given by

$$\text{FWHM} = 2.355\,(Ew)^{1/2}. \qquad (7.20)$$

(Note that if E and w are expressed in electron volts the FWHM will be in the same units.) When we put numerical values into this equation, we immediately see an additional bonus for resolution in semiconductor detectors, because as we know, w is only about 3 eV for a germanium device and less than 4 eV for silicon. Putting E then equal to 5.9 keV and using a value of 3 eV for w, we obtain a calculated fractional FWHM of 5.3 per cent or a FWHM of 0.3 keV. (At 662 keV we expect the fractional FWHM to be 0.5 per cent and the FWHM 3.3 keV.) With figures like these it is no wonder that semiconductor detectors are the preferred devices where resolution is the prime consideration.

We have two closing remarks to make for this section. The first is a note of caution. As no internal multiplication occurs in semiconductor detectors, some external amplification must be provided before the pulses from the detector can be analysed and recorded. As we noted earlier in this chapter, all amplifiers add some noise, and therefore linewidth, to the signal. Consequently a great deal of care must be taken in their design to minimize degradation of the excellent linewidth provided by semiconductor detectors. We shall be taking up this topic of amplifier design later in this chapter. We will merely note now that, as the amplifier noise turns out to be a constant (in keV terms), i.e. does not reduce with decreasing photon energy, it will obviously become increasingly important as the energy decreases, and will ultimately become the dominant factor in determining the linewidth.

Our second remark is similar to the one we made at the end of the discussion in the previous section on proportional counters. If we make an experimental determination of the FWHM using, say, a germanium detector, (and to avoid confusion let us assume that we have an amplifier sufficiently good for its noise contribution to be negligible at the energy in question) then we find that our measured value for the FWHM will be very much less than its value computed from equation (7.20). Clearly there is no way to explain the discrepancy in this case, except through some defect in our calculation of the relative variance $\mathcal{V}_1$ for the production of the charge carriers, and we proceed immediately to make some remarks on this problem.

7.3.6 *The Fano factor*

The calculation we made in the preceding section on the relative variance in the number of charge carriers was based on Poissonian statistics. We now examine this assumption more closely, first of all for the case of a semiconductor detector. The energy to produce a hole–electron pair in silicon or germanium is approximately three times the respective band gap, and indeed as we noted in Chapter 4 this ratio holds reasonably well for a whole range of semiconductors. Clearly the major part of the energy of an incident particle or photon is thus taken up by then semiconductor lattice. Now if it were possible for all the radiation to be converted into carriers so that the number produced would be E/E_G, where E is the incoming energy and E_G is the band gap, then this number would not change from event to event, and the variance, and therefore the FWHM, would be zero. In practice of course the energy is shared statistically between the lattice and carrier production, and it is the small variation in sharing from event to event which produces the linewidth. The theory underlying Poissonian statistics implies that they should hold in the present situation only if a very small part of the incoming energy went on average into carrier production—as is clearly not the case. Consequently we now expect the relative variance to lie somewhere between the two extremes of zero variance and Poissonian variance.

We express this formally by writing

observed relative variance $= F \times$ Poissonian relative variance

$$(7.21)$$

where F is called the Fano factor and is of course less than unity. Replacing $\mathscr{V}$ then in equations (7.18) by $F\mathscr{V}$ means that equations (7.19) and (7.20) become, respectively,

$$\text{fractional FWHM} = 2.355 \, (Fw/E)^{1/2} \qquad (7.22)$$

and

$$\text{FWHM} = 2.355 \, (FEw)^{1/2}. \qquad (7.23)$$

(Note again that if E and w are expressed in electron volts then the FWHM will be in the same units.) Because of the complex nature of the interaction between nuclear radiation and the semiconductor lattice, theories predicting the value of the Fano factor have not been particularly successful, and indeed experimental determinations of F have shown considerable variation. However, both for silicon and germanium the currently accepted values are around 0.1. In our future treatment we will use the value of 0.12 suggested by Goulding and Landis (reference 3). For HgI_2 a value of 0.12 has also been reported.

Similar remarks apply to our treatment of the linewidth using proportional counters. In this case we replace $\mathscr{V}_1$ by $F\mathscr{V}_1$ (but leave $\mathscr{V}_2$ unchanged). Equation (7.15) then becomes $\mathscr{V} = (1/\bar{x}) (F + 1)$ or using the more accurate expression we alluded to later $\mathscr{V} = (1/\bar{x}) (F + 0.5)$ where $\bar{x} = (E/w)^{-1}$ as usual. The fractional FWHM then becomes $2.355 \, [\, (E/w)^{-1} (F + 0.5) \,]^{1/2}$. There has been a wide range of values of F suggested for various gases, but a number have F around 0.17. Using this and our previous value of $w = 25 \, \text{eV}$ with $E = 5.9 \, \text{keV}$ yields a fractional FWHM of 12.5 per cent which is around the experimental value. We found for scintillation counters that theory and experiment agreed reasonably well without allowing for any Fano factor reduction (that is $F = 1$), but this could possibly be the result of a Fano factor actually less than unity balanced out by some line broadening due to crystal imperfections.

7.4 Electronic noise and linewidth

7.4.1 *Introductory considerations*

We have noted earlier that for all three types of detector we have been discussing for use in particle and photon spectrometry — scintillation counter, proportional counter, and semiconductor detector — some amplification of the output signals is needed before they can be processed and analysed. The first two of these detector types feature internal multiplication of charge and the added amplification can therefore be quite modest. For a semiconductor detector, however, the amplification required may be quite large, and in these circumstances the amplifier system will not only increase the size of the signal but may add an appreciable amount of 'noise'. It is therefore in the context of semiconductor detectors, and the possible effect of this noise on the energy resolution, that we discuss the problem.

If we observe (on a fast oscilloscope) the output from an amplifier of sufficiently high gain, with no input, and well shielded from any source of extraneous 'pick-up', we nevertheless obtain on the screen a changing pattern of sharp pulses of random height (see Figure 7.4). This spurious output is known as electronic noise, and because of the progressive amplification by successive stages of the amplifier clearly comes, in the main, from the first stage. What the large amplification has enabled us to see is not of course the effect due to individual electrons, but the statistical fluctuations in the numbers of the electrons which are responsible for the currents and voltages in the system. As such we are looking at something very fundamental, whose effects can therefore be minimized, but not eliminated entirely.

This noise is of practical importance for two reasons. First of all it clearly sets a limit on the smallest genuine signal we can detect. Once the signal is lost in the noise there is no way its presence can be identified. (We pause to interject that we are talking here about signals in the form of individual pulses from a detector: the prognosis for something like a continuous sinusoidal signal buried in noise is much more hopeful, but outside our present brief.) However, long before the signal begins to disappear in the noise there are serious implications arising for spectrometric measurements. If we imagine a series of quite large pulses all of equal height superimposed on the noise, then because the baseline is fluctuating at random, a measurement of the pulse height will differ somewhat depending on whether the pulse is sitting on a 'spike' or a 'hollow'. The resulting spectrum of heights will be broadened out from the ideal line by an amount depending on the signal-to-noise ratio. It is methods of maximizing this signal-to-noise ratio that we are now concerned with, and we start by outlining the facts concerning the basic physical attributes of noise.

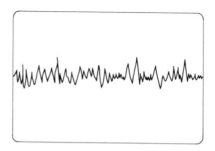

Figure 7.4 Noise displayed on an oscilloscope.

7.4.2 *Thermal noise*

Thermal noise and shot noise are the two basic types of noise occurring in electrical circuits. Thermal noise refers to noise occurring in resistors even in the absence of current flow, while shot noise is associated with a flow of current. We start with a discussion of the former. Thermal noise (or Johnson noise as it is sometimes called after its first experimental investigator) manifests itself as a small fluctuating voltage appearing across a resistor, and can be thought of as due to the thermally driven random Brownian motions of the electrons in the resistor. Clearly we expect no steady voltage in any one direction, so the average voltage appearing, $\bar{v}$, must equal zero. However, the mean square voltage $\overline{v^2}$ need not be zero, and its value was first deduced by Nyquist in 1928 (reference 4). He found that the mean square voltage $\overline{v^2}$ developed by a resistor in any frequency interval df was given by

$$\overline{v^2} = 4kTR \, df \tag{7.24}$$

where R is the value of the resistor, and the thermal effect comes in through T the absolute temperature and Boltzmann's constant k. (The purist might prefer to write equation (7.24) as

$$d(\overline{v^2}) = 4kTR \, df$$

to emphasize that the left-hand side refers to the noise in a narrow frequency band. However, the convention is to write it simply as $\overline{v^2}$, and we continue this practice for other noise quantities introduced in the text right up to equation (7.34). In this equation because we will have just integrated the noise over the amplifier bandwidth, the quantity $\overline{v^2_{\text{total}}}$ will not in any case need to appear in differential form). What equation (7.24) tells us, in other words, is that the resistor R can be considered as a noise source of voltage $\overline{v^2}$ (as given by equation (7.24)) in series with an ideal noiseless resistor of size R. Sometimes it is more appropriate to think of the real resistor as being made up of an ideal current source in parallel with an ideal resistor R, in which case it is not hard to see that the mean square value of the noise source of current, $\overline{i^2}$, is given by

$$\overline{i^2} = 4kT \, df/R \tag{7.25}$$

with the same notation as before. Either of these statements can be shown to be equivalent to the statement that the available power P (that is the power which could be delivered in the frequency interval df by the resistor to a load resistor of similar size) is given by

$$P = kT \, df. \tag{7.26}$$

It was this last formula that Nyquist deduced through an ingenious 'thought experiment' in which he imagined two equal resistors, each of size R, connected together by a long transmission line, with noise energy flowing in both directions from one resistor to the other. It would, however, take us too far afield to discuss the matter in full, but interested readers are referred to the original Nyquist paper or to reference 5.

7.4.3 Shot noise

Shot noise, as we have already noted, concerns the fluctuations in a current: these are due to the fact that the current consists of a flow of discrete electrons. Schottky first investigated this problem in connection with current fluctuations in thermionic devices. We can in fact deduce the required result in a simple way as follows.

Suppose we were able to measure a current by counting the number of electrons passing in successive intervals of time t_0. If $\bar{N}$ is the average number of electrons passing in this interval, then the average current $\bar{I} = \bar{N}e/t_0$ where e is the electronic charge. The number N found in successive measurements will fluctuate statistically in a Poissonian fashion around $\bar{N}$, with a standard deviation given by $(\bar{N})^{1/2}$. Expressing this in terms of $\bar{I}$, the fluctuation in number is given by $(t_0\bar{I}/e)^{1/2}$. Multiplying this by e and dividing by t_0 gives the current fluctuation as $(e\bar{I}/t_0)^{1/2}$, and squaring implies that the mean square current fluctuation is $e\bar{I}/t_0$. We would like to express this result in terms of frequency in order to make it comparable with the result for thermal noise, and we do so again by a plausible argument. Clearly the fluctuations we are concerned with are the ones with a period of the order t_0 and greater, as those more rapid than this will be smoothed out over the measurement time t_0. So if we write $t_0 = 1/f$ then $e\bar{I}/t_0$ becomes $e\bar{I}f$, and we are talking of all those fluctuations with frequency f or less. That is, we have a statement about the integral of the noise from f to zero. In the small frequency interval df then, the mean square current fluctuations can plausibly be written $\overline{i^2} = e\bar{I}\,df$. Since our argument has been a very simple one we would not be surprised to be out by a numerical factor, and in fact the correct answer for the shot noise turns out to be

$$\overline{i^2} = 2e\bar{I}\,df. \tag{7.27}$$

How are we to apply this result to noise in semiconductor devices? Remembering that Schottky originally obtained this formula for the flow of electrons in the vacuum of a thermionic diode, we might expect that the result could be applied to the small minority carrier currents flowing at semiconductor junctions—that at the gate of a FET for example—and this proves to be the case. Majority carrier currents require a different approach. We have already pictured the electrons in a resistor as moving around rapidly and randomly due to thermal agitation. If in addition a current is flowing through the resistor, all that happens is that a small overall drift velocity is superimposed on the random motion. Any shot noise effects are thus swallowed up in the larger thermal noise. We will apply these considerations shortly to the channel of a FET, which of course acts as a majority carrier resistor.

7.4.4 *The role of the amplifier bandwidth*

Thermal noise and shot noise are both uniformly distributed across the frequency spectrum (and are thus often loosely referred to as 'white noise'). However, not all the thermal and shot noise from the important first stage of the preamplifier will reach the output of the main amplifier, because of the latter's finite bandwidth. Indeed, as we shall see, we will wish to wilfully restrict the bandwidth in order to reduce the noise output, but before we discuss this matter we must first recall a few simple facts about amplifier response. A general-purpose wide-band amplifier will have a voltage amplification factor A_0, say, that will be constant over a range of frequencies which may extend from a few hundred hertz to many tens of megahertz. Eventually at a sufficiently high frequency, f_U say, the amplifier response starts to fall off significantly due to the effect of stray capacitance to ground. Similarly at some low frequency, f_L, determined by the largest convenient size for the capacitors used to couple the various stages, the response again falls off significantly. The bandwidth is thus defined by the quantities f_U and f_L, where the subscripts refer to the upper and lower fall-off in response.

We can look at the matter in an alternative manner. We know that the response of an amplifier to an idealized voltage step input is a signal with a rise-time T_R (usually quite small so as to reproduce the rising edge of the input signal) followed by a longer fall time T_F. The two points of view are of course closely related, and it can be shown that ω_U ($= 2\pi f_U$) is equal to $1/T_R$ and ω_L ($= 2\pi f_L$) is equal to $1/T_F$. What then can we do to reduce the noise — or to be more precise increase the signal-to-noise ratio? Clearly the answer is to reduce the noise reaching the output by reducing the bandwidth — that is by bringing ω_U and ω_L closer together, or what is the same thing by bringing together T_F and T_R. This will normally be achieved using resistance–capacitance filter circuits in the main amplifier to increase T_R and reduce T_F. While this technique certainly reduces the noise output, unfortunately it also reduces the output size of a pulse from a genuine event, since if T_R and T_F are not too different, the signal gets 'cleared away' before it can reach its expected value (the ballistic deficit of Chapter 6). None the less the overall result is, on balance, favourable and the signal-to-noise ratio increases — and continues to do so as T_R and T_F are brought closer and closer together. However, when T_R and T_F reach equality it can be shown that the maximum value of signal-to-noise ratio has been achieved—a result indeed that we might have intuitively foreseen. This common value of T_R and T_F, which we will call τ (not to be confused with the τ used for the detector pulse fall time in Chapters 4, 5, and 6), is still at our disposal, and we will explain a little further on how a numerical value, optimum from the point of view of signal-to-noise ratio, can be selected for it.

The reader must not think that because we have made T_R equal to T_F, and thus ω_U equal to ω_L, that the bandwidth has been reduced to zero! The precise definition of ω_U and ω_L (which we have not attempted here) looks after that. What is of more importance to us now is an expression for the magnitude of the amplification A in terms of the quantity ω ($= 2\pi$ times the frequency), and this can be shown to be (reference 6)

$$A = A_0 \left(\frac{\omega\tau}{1 + \omega^2\tau^2} \right) \tag{7.28}$$

where τ is the quantity we met a moment ago, the common value of T_R and T_F, and A_0 is the midband gain of the system when ω_U and ω_L are far apart. We will use this result for A very shortly.

7.4.5 *Noise calculations for the field-effect transistor*

We are now in a position to discuss the total noise emanating from an amplifier system. This we will attempt to do in general terms and without undue mathematical complexity. More detailed discussion can be found, for example, in reference 7. As we know, this noise comes almost entirely from the first element of the system, and we choose this element to be a junction field-effect transistor, rather than a MOSFET or bipolar transistor, for reasons which will be apparent later. Figure 7.5 shows the set-up at the FET input. The semiconductor detector providing the signals to be amplified is shown as a reverse biased diode: R_g acts both as the detector load and the gate resistor of the FET as we saw in Chapter 6. I_D is the leakage current in the reverse biased detector diode, while I_G is the FET gate current, again from a reverse biased junction, and usually much smaller than I_D. The symbol labelled C_{in} (called C_s in a slightly different context in Chapter 4) indicates what is known as the 'stray input capacitance'. It represents the unavoidable capacitance between the FET input and ground, and includes the detector capacitance C_d (called simply C in Chapters 3 and 6), the effective capacitance between the gate and the other electrodes of the FET (the C_i of Chapter 3), together with whatever capacitance is due to wiring, feedthroughs, etc. We shall see that C_{in} plays a crucial role in what follows. We note in passing that we are using a voltage-sensitive preamplifier in Figure 7.5 for our discussion rather than the charge-sensitive one, shown in

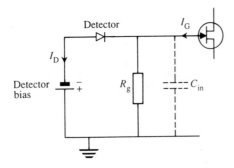

Figure 7.5 Circuit for discussion of noise in the FET.

Figure 7.6 (and earlier in Figure 6.6), which we might reasonably have expected. We do this for simplicity of treatment: it is not difficult to show that changing over to the charge-sensitive configuration will make negligible difference to our answer for the signal-to-noise ratio (see the Appendix to this chapter). The only change in fact is that we must include the small feedback capacitance C_f with the much larger contributions to C_{in} that we have already mentioned. C_f on its own will not appear in the result. The role of R_g of course will be played in the charge-sensitive case by the feedback resistor R_f. Figure 7.7 gives the corresponding picture from the point of view of noise. To represent the thermal noise from the gate resistor we have chosen to show it as a current source $\overline{i_R^2} = 4kT\,df/R_g$ in parallel with R_g. The shot noise in I_D is shown as a current source of

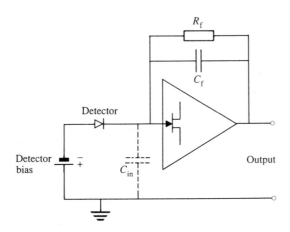

Figure 7.6 Detector with charge-sensitive preamplifier.

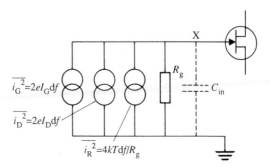

Figure 7.7 Current noise sources for the FET.

size $\overline{i_D^2} = 2eI_D\,df$, and similarly for I_G. If we look at the expression for thermal noise it is clear that we can minimize it by making R_g as large as we can compatible with the correct operation of the FET. We can in fact make it so large that the value of its resistance can be ignored compared with the much smaller impedance of C_{in}, which is in parallel with it. Consequently all noise currents shown in the diagram will flow almost entirely through C_{in}. In order then to find the mean square voltage at the FET input due to the thermal noise in the resistor we simply multiply the value of $\overline{i_R^2}$ ($= 4kT\,df/R_g$) by $(1/C_{in}\omega)^2$, that is the square of the impedance of C_{in}. (We use the square of the impedance because we are talking about mean square values of current and voltage.) This gives us for $\overline{v_R^2}$, the mean square noise voltage at the input due to R_g,

$$\overline{v_R^2} = \left(4kT\frac{df}{R_g}\right)\left(\frac{1}{C_{in}\omega}\right)^2$$

or remembering that $f = \omega/2\pi$, and therefore $df = d\omega/2\pi$,

$$\overline{v_R^2} = \frac{2kT}{\pi\omega^2 R_g C_{in}^2}\,d\omega. \tag{7.29}$$

Similarly for the mean square noise voltage $\overline{v_D^2}$ due to the detector shot noise we multiply $2eI_D\,df$ by $(1/C_{in}\omega)^2$, and eliminating df obtain

$$\overline{v_D^2} = \frac{eI_D}{\pi\omega^2 C_{in}^2}\,d\omega. \tag{7.30}$$

The expression for the shot noise voltage from the FET gate current will have the same form as that for the detector noise, so we can incorporate this small contribution into equation (7.30) by considering I_D in that equation to represent both I_D and I_G. (Note that since shot noise currents are non-directional, the contributions from the detector and the FET gate always add, whatever the relative directions of the original currents I_D and I_G.)

We have one more important noise source to deal with — not yet shown in Figure 7.7 — that from the conducting channel of the FET. We know from our discussion on majority carriers at the end of Section 7.4.3 that this noise source will be predominantly thermal noise from the channel considered as a resistor. Thus we can write

$$\overline{i_c^2} = 4kT \frac{df}{R_0} \tag{7.31}$$

where R_0 represents the effective channel resistance. Now this noise contribution is rather different from the previous ones we discussed inasmuch as it is not directly across the FET input, and we would like to put it on an equal footing with them. To do this we use the transconductance g_m which is the parameter which links current through the FET with voltage at its input, and whose dimensions are amperes volt^{-1}. By dividing our previous result in equation (7.31) by g_m^2 we can obtain the equivalent mean square voltage at the input. Thus the channel noise $\overline{v_c^2}$, is given by

$$\overline{v_c^2} = 4kT \frac{df}{R_0 g_m^2} \, .$$

Now it turns out from the theory of the FET that, to a reasonable approximation, R_0 may be taken as equal to $1/g_m$ so

$$\overline{v_c^2} = 4kT \frac{df}{g_m} \, . \tag{7.32}$$

Finally changing over to ω we have

$$\overline{v_c^2} = \frac{2kT \, d\omega}{\pi g_m} \, . \tag{7.33}$$

(If we wish to include this noise source in our previous figure 7.7, it can be shown as a voltage source of size given by equation (7.33) inserted between the point X and the FET gate.)

The three results we have obtained in equations (7.29), (7.30), and (7.33) for the resistor, detector (plus gate), and channel noise contributions are for the noise at the input of our preamplifier–amplifier system, and for the small frequency range corresponding to $d\omega$. To obtain the total noise at the output we must multiply each of these expressions in turn by A^2 (where A is given by equation (7.28), and A^2 is used because we are concerned with mean square voltages) and integrate from $\omega = 0$ to $\omega = \infty$. The integrations are straightforward and having done them we can then add the three mean square expressions together (as is appropriate for random uncorrelated contributions) to obtain finally for the total mean square noise at the output

$$\overline{v_{\text{total}}^2} = A_0^2 \left(\frac{kT}{2R_{\text{g}}} \frac{\tau}{C_{\text{in}}^2} + \frac{eI_{\text{D}}}{4} \frac{\tau}{C_{\text{in}}^2} + \frac{kT}{2g_{\text{m}}} \frac{1}{\tau} \right). \tag{7.34}$$

The individual contributions to $\overline{v_{\text{total}}^2}$ appear in the same order as we discussed them — that is resistor, detector plus gate, and channel noise.

We know of course that the calculation of the total noise is not the whole story — a low noise level will not itself be advantageous if the corresponding signal level is equally low. What we must consider is something that gives a measure of the signal-to-noise ratio, and in the context of an amplifier for a radiation detector, where the input signal is basically a charge, we do so by computing the 'equivalent noise charge', usually called Q or simply ENC. This is defined as the genuine charge signal which will produce the same output as the root mean square (RMS) noise, and is the valid parameter for comparison of two amplifier set-ups. The smaller the value of Q, naturally, the better the amplifier noise performance. We deduce an expression for Q as follows. If a charge Q is deposited on the input of the preamplifier then the corresponding input voltage step is Q/C_{in}, and with a normal wide band amplifier one expects the output to be $A_0 Q/C_{\text{in}}$ (where the quantities have the same meaning as previously). However, since T_{R} and T_{F} have been made equal the peak pulse-height voltage will be less than this, and, as has been noted in Section 6.4.2 will be down by a factor of $\exp(1)$ to $A_0 Q/C_{\text{in}}\exp(1)$. If Q is to represent the equivalent noise charge then the square of this quantity, that is $(A_0 Q/C_{\text{in}}\exp(1))^2$, must be put equal to the value of $\overline{v_{\text{total}}^2}$ as given by equation (7.34). Solving for Q^2 we have

$$Q^2 = (\text{ENC})^2 = [\exp(2)] \left(\frac{kT}{2R_{\text{g}}}\tau + \frac{eI_{\text{D}}}{4}\tau + \frac{kTC_{\text{in}}^2}{2g_{\text{m}}\tau} \right) \tag{7.35}$$

and Q of course can be obtained by taking the square root of the right-hand side.

7.4.6 *The ENC and gamma linewidths*

The equivalent noise charge, being the genuine signal which will produce the same output as the root mean square noise, represents the lower limit that can be seen in an amplifier output. More importantly for our discussion here, it also determines the fluctuations appearing in the output when a succession of equal inputs are applied to the system — even when these inputs are well above the detectability limit. In fact the ENC gives the standard deviation in the size distribution of full-energy gamma peaks from a detector due to electronic noise, and this spread is independent of, and in addition to, the statistical spread we discussed at the beginning of the chapter.

In order to obtain an appropriate measure of this electronic spread, we must first express the ENC in energy terms, and this we do as follows. The number of electrons arriving at the preamplifier input corresponding to the equivalent noise charge is ENC/e, where $e = 1.6 \times 10^{-19}$ C and ENC is of course also in coulombs. This in turn corresponds to an expenditure of energy in the detector of $(ENC/e) \times w$, where w as usual is the energy in eV to produce an electron–hole pair there ($= 2.96$ eV for germanium). We now have the standard deviation in electron volts and just as in Figure 7.3(b) we must multiply it by 2.355 to get the FWHM. Thus the FWHM in eV is given by $2.355(ENC/e)w$, or one-thousandth of this value if, as is normal, we express FWHMs in keV. So finally

$$FWHM(keV) = (2.355 \times 10^{-3})(ENC/e)w. \qquad (7.36)$$

Putting $w = 2.96$ eV, as appropriate to a germanium detector at around 77 K (where it must always be used, as we have seen earlier), we obtain

$$FWHM(keV\ germanium) = (4.4 \times 10^{16})(ENC). \qquad (7.37)$$

Similarly if we were using the amplifier in conjunction with a silicon detector at around 77 K, one would put $w = 3.76$ eV to obtain

$$FWHM(keV\ silicon) = (5.5 \times 10^{16})(ENC). \qquad (7.38)$$

So our FWHM is directly proportional to the equivalent noise charge.

Let us return then to equation (7.35), talk a little about the individual terms to see how their noise contribution can be minimized, and then put some typical numbers in to decide what FWHMs we might reasonably expect in practice. Our first remark is concerned with a basic difference between the first two terms in equation (7.35) (i.e. those due to R_g and I_D) and the third. The former are independent of C_{in} and thus their contributions to the FWHM are also independent of it. They are often referred to as parallel noise sources as they are applied to the FET input in the same way (i.e. in parallel with) the genuine signal. Thus if C_{in} is for some reason made larger, then (at least as far as these two terms are concerned) both the signal and the noise decrease in the same way, and the ENC (and FWHM) remain unaffected. The third term ('series noise') obviously increases with C_{in}: one of our aims then will be to reduce C_{in} as far as possible, a topic we will develop more fully a little further on.

Our next point concerns the dependence of the ENC on τ, the common rise and fall time of our amplifier. As the first two terms increase with increasing

τ, and the last term decreases with increasing τ, clearly an optimum value for τ exists, which is obtained by differentiating the right-hand side of equation (7.35) with respect to τ, and setting the result equal to zero. This optimum value of τ which minimizes the ENC is then given by

$$\tau_{\mathrm{opt}} = \left(\frac{kT/2g_{\mathrm{m}}}{(kT/2R_{\mathrm{g}}) + (eI_{\mathrm{D}}/4)} \right)^{1/2} C_{\mathrm{in}}. \tag{7.39}$$

The minimum value of $(\mathrm{ENC})^2$ can then be written

$$(\mathrm{ENC})^2_{\mathrm{min}} = [2\exp(2)] \left[\left(\frac{kT}{2R_{\mathrm{g}}} + \frac{eI_{\mathrm{D}}}{4} \right) \left(\frac{kTC^2_{\mathrm{in}}}{2g_{\mathrm{m}}} \right) \right]^{1/2}. \tag{7.40}$$

(Incidentally it is instructive to write equation (7.39) as

$$\frac{kT}{R_{\mathrm{g}}} \tau_{\mathrm{opt}} + \frac{eI_{\mathrm{D}}}{4} \tau_{\mathrm{opt}} = \frac{kTC^2_{\mathrm{in}}}{2g_{\mathrm{m}} \tau_{\mathrm{opt}}} \tag{7.41}$$

The equation in this form tells us that the optimum time constant is the one which makes those terms that increase with increasing τ equal to the term which decreases with increasing τ — a result we might have expected anyhow.) The value for τ_{opt} usually works out, as we shall see, at a few microseconds for the sort of high-resolution detectors we are talking about. (However, it is worth noting that if, for example, the prime requirement of our system is to accurately determine the time of arrival of a particle or quantum, then we may be forced to use a value of τ very much less than the optimum. In this case our equation (7.35) clearly indicates that the dominant contribution will come from the third term, and our value for $(\mathrm{ENC})^2$ will be given by $(\mathrm{ENC})^2 \approx [\exp(2)](kTC^2_{\mathrm{in}}/2g_{\mathrm{m}}\tau)$ for $\tau \ll \tau_{\mathrm{opt}}$. We have clearly sacrificed some energy resolution for better timing ability.)

Having agreed then that for best resolution we choose our time constant to be τ_{opt}, we still have to look at the individual terms in our equation (7.40) for $(\mathrm{ENC})^2_{\mathrm{min}}$ to see how we might reduce them. We have already noted that R_{g} must clearly be as large as possible, and in practice this means about $10^{10}\,\Omega$. Beyond this value it becomes difficult to make a real resistor behave as a pure resistance. Secondly we must ensure that I_{D} is as small as possible. We recall that I_{D} contains both the gate current of the FET, I_{G}, and the leakage current of the detector. This 'leakage' current includes both the usual reverse current expected in a back biased diode, as well as any genuine surface leakage current between the detector electrodes. For room-temperature operation of silicon detectors the former contribution will usually predominate. Germanium detectors because of their smaller band gap have an unmanageably high reverse diode current at room temperatures and must always be cooled to liquid-nitrogen temperatures. The reverse diode current is then essentially eliminated, and the remaining current represents surface leakage only. This current might vary from less than 1 pA for a small detector to many tens of picoamperes for a large one. The gate currents of selected FETs suitable for this type of work can have values considerably lower than 1 pA.

Turning finally to the last term in equation (7.40) we clearly would like to have C_{in}^2 (the square of the stray capacitance) as small as possible, and g_m as large as possible. However, things are not quite as simple here as they might appear. We know that C_{in}, the total capacitance at the FET input, is made up of the input capacitance of the FET, C_i, plus the detector capacitance, C_d, as well as residual capacitance due to wiring, feedthroughs, etc. So we can write $C_{in} = C_i + C_d$, including for convenience the remaining residual capacitance with C_d. Now naturally we will try to make C_d as small as possible – that is to reduce the area of our detector. However, the minimum size of the detector will be determined by other factors – such as the minimum value of the absolute detection efficiency we can tolerate in the work in question. So C_d is then fixed for us, and we will make the contribution from the residual capacitances also as small as possible by good design. That leaves us with C_i. Unfortunately due to the characteristics of the FET if we try to reduce C_i by changing the appropriate physical dimensions of the device, we also reduce g_m proportionately. In fact for a tolerably good FET the value of g_m/C_i is approximately $10^9 \, \mathrm{s}^{-1}$, that is we can have a device with a g_m of 10 mA V^{-1} and a C_i of 10 pF or, if we like, one with C_i only 5 pF, but then g_m will be only 5 mA V^{-1}. We must not take these exact figures too seriously, however, because of the differences which normally occur from transistor to transistor – but the general trend is undoubtedly valid.

Returning then to the problem of minimizing C_{in}^2/g_m we see that this is equivalent to minimizing $(C_i + C_d)^2/C_i$. A simple differentiation shows that the optimum situation is when we choose C_i to be equal to C_d. So if our detector has a small capacitance we choose a FET with a correspondingly small C_i, while if the detector capacitance is large we choose a FET with a large capacitance (but of course with a large g_m). Indeed for large-capacitance detectors it may be possible and profitable to use two FETs in parallel. (Note of course that with C_i put equal to C_d, and g_m proportional to C_i, the noise term C_{in}^2/g_m is proportional to C_d, so a smaller detector still means less FET noise.)

7.4.7 *A numerical example on noise*

Let us now see how far our previous theoretical calculations work out in practice, and we take some figures from actual data accompanying a commercial Ge(Li) detector. This coaxial detector is quite large with a volume of over 100 cm³. Its capacitance when biased is 15 pF, and its leakage current (when cooled of course) is specified as less than 50 pA. The FWHM is quoted as 0.78 keV for the standard gamma ray of 122 keV, using a 4 μs shaping time constant (our τ). The only information available on the preamplifier relevant to our calculation is that the gate resistor is $2 \times 10^9 \, \Omega$. Manufacturers are naturally reluctant to reveal the type and selection methods used for their FETs.

What would our design criteria be? Since the detector capacitance is 15 pF we should, in line with our previous considerations, choose a FET with a similar capacitance (assuming for simplicity that any further stray capacitance is small). This allows us to assume a transconductance of 15 mA V^{-1}. As the detector leakage current is given as 'less than 50 pA', and the FET gate current will probably be less than 1 pA, we will assume a value for I_D of 50 pA exactly. For room temperature, T (in kelvin) is close to 300 K, and thus kT is 4.14×10^{-21} J. Now C_{in} is of course $15 + 15 = 30$ pF, R_g has been fixed for us at $2 \times 10^9 \, \Omega$, and $e = 1.6 \times 10^{-19}$ C. Substituting these values into equation (7.39) gives us for τ_{opt} a value of 6.4 μs — rather far from the value used by the manufacturers — but we will come to that later. The value for ENC^2_{min} can be predicted from equation (7.40), and the corresponding value for the FWHM from equation (7.37). The latter turns out to be 0.75 keV(Ge) which at first sight looks quite close to the figure quoted for the system. However, the agreement is illusory because we remember from Section 7.2 that the total FWHM must be made up by combining the contribution from the noise in the preamplifier and the statistical broadening in the detector. This we do by adding the squares, as we did for the mean square voltages, giving, using an obvious notation

$$FWHM^2_{total} = FWHM^2_{FET} + FWHM^2_{det}. \tag{7.42}$$

So we need to predict a FWHM$_{FET}$ well below FWHM$_{total}$ to allow us to include the the contribution of the detector. So how can we reduce the FET noise contribution?

The most obvious candidate is the temperature T in equation (7.40). By its very nature thermal noise is reduced at lower temperatures: thus by cooling the FET we can significantly reduce the terms in T in equation (7.40). (I_G the FET gate current will also go down, but as it was already almost negligible compared with the detector current, this does not help us significantly.) As we will already have a cooling system provided for the germanium detector (whether hyperpure or lithium-drifted), all we need to do is to arrange for the input FET of the preamplifier to be placed in the cryostat. (Actually if we cool the FET down to near liquid-nitrogen temperature (77 K), the thermal energy available, unlike the situation described in Section 5.2, becomes insufficient to ionize all the dopant in the channel. The resulting reduction in the charge carrier density, or carrier 'freeze-out' as it is called, reduces the FET conductance — and hence g_m —

thus increasing the noise. It is necessary then to arrange to have the FET temperature maintained at around 120–130 K.)

Changing the value of T we have used in equations (7.39) and (7.40) from 300 K to 120 K gives us a new value for τ_{opt} of 4.5 μs and a new FWHM of 0.56 keV. What else can we look to for further reduction? Strangely enough it is to the term exp(2) which appears in front of the whole expression in equation (7.40). We remember it arose because of the type of shaping we imposed on the pulses in the main amplifier. In the case we discussed we had the pulse shape defined by two time constants T_R and T_F, which we agreed should be made equal for best noise performance. However, as we noted at the end of Section 7.4.4, making T_F equal to T_R did cause a drop in the pulse height by a factor of exp(1), which is where the factor exp(2) derived from. Figure 7.8(a) (which is a simplified version of Figure 6.18) shows schematically how this pulse shaping is done. The *CR* differentiating circuit at the left of the diagram defines T_F, and the *RC* integrating circuit on the right defines T_R. (A buffer amplifier is shown between them to indicate that the second circuit should not load the first.) As we wish T_F and T_R to be the same, the time constant *RC* is identical in both sections, and equal to

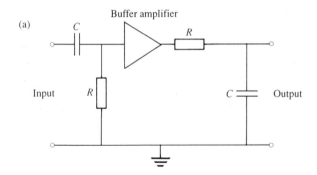

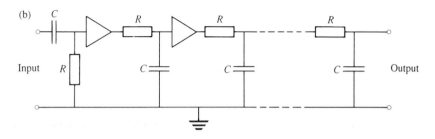

Figure 7.8 (a) CR–RC pulse shaping. (b) Gaussian pulse shaping.

their common value of τ. The circuit thus converts a step function of voltage (or a slowly decaying one) into the required shape.

The question naturally arises as to whether other forms of pulse shaping might reduce the ENC below that obtaining with our present relatively simple circuit. The answer is yes, but the full mathematical analysis would unnecessarily prolong the present discussion. Instead we will simply summarize the main points, and the interested reader can consult, for example, reference 8. The pulse shape with optimum ENC can be shown to be the cusp-shaped one of Figure 7.9(a). Other good shapes are the triangular pulse, Figure 7.9(b), which has an ENC only 7.5 per cent greater than the cusp-shaped pulse, and the Gaussian pulse of Figure 7.9(c), which has an ENC greater than the optimum by only 12 per cent. (Note that the pulses are all symmetrical about a vertical line through their peak: this is a necessary condition for good ENC.) By comparison our $CR-RC$ circuit gives an ENC that is about 36 per cent greater than the optimum case, so we have plenty of room for improvement. However, the possibility of such improvement must be balanced against the complexity of the circuits necessary to produce these better pulse shapes—particularly one resembling the optimum of Figure 7.9(a). Consequently most attention has been focused on triangular or Gaussian-shaped pulses. The latter in particular can be produced quite simply by adding further RC sections to our earlier circuit of Figure 7.8(a), as shown in Figure 7.8(b). In theory we need an infinite number of extra sections to produce an exactly symmetrical Gaussian: practical circuits with four or even two RC sections (in total) produce an appreciable improvement over our simple circuit with one RC section. With four RC sections one obtains a calculated ENC of 1.165 times the ideal one: with two sections the corresponding value is 1.2155. These figures are equivalent to 85.7 per cent and 89.4 per cent, respectively, of the ENC for the simple $CR-RC$ case, and represent a worthwhile reduction in the ENC for little additional circuit complexity. For even more sophisticated shaping methods the interested reader can consult reference 9.

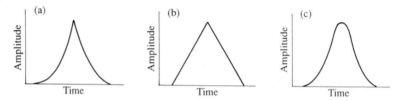

Figure 7.9 (a) Optimum pulse shape. (b) Triangular pulse. (c) Gaussian pulse.

Returning then for a last shot at our figure for the FWHM in the numerical case we were considering, we see that using an amplifier with one CR and two RC shaping circuits we should expect an ENC of 89.4 per cent of our previous figure of 0.56 keV, that is 0.50 keV FWHM(Ge). To see what overall linewidth we would get in practice with such an amplifier we make use of equation (7.42) to obtain the total full width at half maximum. The contribution from the statistical processes in the detector is (from equation (7.23)) FWHM (in eV) = 2.355 $(FEw)^{1/2}$, where E, the energy of the gamma ray is here 122 keV, $w = 2.96$ eV for germanium, and we take F to be 0.12. This gives 490 eV or 0.49 keV, which combined quadratically as in equation (7.42) with our previous value of 0.50 keV for the FET contribution, gives an expected value for the overall linewidth of 0.70 keV FWHM. This is to be compared with the experimentally determined result of 0.78 keV FWHM under similar operating conditions. The agreement must be considered satisfactory, remembering that in practice it would not be possible to match the FET capacitance with that of the detector as perfectly as we have assumed, nor indeed would the amplifier, with a limited selection of shaping times, be able to provide an exact match for the optimum value of τ.

7.4.8 *Further sophistications and the current state of the art*

If we are pushing for the ultimate in resolution, then first of all we will clearly opt for small planar detectors with low capacitance (≈ 1 pF) and small leakage currents (≈ 1 pA). (Of course we may not find FETs with stray capacitances low enough to match the capacitances of these detectors, but we will simply use those with the lowest available C_i.) Having reached this point, about all that can be done to produce a final small reduction in linewidth is to look again at the resistor R_g (or rather its equivalent in the charge-sensitive configuration R_f). We have already noted that although we would like to make this resistor as large as possible, it is not technically feasible to make satisfactory resistors much above $10^{10}\,\Omega$. Let us ask ourselves therefore what would happen if we took the rather dramatic step of removing it altogether from the circuit, thus effectively making it infinite, and eliminating the corresponding term entirely from the equivalent noise charge equation (7.40). From the circuit point of view the role of R_f is to discharge the feedback capacitor C_f (Figure 7.6) after it has been charged following an event in the detector, and thus prepare it for the next event. This it does in an exponential fashion with a time constant $R_f C_f$. If the resistor is absent then in place of a series of pulses with exponential tails we have a 'staircase' of pulses appearing at the output of the feedback stage, as shown in Figure 7.10. (The pulses will of course appear randomly in time, and the rising edge of the steps will in general be of different heights: we have shown

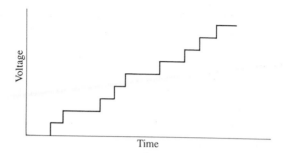

Figure 7.10 'Staircase' output with R_f absent.

them fairly closely of the same height, as would occur for the case we are interested in — that of a narrow gamma-ray line.) The fact that the pulse train has this staircase form need not worry us unduly, as we know that when it reaches the shaping section of the main amplifier the individual pulses will be separated out with the critical information on their relative heights retained. (Indeed the flat tops on the steps will actually make pulse processing in the main amplifier easier, in that it avoids the necessity for pole-zero cancellation.) What does concern us is the steadily increasing d.c. level at the output, which must not be allowed to reach a value where it will upset the operating point of any transistor connected there. (Even this problem is not entirely new, as for our previous case of R_f finite but large, we must also be careful with high counting rates that the d.c. level does not reach an unacceptable value — as we saw in Chapter 6. However, with R_f removed an unacceptable value will always be reached sooner or later, even for low counting rates.)

One solution to this problem is shown schematically in Figure 7.11(a). The action of the circuit is as follows. As the negative pulses from the detector arrive at the gate in turn, the amplifier feedback action, as we have seen in Section 6.3.2, draws almost their entire charge on to C_f, and produces a positive-going staircase at the output, as shown in Figure 7.11(b). When the output level reaches a predetermined value we need to rapidly reverse this process and 'reset' the preamplifier by returning the output to its original state. This is done by means of an optoelectronic link between the LED shown in the diagram and the FET gate. What happens is that when the discriminator circuit of Figure 7.11(a) trips at the upper level, an electrical pulse is applied to the LED which in turn illuminates the FET gate–drain diode and renders it conducting. A current flow to the gate thus ensues, lowering the output voltage and discharging C_f. When the output reaches a predetermined lower level the LED is turned off and the whole process resumes (Figure 7.11(b)). The discharging of C_f takes a few microseconds

(a)

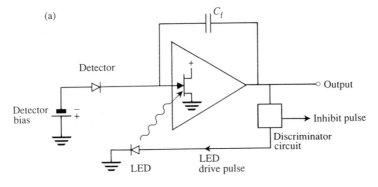

(b)

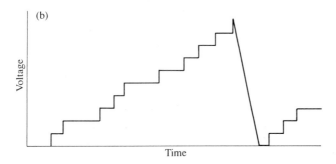

Figure 7.11 (a) Preamplifier with optoelectronic reset using light-emitting diode (LED). (b) Output from circuit of Figure 7.11(a).

and during this time another pulse from the discriminator circuit is used to inhibit later parts of the recording system, and prevent them being affected by the large negative-going signal produced at the preamplifier output during reset.

The values of the lower and upper discriminator levels depend on the particular circuit design, but the range through which the voltage is allowed to change is typically a few volts. The actual duration of the staircase between resets depends of course on the rate of arrival of the radiation at the detector and its energy. For gammas of 122 keV, say, each pulse will produce on average in a germanium detector a charge of $(122 \times 10^3/2.96)(1.6 \times 10^{-19})$ C or 6.6×10^{-15} C. If C_f is 1 pF this charge produces a voltage of 6.6 mV across it. With a count rate of, say, 5000 s^{-1} and the discriminator levels 2 V apart, the duration of the staircase between resets will then be, on average, about 60 ms. Even in the absence of radiation the output voltage will increase slowly due to the detector leakage current

(Section 6.3.2). For example if the leakage current is 1 pA (and the FET gate current negligible) then with a 1 pF feedback capacitor as before, the voltage rises at 1 V s^{-1} and the time between resets with little or no radiation present will thus be 2 s, for the same 2 V range at the preamplifier output.

With this new technique and a small detector let us make a calculation to see how low we can push the linewidth. We assume of course that the detector and FET are appropriately cooled, and that the leakage current can be taken as 1 pA. The detector capacitance for a small detector will be about 1 pF, but we must add to this the FET capacitance of say 4 pF — and also at this level take account of any residual stray capacitance, which might be about 2 pF. This gives us a total of 7 pF. (Note that we have chosen a FET with a capacitance about as low as commercially available, in order to match as best we can the small detector plus stray capacitance, 3 pF.) The value we have quoted, 4 pF, implies as explained earlier, that the transconductance we assume will be 4 mA V^{-1}. With these parameters and a FET temperature of 120 K as before, the optimum time constant is 16 μs, and assuming a main amplifier shaping network of one CR and four RC sections, the FWHM(Ge) comes out as 0.116 keV or 116 eV. This represents the linewidth introduced by the electronic noise of the system, and is incidentally the result we would get if, as is common practice, we checked the system by injecting test signals, all of the same size (i.e. with zero spread) into the preamplifier input from a pulse generator (Section 6.3.6). Photons from say a ^{55}Fe source with energy 5.9 keV would produce signals with their own FWHM due to detector statistics of $2.355(FEw)^{1/2}$, as we saw in Section 7.3.6. F and w have the same values as before while E is here 5.9 keV. Putting in these values gives a FWHM of 108 eV for the detector statistics, which when combined quadratically with the electronic noise value of 116 eV gives an overall value of 158 eV FWHM(Ge).

Incidentally if we repeat the calculation for resistive discharge of the feedback capacitor we obtain a value of 138 eV for the electronic noise, using a resistor of $2 \times 10^{10} \, \Omega$ (which is close to the maximum feasible value). The corresponding figure for the overall FWHM for ^{55}Fe photons is 175 eV, which gives us an idea of the sort of improvement obtainable by the removal of the discharge resistor. Although obviously we can expect some further improvement if we postulate a lower leakage current in the detector, a value of around 150 eV FWHM(Ge) for 5.9 keV photons represents about the lower limit for commercially available germanium systems, although values below 120 eV have been reported using more sophisticated pulse-shaping methods and very small detectors.

Just how far we have been able to go in reducing the electronic noise contributions even to 116 eV can be appreciated by rewriting equation (7.36) to read

$$\text{ENC}/e = \frac{\text{FWHM (keV)}}{(2.355 \times 10^{-3})w}. \tag{7.43}$$

Now the left-hand side of this new equation represents, as we noted in Section 7.4.6, the equivalent noise charge expressed in electron numbers. Putting in the value for the FWHM of 0.116 keV gives us a value for ENC/e corresponding to fewer than 17 electrons! The ability of such an electronic system to process signals while adding only minimal amounts of noise is thus underlined. In practice this means two things. Firstly for gamma rays of even moderate energy (say around 100 keV) the linewidth will be due almost entirely to statistical processes. Secondly even at the lower end of the energy scale (where the electronic noise predominates) an X-ray line at say 1 keV can be very well separated from another at 1.5 keV. We will return to an important application of this latter remark when we discuss silicon detectors.

At this stage we seem to have exhausted all the stratagems available for the reduction of preamplifier noise, but one final possibility should be mooted. If it were possible to design a small unconventional FET directly integrated with the detector, then we might be able to eliminate the stray capacitance associated with the normal detector–FET connection, and have also an FET input capacitance better matched to that of a small detector (or to that of a moderate-sized drift detector – see Section 7.4.11). The resistivity of normal FET material is very much lower than that of detector material, and this presents technical problems. However, such integration on silicon has been achieved (reference 10), although the relatively large gate currents reported mean that applications will probably be with room-temperature detectors or at short shaping times.

7.4.9 *Noise with other detector types*

In all our previous discussions in this chapter we have concentrated on one particular detector–transistor combination, that of a germanium detector coupled to a field-effect transistor. Although the principles we have already developed for this system will apply equally to other combinations, the performances achievable will not necessarily be the same. We will therefore briefly discuss some alternative detectors in this sub-section, and alternative transistors in the next.

Let us start then with Si(Li) detectors and ask ourselves how we would expect their properties to differ from their germanium counterparts. Clearly the first important difference is the larger value in silicon for w, the energy (3.76 eV at 77 K) required to produce an electron–hole pair, which of course arises directly from the larger band gap in silicon (≈ 1 eV). This results, naturally, in fewer current carriers being produced in silicon for a given energy input, which, both from common sense and from our formula FWHM = 2.355 $(FEw)^{1/2}$, means rather worse statistics and linewidth, as compared with germanium. (We remember that the accepted value of the Fano factor F for silicon is the same as that for germanium.) A second effect from the lower number of carriers produced by a given energy deposition in silicon is that the noise produced by the FET will appear relatively worse. Again we can see this formally by comparing equations (7.37) and (7.38), where the larger value for w in silicon increases the FWHM. Both these effects degrade the noise performance of silicon relative to germanium.

However, there is a counterbalancing factor. A parameter which featured prominently in our FET noise calculations was the capacitance of the detector, which of course depends directly on the dielectric constant of the material. As this has a value of 12 for silicon compared with 16 for germanium, the consequent reduction in capacitance for a silicon detector will lead to better linewidth — particularly at low energies where electronic noise is dominant.

With these various gains and losses for the linewidth of a silicon detector compared with its germanium counterpart, it is hardly surprising that the FWHM for the best commercial Si(Li) detectors is comparable with that quoted previously for germanium detectors — although again better figures have been reported in the research literature. For reasons noted in Chapter 5, silicon detectors are particularly suited for the analysis of low-energy photons — that is X-rays. One important application is their use in X-ray fluorescence spectrometry. Here a sample — say a metal alloy — is bombarded with X-rays, producing the re-emission of characteristic X-rays from the sample. These X-rays, for low- and moderate-atomic-number materials, have energies of the order of 1 keV, and the spectra obtained can thus be analysed with good resolution and high efficiency by a cooled Si(Li) system. This would allow us, in the example above, to rapidly detect very small quantities of constituent metals in the alloy.

Turning finally to HgI_2 detectors, we remember that their outstanding advantage is their ability, because of their large band gap (2.14 eV) to operate successfully at or near room temperatures. The other side of the coin is, as we mentioned when discussing silicon, that such a large band gap implies a large value for the energy, w, to produce an electron–hole pair. This is in fact about 4.2 eV for HgI_2, and results in rather worse statistics in the signal production and a higher FWHM from the FET. We need not expect then quite as good linewidth with HgI_2 detectors, but in fact values down to 175 eV FWHM have been reported for 5.9 keV photons — as mentioned in Section 5.12.

7.4.10 *Possible use of other transistor types in low-noise preamplifiers*

We have concentrated solely on the use of junction field-effect transistors in our discussion of low-noise preamplifiers: we now explain our reasons for this. Let us first consider the possibility of using a bipolar transistor as the input device. Here the difficulty is that the current at the input electrode, that is the base, is that due to a forward biased diode. It is thus many orders of magnitude greater than that for a junction FET with its reversed biased diode gate. The term in I_D in our noise equation (7.35) which there included the small contribution from the gate current will now contain a much larger and dominant term from the base current. This new large value of I_D has two effects. First of all τ_{opt} (equation (7.39)) is moved down to the nanosecond region, and secondly even here ENC_{min} (equation (7.40)) for the bipolar transistor is markedly inferior to that for the FET at its own microsecond τ_{opt}. However, one should add that if for timing or count rate reasons we are compelled to work in the nanosecond region, then the FET, now far from its τ_{opt}, is somewhat inferior as regards noise to the bipolar device. But in this case of course we are primarily seeking speed of response, not optimum energy resolution.

The other potential competitor to the junction FET is of course the MOSFET. Here we can not plead that the MOSFET's input current is greater than that of the FET—the contrary is true. So why do we not choose it for low-noise use? To answer this we must digress briefly to talk about '1/f noise'. The contribution of this noise is, as its name implies, inversely proportional to frequency (at least approximately), so this is not 'white noise'. We will allude to possible origins for this noise a little later. For the moment let us, as we did earlier for the two other noise sources, express the contribution of 1/f noise in terms of an equivalent noise source at the input of our device as follow

$$\overline{v_f^2} = A_f (1/f) \, df \qquad (7.44)$$

where A_f is a constant defining the amount of noise present in a particular device. Integrating this over the amplifier bandwidth leads to a fourth term inside the brackets in equation (7.35) equal to $(A_f/2)C_{in}^2$, which we note incidentally is independent of τ. The nature of this $1/f$ noise is not well understood, but it is generally attributed to crystal imperfections and surface defects in devices. In junction FETs it is usually negligible with respect to the other terms, but this is not so with MOSFETs. The significant contribution from $1/f$ noise in these devices is usually considered to be due to effects at the semiconductor–oxide interface. In any case it renders them unsuitable for use in preamplifiers where the very best energy resolution is required. We conclude then that the junction FET is still the best input device in low-noise preamplifiers for nuclear radiation spectrometry.

7.4.11 *Drift detectors*

We have already met drift detectors in Chapter 5, where they were introduced primarily as position-sensitive devices. In addition the dramatic drop
in capacitance (and leakage current) which occurs when these detectors are
appropriately biased suggests the possibility of a substantial improvement in
energy resolution over their conventional counterparts. Let us illustrate this
with a numerical example.

The germanium detector described by Luke (reference 11) had an area of
5 cm^2 but only a capacitance of 1.4 pF. (A normal planar device with the
same area and a depletion depth of 5 mm would have a capacitance of about
ten times this.) An optoelectronic preamplifier was employed, and a main
amplifier with a shaping time of 17 μs, presumably the optimum one. As we
remember from our earlier discussions in this chapter, it is always desirable
to match the FET capacitance as closely as possible to that of the detector.
Clearly no exact match is possible with ordinary FETs, so the best that can
be done is to pick one with as low a capacitance as possible. No details are
given in the paper, so let us assume the FET capacitance to be 4 pF, and that
strays amount to 2 pF, which together with the detector capacitance gives a
total of 7.4 pF. Our choice of FET capacitance then allows us to assume a
value for the transconductance of 4 mA V^{-1}.

We are now in a position to make use of equation (7.35) to determine the
equivalent noise charge, if we remember the result we deduced in equation
(7.41). This was that, with the optimum time constant τ_{opt}, those terms
which increase with τ are just equal to the term which decreases with τ. Equation (7.35) may thus be written

$$\text{ENC}^2 = [\exp(2)]\left(2 \times \frac{kTC_{in}^2}{2g_m\tau_{opt}}\right)$$

and all the quantities are now known, if we take, as usual, the temperature
to be 120 K. Converting the ENC to FWHM and assuming that more
sophisticated Gaussian shaping reduces this to 85.7 per cent of our calculated
value gives an answer of 118 eV FWHM(Ge). Combining this with a value
of 108 eV for the statistical spread for the 5.9 keV photons from a ^{55}Fe
source yields a calculated overall FWHM of 160 eV at this energy. This compares well (but perhaps a trifle fortuitously) with the reported experimental
figure of 162 eV. This experimental value of 162 eV FWHM is considerably
less than would be obtained with a conventional detector of the same area —
in fact only about half as much. However, this improvement hardly seems
quite as good as what one might originally have hoped for from such a
low-capacitance device, particularly as it must be balanced against the more
complex fabrication procedures involved.

The difficulty (and it is one that we met even with conventional detectors in Section 7.4.8) is the mismatch in capacitance between that of the detector and that of the lowest capacitance FET available. As we mentioned there, FETs with very low input capacitances and integrated with the detector are being developed. If low gate currents could be achieved, a further significant reduction in linewidth for the drift detectors we are discussing would be possible.

7.4.12 *Resolution and count rate*

One last caveat must be entered if we are striving for the very best in spectral resolution, and that is that the overall counting rate in the detector must not be too high. This constraint arises for two reasons. Firstly we know that if we are using resistor discharge of the feedback capacitor in our preamplifier, then this resistor R_f must be large to minimize noise, with a consequent large time constant $R_f C_f$ for the preamplifier output pulses. These pulses therefore may 'climb on each other's tails' (as we saw in Chapter 6), and while they will be separated out again in the shaping section of the main amplifier, may in the mean time have occasionally reached the overload level of the preamplifier. This effect will clearly become worse as the count rate increases, producing spectral distortion and a drop in resolution. If we wish to prevent this dependence of resolution on count rate then we can reduce the value of R_f and accept a worse value for the resolution as the price we must pay for its greater constancy. The optoelectronic system we discussed earlier clearly has the additional advantage from the point of view of our present discussion of keeping the preamplifier output within operational limits at high count rates, with the reset mechanism merely operating more often.

Quite apart from the preceding problem, however, we may still have difficulties in the main amplifier due to the very long shaping times (up to some tens of microseconds) which we have seen may be needed for the best linewidth figures. Long-time-constant unipolar pulses, such as the Gaussian pulses we discussed earlier, can produce shifts in the baseline at high counting rates (as we saw in Chapter 6) because of any CR interstage couplings in the main amplifier, with consequent spectral distortion. Baseline restorer circuits can be used to counteract this, but no restorer is perfect, and some may even inject a little of their own noise into the signals. Reduction of shaping time will of course of itself guarantee constancy of the resolution to higher count rates, but again only at the expense of a reduced value for this resolution. For the very best results, therefore, the count rate should be kept low — say to a few thousand counts per second — but it must be added that if count rates even in the hundred thousand counts per second range are needed (say for example in X-ray fluorescence analysis) then modern equipment can, under these conditions, typically produce linewidths for low-energy photons less than a factor of two worse than the optimum. (Reference 12, for example, although it uses a different type of FET reset to the one we described earlier, is worth consulting if details of a high-count-rate gamma spectrometry system are of interest.)

Appendix: Noise calculations for the charge-sensitive preamplifier

While it is possible to predict from very general considerations that the expression for the equivalent noise charge will remain almost unaltered when we change from the voltage-sensitive to the charge-sensitive configuration (but of course with R_f taking the place of R_g) it may be helpful to look more directly at what is happening. In the charge-sensitive mode (Figure 7.6) the roles of R_g and C_{in} in Figure 7.5 would appear to be played by R_f and C_f, respectively. Therefore we might expect the first two terms inside the brackets in the expression for $\overline{v_{total}^2}$ in equation (7.34) to become $(kT/2R_f)$ (τ/C_f^2) and $(eI_D/4)(\tau/C_f^2)$. Note that since R_f and C_f have their left-hand ends connected to a 'virtual earth' (i.e. a place held by feedback action close to ground) and their right-hand ends to the output, these two noise terms now represent noise at the *output*. The third term in the brackets in equation (7.34), although it does not contain any quantities to be changed, specifies the channel noise referred to the *input*. We must now find its equivalent value when referred to the output. We do this as follows. We indicated earlier that channel noise could be represented symbolically by a voltage source between the point X (Figure 7.7) and the input. So let us first see what the effect of introducing a general voltage source v in this position would be at the output of a charge-sensitive loop (Figure 7.12). We assume the impedance of R_f is large with respect to that of C_f, just as we did with R_g and C_{in} in the voltage-sensitive case, and so we do not show R_f in the diagram.

The feedback action (if the gain is large) is again to reduce the voltage at the input almost to zero with respect to ground. So a voltage must appear across C_{in} in the opposite sense to that produced by our voltage source, but of the same numerical size. This voltage is provided by v_{out} and the potential divider action of C_f and C_{in}. In fact $v = v_{out}\,[Z_{in}/(Z_{in} + Z_f)]$ where Z_{in} and Z_f are the impedances of C_{in} and C_f, respectively. Now Z is proportional to $1/C$ for a capacitor, so putting in the appropriate values and solving for v_{out} gives

$$v_{out} = v\left[1 + \frac{C_{in}}{C_f}\right] = v\left(\frac{C_{in} + C_f}{C_f}\right).$$

Returning now to the case in equation (7.34) where we are talking of

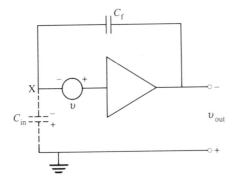

Figure 7.12 Diagram for channel noise calculations.

mean square voltages, we see we must multiply the third term by $[(C_{in} + C_f)/C_f]^2$ to give for the noise at the output of the whole amplification chain the expression

$$\overline{v^2_{total}} = A^2_o \left[\frac{kT\tau}{2R_f C^2_f} + \frac{eI_D\tau}{4C^2_f} + \frac{kT}{2g_m\tau} \left(\frac{C_{in} + C_f}{C_f} \right)^2 \right].$$

Here A_o represents the gain of the main shaping amplifier which follows the feedback loop.

If as usual Q denotes the equivalent noise charge, then the output from the feedback loop is Q/C_f, and after passing through the main amplifier becomes $A_o Q/[C_f \exp(1)]$ for the usual CR–RC shaping. Equating the square of this to the expression just given for the mean square noise we notice that C^2_f cancels out all across (as well of course as A^2_o) to leave us with

$$Q^2 = \text{ENC}^2 = [\exp(2)] \left[\frac{kT}{2R_f}\tau + \frac{eI_D}{4}\tau + \frac{kT(C_{in} + C_f)^2}{2g_m\tau} \right].$$

So our expression for the ENC of a charge-sensitive amplifier is obtained from the analogous one for the voltage-sensitive case by replacing R_g by R_f, but retaining C_{in} (only marginally altered by the addition of C_f). The value of C_f, although determining the gain for the charge-sensitive preamplifier, thus plays no major part in equivalent noise charge considerations.

References

1. Breitenberger, E. (1955). Scintillation spectrometer statistics. *Progress in Nuclear Physics*, **4**, 56–94.

2. Knoll, G. F. (1989). *Radiation detection and measurement*, (2nd edn), pp. 174 ff. Wiley, New York.
3. Goulding, F. S. and Landis, D. A. (1988). Ballistic deficit corrections in semiconductor detector spectrometers. *IEEE Transactions on Nuclear Science*, **35**, 119–24.
4. Nyquist, H. (1928). Thermal agitation of electronic charge in conductors. *Physical Review*, **32**, 110–13.
5. Delaney, C. F. G. (1980). *Electronics for the physicist*, p. 254. Ellis Horwood, Chichester.
6. Ibid., p. 46 and p. 260.
7. Ibid., pp. 257 ff.
8. Ibid., pp. 292 ff.
9. White, G. (1988). Pulse processing for gamma-ray spectrometry: a novel method and its implementation. *IEEE Transactions on Nuclear Science*, **35**, 125–30.
10. Radeka, V., Rehak, P., Rescia, S., Gatti, E., Longoni, A., Samprieto, M., Bertuccio, G., Holl, P., Strüder, L., and Kemmer, J. (1989). Implanted silicon JFET on completely depleted high-resistivity devices. *IEEE Electron Device Letters*, **10**, 91–4.
11. Luke, P. N. (1988). Low noise germanium radial drift detector. *Nuclear Instruments and Methods*, **A271**, 567–70.
12. Simpson, M. L., Becker, T. H., Bingham, R. D., and Trammell, R. C. (1991). An ultra-high-throughput, high-resolution, gamma-ray spectroscopy system. *IEEE Transactions on Nuclear Science*, **38**, 89–96.

Further reading

Buckingham, M. J. (1983). *Noise in electronic devices and systems*. Ellis Horwood, Chichester. A general treatment of noise sources in electronics.

Goulding, F. S. and Landis, D. A. (1982). Signal processing for semiconductor detectors. *IEEE Transactions on Nuclear Science*, **NS-29**, 1125–41. Part of a special tutorial issue containing a short course on semiconductor detectors.

van der Ziel, A. (1970). *Noise sources, characterization, measurement*. Prentice Hall, Englewood Cliffs, NJ. A classic text.

8
Miscellaneous detectors

8.1 Introduction

All the detectors considered so far in this book produce an electrical signal, either directly through the collection of ionization charges or through the conversion of a scintillation of light into a signal with, for example, a photomultiplier. In this final chapter we shall consider some detectors whose basic operation can depend on non-electrical techniques, such as the formation of a visible track in a plastic material or a photographic emulsion, the emission of light in a previously irradiated 'thermoluminescent' material when heated, or the calorimetric measurement of the heat generated as radiation is stopped. Detectors of this type may be described as electrically passive. Some miscellaneous detectors are also considered which rely on a variety of detection methods and which are conveniently grouped together in this chapter. These include Cherenkov, gas scintillation, and liquid ionization detectors.

In general, as throughout this book, we restrict our attention to the detection of radiation of relatively low energy. One or two exceptions are, however, made—for example, we discuss Cherenkov detectors, partly because we have already mentioned the effect in Section 4.4.4, and also because low-energy applications are in fact possible as well as the commoner high-energy ones. We shall consider these detectors first.

8.2 Cherenkov detectors

We have already alluded briefly to the Cherenkov interaction in Chapter 4, where we noted that it was responsible for light production in the glass of photomultiplier tubes. In that case the result was unwanted background counts, but the Cherenkov effect can be used as a method of radiation detection in its own right, and it is this aspect that we are considering briefly here.

The basic principle of this effect (discovered experimentally by P. A. Cherenkov in 1934) can be easily understood from elementary optical considerations. A charged particle passing through a transparent material excites the atoms of the material, and if the velocity v of the particle exceeds

Radiation Detectors

the phase velocity of light in the material (i.e. $v > c/n$, where c is the velocity of light *in vacuo* and n is the refractive index) then the light produced appears as a coherent wave.

This is shown schematically in Figure 8.1. P is the position of the particle (velocity v) at a particular time. At a time t seconds earlier the particle was at the point R, a distance vt to the left of P. The light produced near the point R has spread out in a spherical wavelet of radius $(c/n)t$, as shown. Huygens' construction states that the wavefront is the envelope of all the various wavelets, and as the wavelet at P has not yet developed, the wavefront in this two-dimensional diagram is thus the two tangents (shown as broken lines) to the large circle. It is easy to see that any other wavelet—such as that centred on the point Q—will also touch the broken lines.

The wavefront and the emerging light is usually defined by the angle θ where

$$\cos \theta = \frac{(c/n)t}{vt} = \frac{c}{nv} = \frac{1}{n\beta}$$

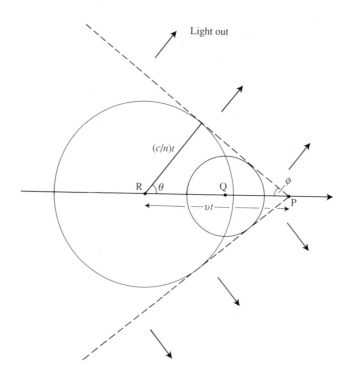

Figure 8.1 The production of Cherenkov radiation.

(where $\beta = v/c$). The wavefront thus forms a cone of half angle

$$\phi = (\pi/2) - \theta$$

in three dimensions.

This is an extraordinarily simple result: the cone angle depends only on the refractive index and not, for example, on the density or atomic number of the material — although these may, of course, be important for other reasons. Note also that our formula indicates a threshold for the production of Cherenkov light. Only if $n\beta > 1$, that is if $v \geqslant c/n$ (where c/n is the velocity of light in the material) does a real value for θ exist. The threshold velocity is thus gives by c/n, and as v is increased from this value θ increases from zero, that is, ϕ decreases from $\pi/2$. So for greater velocities the cone angle shrinks and the light is emitted in a more perpendicular fashion to the particle path. A maximum value of θ (minimum value of ϕ) is reached when $v \approx c$, in which case

$$\theta_{max} = \cos^{-1}(1/n).$$

For an electron in a medium of refractive index 1.5 (a glass or plastic, say), the threshold velocity is $2 \times 10^8 \text{m s}^{-1}$ and the corresponding energy is 175 keV. For protons the threshold energy is 320 MeV. The value of θ_{max} for $n = 1.5$ is 48°. The corresponding threshold values for a gas Cherenkov detector with $n = 1.0003$ (nitrogen at atmospheric pressure, say) are about 20 MeV for electrons and 40 GeV for protons.

(Note that our assumption in the above calculation of a unique value for n in each case is only an approximation. Because Cherenkov light is, as we shall see in more detail in a moment, emitted as a continuous spectrum, and because n is a function of wavelength, the angle θ will not be unique but will have a small spread of values. For a detailed treatment of this problem see reference 1 listed at the end of the chapter.)

The existence of a detection cut-off for particles below a certain velocity makes the Cherenkov detector of obvious value in areas like high-energy particle physics, where attention may be centred on relatively few high-energy particles in the presence of many more particles of lower velocity. Cherenkov detectors also have the advantage of very rapid response. The light output occurs only while the particle is passing through, or (in the case of a solid or liquid detector) until it has lost enough energy to drop below the light emitting threshold. The duration of the light signal will therefore be comparable with, or in many cases very much less than, the response time of the photomultiplier detecting the light.

On the debit side the actual number of photons produced per unit length of path is far less for a Cherenkov detector than for its scintillation counterpart. Theory shows that the number N of photons emitted per unit path length is given approximately by

$$\frac{dN}{dx} \propto \sin^2\theta \int \frac{d\lambda}{\lambda^2}$$

where θ is the usual Cherenkov angle, and λ the wavelength of the radiation emitted. We can write this even more crudely as

$$\frac{dN}{dx} \propto \sin^2\theta \frac{\Delta\lambda}{\lambda_m^2}$$

where $\Delta\lambda$ is the band of wavelengths around a mean value λ_m in which the light is collected. Writing it in this latter form shows how important it is to collect the light well into the ultraviolet, but of course this creates difficulties with the transmission of the radiation through the radiator and any focusing lenses, as well as requiring detecting devices sensitive in the ultraviolet region. In practice for a range of 400–700 nm we can write the number of photons per millimetre numerically as $50\sin^2\theta$ approximately.

For a 1 MeV electron entering Perspex ($n = 1.5$) the Cherenkov angle is approximately 45°, so $\sin^2\theta = 1/2$, and therefore initially photons are being produced at the rate of 25 mm^{-1}. Now the range of such an electron in Perspex is only about 3 mm, so therefore the total number of photons produced as it slows down and θ decreases can hardly be as much as 50. Compare this with the number of photons produced by a 1 MeV electron even in a plastic phosphor: there about 90 eV are required per photon, so we will produce about 1.1×10^4 quanta, a number more than two orders of magnitude greater than the Cherenkov case. With such low Cherenkov yields one must be careful that any radiator used must not confuse the issue by being a feeble scintillator!

A Cherenkov detector can be made very simply by coupling a radiator, either solid or gaseous, to a photomultiplier. Depending on the energy of the particle and the dimensions and nature of the radiator, the particle will either be stopped and detected, or else will pass through the radiator, possibly for further analysis, producing a Cherenkov signal which can be used for a variety of gating purposes. However, with this simple set-up we are not using to the full all the attributes of Cherenkov radiation. Figure 8.2 shows a very elegant device, the ring image detector, used with high-energy particles in a gas radiator (references 2 and 3). The particle, entering from the left, produces Cherenkov radiation as shown. If we assume that the particle loses only a small fraction of its initial energy in the gas radiator the angle θ will remain almost constant. The resulting parallel rays are brought to a focus (in this two-dimensional diagram) at two points P and Q in the focal plane of the lens. In three dimensions of course the result is a ring focus, and hence the detector name.

The ring radius, r, can be seen to be given by $f\tan\theta$, where f is the focal length of the lens, and so its magnitude allows us to determine θ

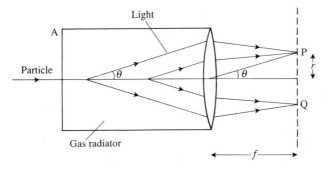

Figure 8.2 A ring image detector of Cherenkov radiation.

and hence the particle velocity. If the particles have, as is usual, already been pre-selected for a particular momentum, then different particles — say a pion and a proton — may be identified by their ring radii. In practice, because of the necessity to collect light as far into the ultraviolet as possible, the lens is replaced by a concave mirror, with the rings being formed just outside the end A, now made of material transparent to ultraviolet radiation.

Obviously from all the preceding discussion, the applications for Cherenkov detectors lie mostly in the high-energy region and are thus largely outside our area of interest in the present book. The Cherenkov effect itself of course is easily observable at lower energies, as evidenced by the very striking blue glow produced, for example, by fast electrons from spent reactor fuel elements in their cooling pool water. There has, however, been at least one example involving the use of Cherenkov radiation in radioisotope work. This involves the detection of tumours in the back region of the eyeball which have been labelled with ^{32}P. The betas from this isotope are totally absorbed in the tissue of the eye, and thus undetectable by normal means; however, the Cherenkov light formed in the transparent media of the eyeball can be observed externally, and this provides a viable alternative detection method (reference 4).

8.3 Special detectors using noble gases

8.3.1 Scintillation detectors

We saw in Chapter 4 that scintillators for radiation detection normally use solid organic or inorganic materials, or else are made in liquid solution form. It is also possible to use certain high-purity gases as scintillators, such as the noble gases helium, argon, krypton, or xenon. Here the incident radiation leaves a track of excited gas atoms, which then radiate their excess energy as photons in the visible or near-visible region. In many cases much of the emission is in the near-ultraviolet rather than the visible region. The photomultiplier or other light-detecting element must therefore be specially designed to respond at those wavelengths. Alternatively some form of wavelength shifter may be added, such as a second gas or a coating of a suitable material on the walls of the container. When struck by ultraviolet radiation the shifter emits visible photons, to which conventional photodetectors can respond.

Other modes of de-excitation compete with light emission, and the light yield obtained for a given energy loss is in fact an order of magnitude or more worse than with NaI(Tl). However, these gases share with liquid scintillators the advantage of easily variable size and shape, and in addition are not flammable. Furthermore, different stopping powers dE/dx can be obtained if the gas pressure is varied. The response can also be very fast, typically in the nanosecond region. Applications include the detection of heavy ions such as fission fragments with gases at high pressures.

8.3.2 *Gas proportional scintillation detectors*

A variant of the last type of detector is the gas proportional scintillation detector (reference 5). In this a proportional counter structure incorporates a noble gas such as xenon (Section 3.3.3), which also acts as a scintillant. Incident radiation produces a 'prompt' scintillation response in the way just described, which can be used for fast-timing measurements. The electrons also produced then drift towards the high-field region around the anode wire, where they themselves excite the gas and cause further scintillations. Since the excitation potential is less than the ionization potential, conditions can be such that a large amount of light can be produced with little or no charge multiplication. This 'secondary' response is on a much slower time-scale of microseconds, and can be used for energy measurements. In practice the number of photons produced is such that the contribution of the photomultiplier to the energy resolution is much smaller than the contribution of the large amount of charge multiplication to the resolution in an ordinary proportional counter. For example, the energy resolution of 5.9 keV X-rays can be nearly twice as good as with conventional proportional counters, and this is one of their main advantages. Continuous gas flow is, however, often needed because the gases must be very pure; any impurities tend to degrade the performance of these detectors. Because of this and their general complex nature their use has not become widespread.

8.3.3 *Ionization chambers and proportional counters*

Some of the noble gases, when cooled to form liquids, have also been observed to act as materials for ionization chambers and proportional counters. Of these xenon has the highest atomic number (54), and it therefore has particular potential as a gamma-ray detector as efficient as NaI(Tl) but with better energy resolution. In general one of the main problems is that electronegative impurities such as oxygen can trap the electrons produced by ionization, and purity levels of parts per million or better are necessary to ensure that the electrons can drift through distances long enough to ensure reasonably large detector volumes. Another experimental difficulty is, of course, that cryogenic techniques are necessary. Incidentally, a few organic liquids in which high enough purity levels can be attained have also been investigated for these purposes. Other possibilities which have been studied include the use of solidified noble gases in ionization chambers and the use of liquid helium and liquid argon as scintillators.

8.4 Detectors with event storage properties

8.4.1 *Introduction*

In all the detector systems we have discussed up to now, the detection and recording systems are quite separate. By this we mean that the detector itself produces an electrical or optical pulse of a transient nature, but before it disappears it is electronically recognized and subsequently stored as an event in the memory of a scaler or MCA.

The five detectors with which this section is concerned — nuclear emulsions, solid-state nuclear track detectors (SSNTDs), activation foils, thermoluminescent detectors (TLDs), and superheated drop detectors — allow events in the detectors to be stored there in a latent form either permanently or at least for a period of time, and thus to be read out some time after the incidence of the radiation. Except for TLDs, the data can be read out more than once. Chemical processing is required in the case of the first two detectors before the information can be read out, whereas the last three detectors need no such processing, and can in fact be used more than once.

We now proceed to discuss briefly each of these detector types.

8.4.2 *Nuclear emulsions*

Emulsion methods can claim to be the oldest of nuclear detection techniques, inasmuch as it was the effect on photographic plates by uranium salts left in their vicinity which led Henri Becquerel to the discovery of radioactivity in 1896. Ionizing radiations have the same effect on photographic emulsions as does ordinary light — a sensitization of the silver halide grains in the emulsion, so that subsequent chemical development reduces them to metallic silver. Fine-grain films are of course used routinely in medical X-ray work and in film badges in health physics. In the latter application a small piece of film (in a suitable light-tight holder) is worn for a fixed period (usually a month) by a person working with ionizing radiations. The density of the film blackening (after development) is a measure of the integrated dose received, and as such can serve as an essential check on dose limits.

However, in employing as parameter the overall blackening we are using the method in a very crude way. Much more information could be obtained if we were able to see (with a microscope of appropriate magnification) individual particle tracks in the emulsion — and indeed we can, with specially developed very fine-grain nuclear emulsions available with thicknesses up to about 500 μm. When first introduced these plates were sensitive only to heavily ionizing particles like fission fragments or alphas, but emulsions are now able to record fast electrons. In order to give some numerical feel to the situation one might note that the range of a 2 MeV proton in emulsion would be about 40 μm, which of course from our discussion in Chapter 2 is also what we would expect for an 8 MeV alpha particle.

Nuclear emulsions have had their widest and most spectacular applications in making visible tracks and events involving high-energy particles produced both by cosmic rays and accelerators. For those who have not seen the beautiful micropictures of the 'stars' formed in the emulsion by the break-up of nuclei struck by high-energy particles, references 6 and 7 are well worth looking at. Measurements on such track parameters as range, grain density, and scattering allow a quantitative determination of the energy and charge of the particle producing the track.

At the lower energies with which the present book is concerned, applications of nuclear emulsions tend mostly to take advantage of their 'integrating' property to facilitate measurements of feeble amounts of radioactivity. For example, to examine a rock specimen suspected of containing small amounts of naturally alpha-radioactive inclusions, one would simply place a previously cut and polished face close to the emulsion of a nuclear plate. Depending on the estimate of the activity of the specimen the plate could be left for a period of up to weeks before development. A microscopic examination would then reveal not merely the locations where the activity occurred, but, by an examination of the ranges of the alpha-particle tracks, information on the relative amounts of thorium and uranium present. In the case in question a visual examination of the plate area would not be too time consuming, but in more complicated applications, particularly in high-energy physics where rare events are being looked for, automated track searches using optical scanning are essential.

8.4.3 *Solid-state nuclear track detectors*

In the late 1950s and early 1960s it became clear that the trail of damage caused by the passage of a heavily ionizing particle through a variety of dielectric materials (including glasses and many common transparent plastics) could be rendered visible under ordinary microscopic observation by treatment of the irradiated surface with a suitable etch — usually sodium or potassium hydroxide for the plastics. The etch attacks the sites where the trail damage exists at a much faster rate than the bulk plastic, producing a cone-shaped pit, whose tip eventually rounds off when the etch reaches the end of the track.

The analogy with the sensitization and subsequent development of an emulsion is clear. However, there are obvious differences between the two techniques which make them complementary rather than competitive. The fact that tracks in solid-state nuclear track detectors (SSNTDs) must be etched from the surface rather than developed in the bulk means that they are best suited to the registration of single-particle tracks rather than the complex events often recorded in emulsions. On the other hand track detector materials are cheap, easy to handle, and insensitive to light, and can thus be employed in situations unsuitable for emulsions.

As with nuclear emulsions, heavily ionizing particles such as fission fragments are easiest to detect with SSNTDs. Some commercially available plastics are also sensitive to alpha particles, while specially developed materials can register proton tracks. (This latter facility of course allows us also to use SSNTDs as fast neutron detectors because of the recoil protons produced.) Electrons do not produce etchable tracks. Particle identification, when required, can be made with the help of data from the track etching rate, which is related to the rate of energy loss of the particle. For more information on this and also on the many ingenious methods developed for automatic track counting reference 8 may be consulted.

Among the great variety of applications of nuclear track recording detectors we will mention just three from very different fields — the study of heavy nuclei occurring in cosmic rays, the investigation of trace quantities of uranium present in stainless steel, and the determination of the amounts of radon gas in domestic dwellings. One more in a rather different category is worth adding. Some mineral crystals can act as natural track detectors, and if they contain small quantities of uranium, etchable fission fragment tracks from the very occasional spontaneous fission events in uranium-238 will be present. (The much commoner alphas will not be recorded.) If one determines the density of these tracks in the specimen and combines this with a measurement of the current uranium content then a figure for the age of the mineral is obtainable. (The current uranium content can, for example, be assayed by determining the increase in track density when the crystal is irradiated with neutrons of known number and energy distribution, causing induced fission in the sample.)

8.4.4 *Activation foils*

Neutron activation, in which very small amounts of materials can be identified by the characteristic radiation they emit after being bombarded with neutrons, is a powerful method of trace analysis. (A specialized example of this has just been given at the end of Section 8.4.3 in the assaying of uranium in certain minerals.) In the present sub-section we 'stand the method on its head' in using it to determine fluxes of neutrons (or other particles).

The principle is extremely simple. A foil thin enough not to appreciably perturb the flux to be measured, and made of a material with a nuclear cross-section large enough to produce appreciable activation, is placed (in the most common application) inside the nuclear reactor where the neutron flux is to be determined. Activation takes place usually through (n, γ) reactions, which produce nuclei of a beta- or gamma-emitting isotope of the target element. After a suitable period has elapsed, the foil is removed from the reactor, and the resultant count rate, as determined with a conventional detector system, is a measure of the reactor neutron flux.

If only relative measurements are required — say, a comparison of the flux in different parts of the reactor — then no further information is needed. However, if absolute flux measurements are in question then we must know the target cross-section, the foil dimensions, the efficiency of our detection system and the half-life of the isotope produced (as the radioactive nuclei formed are subject to decay even while more are being produced). Alternatively the system can be calibrated empirically if a known neutron flux is available for a preliminary experiment. Silver, indium, and gold are among the many materials suitable for use as activation foils. The activation method is useful where space considerations, or the inability to handle the large fluxes involved, would preclude the use of more conventional direct reading detectors.

8.4.5 *Thermoluminescent dosemeters*

We saw in Section 8.4.2 that the blackening by radiation of a nuclear emulsion can be used for the measurement of the integrated dose received. At the end of each period for which the dose is to be measured a new photographic film is issued and the old film has to be developed. Thermoluminescent dosemeters, or TLDs, use materials in which light emission occurs when they are heated after being subject to irradiation (normally with gamma rays). Such dosemeters are gradually becoming increasingly popular as an alternative to film badges. No photographic developing of film is required, and in addition we shall find that these dosemeters can be reused many times.

When thermoluminescent materials are irradiated electrons and holes are initially produced by the ionization process in the usual way, that is, by the promotion of electrons across the energy gap from the valence to the conduction band. If the material were, say, silicon or germanium, the carriers would (if not already collected by an applied electric field) recombine with release of energy after a short time (milliseconds or less); furthermore, if the material were instead an ordinary scintillator, such recombination would be accompanied by emission of visible light. Unlike these cases, however, TLD materials have a high concentration of trapping centres within the band gap. The electrons and holes produced are therefore captured by them, but, unlike the case of a scintillator such as NaI(Tl) in Chapter 4, the decay time is very long. Consequently energy is stored, the amount of which is dependent on the quantity of radiation received.

In a suitable TLD material this stored energy (or some fraction of it) can be released by heating it through a few hundred degrees to enable the trapped electrons or holes to escape by thermal activation. This is then followed by recombination of the released carriers, with (as in a scintillator) accompanying emission of light. A photomultiplier viewing the material measures the amount of light released. Provided that the material is heated in a controlled and reproducible manner to a fixed temperature, the total amount of light emission during heating should in practice bear a fixed relationship to the energy initially deposited, that is, the radiation dose received. Calibration of the light output measured against dose received can be made by irradiation of the TLD material with known doses of radiation.

Lithium fluoride, LiF, is one of the most popular materials for TLDs. The necessary trapping centres occur naturally because of the inherent impurities and defects in the crystal. 'Fading', that is, unwanted slow recombination of the trapped carriers at room temperature, is small, unlike in some other materials. This is because the traps do not lie too close to the edges of the electron energy band gap, that is, their ionization energies are not too low. This feature, however, can be shown to decrease the sensitivity of the material, that is, the probability of trapping a carrier. The material is, however, still sufficiently sensitive to be able to measure doses of magnitudes commonly encountered in ordinary work in radiological protection, while also having a wide enough range to enable measurement of large doses such as might occur in an accident situation.

In addition, the average atomic number of LiF is low and not dissimilar to that of body tissue. The significance of this is that the energy deposited by a particular gamma ray changes with absorber atomic number. In order therefore to estimate the dose to body tissue the dosemeter should ideally have an absorption coefficient which varies in a similar way with energy as for tissue. Dosemeters of higher atomic number, however, absorb low-energy gamma rays particularly well, because of the strong increase of photoelectric absorption with atomic number, as we saw in Chapter 2. Allowance must therefore be specially made with such materials for this effect in tissue dose estimations. However, materials of higher atomic number like $CaSO_4$ or CaF_2 (activated with Mn) can be useful. In particular the former can record lower doses than can LiF because the traps in it are shallow, that is, they lie close to the edges of the energy band gap, but it is also therefore susceptible to considerable fading.

LiF TLDs also have a small sensitivity to slow neutrons because of the (n,α) reaction on the naturally occurring isotope ^6Li, as we mentioned on a number of occasions previously. ^6Li is present naturally to about 7 per cent abundance, and this sensitivity can be enhanced or suppressed, if required, by enriching or depleting the lithium in this isotope, although these are costly processes.

TLDs can if desired be mounted in plastic holders somewhat like those used for film dosemeters, or alternatively they can be inserted into thin plastic sachets for wearing over the finger or taping to, say, the forehead. They can be reused many times by annealing out the trapped carriers at a high enough temperature. Dose read-out is a process which can be readily automated and linked to computer-based recording systems. On the other hand any particular dose recorded can be read out only once, since the process destroys the information, whereas a film dosemeter, after being developed, can, of course, be read out as often as desired: neither can TLDs give as much information about the nature of the radiation received as a film can, because with the latter a filter may be used on part of the film to differentiate, for example, between beta and gamma radiations.

It is interesting to quote briefly another rather different application of thermoluminescence, namely, its use in the archaeological dating of pottery. In this the pottery itself is the detector of radiation, though not a very efficient one. The charge carriers trapped in the material of the pottery are produced by naturally occurring radioisotopes such as ^{238}U, ^{235}U, ^{232}Th, and ^{40}K. They can be made to recombine with emission of light by, as before, heating a specimen to a moderate temperature. Starting from no trapped carriers when the pottery was initially fired at high temperatures in a kiln, the number of carriers trapped builds up steadily with the passage of time. Heating a specimen therefore enables an estimate of its age to be made if the dose rate from the radioactivity is found. Again, calibration of the light output against dose received can be made by irradiation of a specimen with a known dose.

8.4.6 *Superheated drop detectors*

A liquid can exist in a metastable, superheated state at a temperature higher than its boiling point without vaporizing. When ionizing radiation passes through certain superheated liquids, vapour bubbles can be formed as a result. In the superheated drop detector the superheated liquid is dispersed in small droplets through a clear holding medium, such as a gel or polymer, which is contained in a plastic vial. Ionization deposited in such a drop can lead to the formation from it of a vapour bubble; thus the incidence of the particle can be detected. By subdividing the liquid into drops we avoid the whole sample being 'used up' by a single ionizing particle. Reference 9, which covers many aspects of these detectors, discusses the factors influencing the vaporization of such drops. These include not only how much ionization is produced and where, but also the thermodynamics of the processes resulting in a microscopic vapour bubble.

Such detectors, if contained in thin-walled vessels, could be used with charged particles, but their most important application has in fact been to neutron detection. In this case fast neutrons can create ionization from the elastic recoil nuclei produced in the drops. Ionization can also be derived from charged particles produced when a neutron induces a nuclear reaction of sufficiently high Q value. In this way thermal (and other slow) neutrons can also be detected. (Compare this with the discussion in Section 5.10.)

Various types of the refrigerant 'freon', for example, have been used for the superheated liquid. At least one of these types is sensitive to thermal (as well as fast) neutrons. This sensitivity is thought to be due to the following reaction undergone by the $^{35}_{17}\text{Cl}$ nuclei contained in the freon:

$$^{35}_{17}\text{Cl} + \text{n} \rightarrow {}^{35}_{16}\text{S} + \text{p}.$$

This reaction has a Q value of 615 keV, which of course, is easily large enough to lead even with thermal neutrons to the creation of ionization by the recoiling products.

If excess pressure is externally applied after the vapour bubbles have been produced, the bubbles can be returned to the liquid phase. Subsequent removal of this excess pressure can make the liquid superheated once again. Reusable superheated drop detectors (called 'neutron bubble dosimeters') are now available (reference 10). Both the superheated drops and the vapour bubbles generated exist for relatively long periods of time. The number of bubbles formed is determined optically, this number being proportional to the neutron dose received. These detectors are insensitive to gamma radiation, which, as we know, often accompanies neutrons in practical situations.

We note in passing that the commonest use of bubble formation by radiation in superheated liquids has been for particle-track visualization in high-energy physics with bubble chambers. The homogeneous liquid is super-heated by the pressure being suddenly dropped below the vapour pressure of the liquid. The trails of bubbles formed in the wake of particles are photographed before the bubbles disappear. The particle energies normally associated with such detectors fall outside the low-energy region covered by this book, but nevertheless reference 7, with its photographs of bubble chamber tracks, is once again well worth looking at.

8.5 Calorimetric detectors

8.5.1 *Introduction*

We finally discuss a radically different type of detector which has been developed in the last few years, one which despite experimental complexities could in principle revolutionize the subject of radiation spectrometry. A calorimeter, of which a calorimetric detector is a specialized example, may in general be regarded as a device for measuring the amount of energy evolved by some means within itself. For example, calorimetry as encountered in the school physics laboratory involves a copper container, appropriate insulation, and a thermometer. This enables measurements to be made of, say, the latent heat released when steam is condensed in water contained in the calorimeter.

In radiation detection the term 'calorimeter' has come to be used for two very different types of detectors which measure the energy of incident radiation. In high-energy physics the term has come to mean a device which enables an assessment to be made of the energy of an incident particle and the ways in which this energy is deposited in the detector. When this energy is, say, many GeV, this task can be complex because of the electromagnetic radiation and the showers of particles produced in the slowing down process, of penetrating nature. Large sandwiches of alternating absorbers and radiation detectors can be used to analyse the processes involved. Somewhat similar complexities can apply at low energies too; we saw in Chapters 4 and 5 that the measurement of gamma-ray energies in a single scintillator or semiconductor detector can require a detailed knowledge of the detection mechanisms involved, and in particular the ways that energy can escape from the detector volume without being detected. However, in all these contexts there is no question of thermometry being used to evaluate the energy evolved as heat, as might at first be supposed by the use of the term 'calorimeter'.

In fact we will not discuss this type of calorimetry here in any further detail, largely because its primary application is to particle energies outside our region of interest. The question can still, however, be asked as to whether the heat evolved and temperature rise produced by nuclear radiation being stopped in an absorber may be used to measure the energy deposited. This would imply a different interpretation of the word 'calorimeter', and one that is close to the original meaning of the word.

The answer to the question is 'yes'. In the earliest days of research into radioactivity it was observed that materials containing radium were warmer than their surroundings because of the radiations stopped within the source. Indeed an extension of 'school-type' calorimetry can be used to measure the heat evolved and hence, for example, the energy released by each disintegration (provided that the half-life and amount of the radioisotope present is known). This is similar to the way in which fluxes of infra-red radiation are measured with bolometers. With nuclear radiation the amount of heat in macroscopic terms is, however, small unless relatively intense sources or fluxes are used.

8.5.2 *Single-particle calorimetry*

In the 1980s much more sensitive calorimetry was developed with small dielectric samples maintained at very low temperatures (although the idea of using low temperature itself was in fact much older). As a result it is now possible to measure the heat and temperature rise produced by a single particle passing through or stopping in a piece of material. In principle this can lead to energy measurements which are fundamentally more accurate than those with conventional detectors which electrically measure the amount of ionization produced. It is this aspect of 'nuclear radiation calorimetry' which represents in many ways a completely new approach to single particle detection, and which we shall now pursue in this sub-section. For more detailed information see, for example, references 11–14.

The operational principle of such a calorimeter (or bolometer, to use the alternative term) is in principle quite simple. Figure 8.3 shows schematically the type of experimental set-up used. In it the absorber is irradiated with photons, such as X-rays, of energy E, each of which produces a small temperature rise ΔT. By making either the mass m of the absorber or its specific heat s small (or both), we can make the absorber heat capacity $C = m \times s$ low enough for this temperature rise $\Delta T = E/C$ to become detectable by a small thermistor in thermal contact with the absorber. The specific heat can be made low by cooling the sample to very low temperatures T far below the Debye temperature. In this region C falls strongly as T falls. For a pure dielectric $C \propto T^3$ (as opposed to C being constant above the Debye temperature). so $\Delta T \propto T^{-3}$. Debye temperatures of order 10^2 K are typical, depending on the material used, so this is one reason (another will become clear in a moment) why temperatures in the region of hundreds

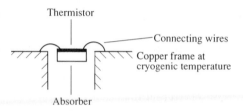

Figure 8.3 Schematic view of a bolometer. (Reference 14, Figure 1.)

of millikelvins or less are used. The detector temperature is measured by applying a steady bias voltage to a series combination of the thermistor and a load resistor. Small variations in the voltage across the thermistor can be measured with a low-noise amplifier with a FET first stage mounted near the detector and operated near liquid-nitrogen temperatures.

Compared with conventional detectors, this calorimetric technique has several advantages which make further investigations into the technique worthwhile despite the very low temperatures which are necessary. One of these is the very good energy resolution possible. With respect to conventional techniques, the resolution achievable is, as we saw in Chapter 7, limited by the statistical fluctuations in the number of the carriers produced (which in turn is associated with availability of alternative routes for energy disposal) and by noise associated with the measurement electronics. In the ideal calorimeter all the incoming energy quickly appears as heat (even if a part is via an initial ionization stage), and the basic source of linewidth is the intrinsic background fluctuations in the number of phonons present in the detector. This is a fundamental effect which follows from the statistical mechanics of the phonon 'gas' in the detector: we shall see below that it can lead in principle to an RMS fluctuation of the order of only 1 eV, independent of the energy of the incoming radiation. Compared with an ordinary semiconductor detector the resolution has thus improved by an even larger factor than when we went from the scintillation counter to the semiconductor detector.

We can estimate the FWHM energy noise due to these fluctuations in a plausible way by first noting that the number of phonons in an absorber is given by its heat content divided by kT. The heat content is CT, and so the phonon number is C/k, with RMS statistical fluctuations of $(C/k)^{1/2}$. The corresponding fluctuations in energy are therefore given by $kT(C/k)^{1/2} = (kCT^2)^{1/2}$. Because the temperature T is well below the Debye temperature, the detector heat capacity C varies as mT^3, m being the detector mass. Hence for small masses and low temperatures this noise is proportional to $T^{5/2}$ and becomes very small. Note that the noise is independent of any other details of the detecting system and, as we stated earlier, of the amount of energy deposited.

There are, of course, other noise contributions which have to be also taken into account. These come primarily from the Johnson noise in the thermistor (given by $\overline{v^2} = 4kTR\,df$, as in equation (7.24)) and from the preamplifier. As regards the first of these, it has been shown that one can, by a suitable choice of the current in the thermistor (and thus its actual working temperature) and by using an appropriate amplifier bandwidth, make the overall fluctuations about $2(kCT^2)^{1/2}$, i.e. with approximately equal contributions from the Johnson noise and the phonon noise. The second contribution, that from the preamplifier, cannot be calculated directly from the results in Chapter 7, because now the optimum bandwidth is quite low and thus the dominant role previously played by the stray capacitance has disappeared. However, with an appropriate choice of the thermistor resistance, it turns out that the preamplifier noise will produce only a small increase in the overall detector noise. This is a rather surprising result at first sight, especially considering that the detector is at a fraction of 1 K while the input FET is near 120 K. However, one must bear in mind the relatively large voltage signals (about 0.5 mV for a 6 keV X-ray) already available from small calorimetric detectors. This should be compared with the equivalent signals from conventional semiconductor detectors, which at best might be an order of magnitude smaller.

We must now see to what extent these predictions for detector resolution have so far been realized. Detector masses and temperatures are limited to the 1 mg and 100 mK regions, respectively, or below, if the noise is to be at all small. In one example published in 1988 (reference 12) a calorimeter had a mass of just over 0.1 mg, and a heat capacity of about $5 \times 10^{-14}\,\mathrm{JK}^{-1}$ at around 100 mK. The calculated RMS fluctuation, $2(kCT^2)^{1/2}$, comes to about 1 eV, and hence the corresponding FWHM to just over 2 eV. Figure 8.4(a) shows an X-ray spectrum from this detector, in which the FWHM resolution, about 17 eV, is already within an order of magnitude of this fundamental limit. In this case excess low-frequency noise and problems with the sensitivity of the thermistor also contributed to the FWHM width. We can see, however, that even 17 eV is nearly an order of magnitude better than the best obtainable with a Si(Li) detector. Figure 8.4(b) shows that with this resolution it is now necessary to include the two components 11 eV apart (labelled $K_{\alpha 1}$ and $K_{\alpha 2}$) which can be shown to make up the K_α 'line', in order to obtain the 'best fit FWHM' of 17 eV. The amplified response to a single 5.9 keV X-ray is also shown in Figure 8.4(a). The absence of any visible fluctuations in the signal (which goes with the excellent energy resolution obtained) and indeed the pulse size before amplification (≈ 0.6 mV) are remarkable achievements for such a small heat input.

The 17 eV FWHM mentioned above represents a twentyfold improvement

(a)

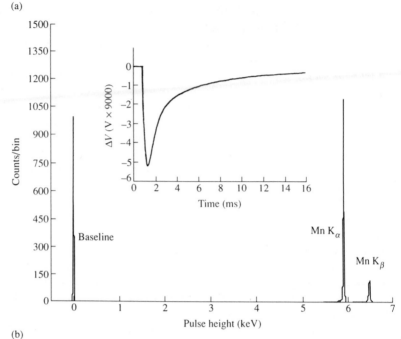

(b)

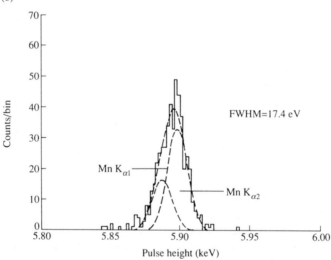

Figure 8.4 (a) The pulse-height spectrum of thermal pulses from a calorimeter used with a ^{55}Fe X-ray source. Inset shows the unfiltered response to a single X-ray. The peak marked 'baseline' is a measure of the system noise. (b) Fine structure in the K_α line. (Adapted from reference 12, figure 2, © 1988 IEEE.)

in resolution over a four-year period – in fact by 1990 this value was reduced still further to 6 eV for a 10 μg device. If resolutions in the eV region could be reached we would then approach that of crystal diffraction instruments but, of course, with none of the disadvantages involved in scanning data one wavelength or energy at a time (as often happens in crystal instruments), as opposed to collecting data at all energies simultaneously.

Masses of below 1 mg are very small compared with that of even a modest surface barrier detector, which is typically at least 100 mg, and are certainly small compared with those of conventional germanium gamma-ray detectors, which can easily reach hundreds of grams. Detection of radiation which is more penetrating than X-rays may not therefore be achievable by calorimeters with low mass, particularly if the atomic number of the detector material is low. However, calorimeters of masses up to 25 g and able to detect gamma rays efficiently have been produced. These have inferior energy resolution (> 10 keV FWHM), although they have other important attributes to be mentioned later (reference 14).

We now consider the question of the materials used in a calorimeter. In order to realize the full potential of the technique when used for high-resolution energy measurements, it is important that all the energy initially deposited in the detector material should appear as heat. If energy is lost by, say, photon emission from the detector, then the statistical fluctuations in this undetected energy would degrade the resolution achievable. Furthermore, the deposited energy should manifest itself as heat within the thermal time constant of the system. Silicon and germanium, for example, have excellent thermal and mechanical properties, but on the other hand at these low temperatures many of the electron–hole pairs generated by incident ionizing radiation can become trapped for quite long times at impurity sites. The fluctuations in these trapping times would then cause a deterioration in the spread of pulse heights. Use of a small-band-gap material such as HgCdTe reduces this problem, although its larger specific heat increases the detector heat capacity and reduces the temperature pulse height obtainable. (The use of metals would eliminate trapping entirely, but in this case the contribution from the electrons to the total specific heat would make the detector heat capacity too high.) The results shown in Figure 8.4 were obtained with a composite calorimeter consisting of HgCdTe mounted on a silicon substrate on which the thermistor was formed. In large calorimeters any problems such as charge trapping are less critical because of the inherently inferior resolution, so a wider range of materials for the absorber, such as germanium, can be employed.

There is no fundamental difficulty in reaching 100 mK temperatures or even lower with a ^{3}He cryostat, provided that the expertise in cryogenic techniques is available, and this, of course, is a crucial limitation. Even lower temperatures (around 10 mK) are often used with the larger detectors in order to mitigate their inferior resolution. One problem with such temperatures is that the various thermal systems in the device (energy absorber/glue/thermistor lattice/thermistor electrons) become thermally more decoupled. This decoupling can lead to very long pulse rise-times, with a consequent reduction of the signal amplitude produced.

Some of the applications envisaged which could use the high resolution and sensitivity of calorimeters include the spectrometry of both laboratory and astronomical X-rays, measurements of beta decay end-points for determinations of neutrino mass, and searches for neutrinoless double beta decay. In addition a unique feature of calorimeters is their sensitivity to non-ionizing as well as ionizing radiation. For example, slow monopoles or weakly interacting 'dark matter' particles (perhaps constituting the 'missing mass' of the universe) could be sought using large calorimeters.

Another case of radiation which ionizes very inefficiently, if at all, is a low-energy heavy ion. We came across this in Chapter 5 when we discussed the pulse-height defect in the response of a conventional detector to such ions. It is significant therefore to note that alpha-particle spectra from a radioisotope source placed close to a calorimeter can show, as well as an expected alpha peak, an accompanying satellite peak about 2 per cent higher in energy. This satellite seems to arise because some of the parent nuclei can be sputtered by the alpha particles out of the source and implanted into the detector. We know from Section 1.5.1 that the daughter nucleus of any alpha decay itself recoils with a small amount of kinetic energy as determined from momentum conservation. For parent nuclei implanted into the detector the daughter as well as the alpha is stopped within the calorimeter, and so in this case the *total* energy released is measured. With conventional ionization detectors the recoil daughter nucleus, being slow and heavy, would create little ionization and so a satellite peak such as this would not appear. The measurement of low-energy recoils such as these or from some types of nuclear reactions are of interest in several areas of study. In the case of heavy ions of higher energy, it would presumably be possible to eliminate the pulse-height defect and the resolution broadening due to the end-of-range atomic collisions described in Chapter 5.

8.5.3 *Calorimetric ionization detectors*

In a conventional ionization detector (gas or semiconductor), particles are detected by the charge produced and collected. However, we ignore entirely the energy Ve involved (where V is the applied voltage) as each ion moves across the chamber, and which is dissipated as heat in collisions with the atoms of the detector material. If we envisage a hybrid detector in which ionization is produced as in a normal semiconductor detector, but measure calorimetrically the total heat evolved, exactly as in our previous calorimetric detector (and at a similar low temperature), then we have a 'magnification' by a factor $(Ve + w)/w$ of the heat produced, where w is, as usual, the energy to produce an ion pair in the material, and which reappears as heat on recombination (reference 15).

This factor $[1 + (Ve/w)]$, which is about 1000 for $V = 3000$ V and $w = 3$ eV, say, means that the thermal noise ($\approx 2(kCT^2)^{1/2}$), and any amplifier contribution, are effectively reduced by the same amount. More pratically, we should in principle have the same signal-to-noise ratio as with a conventional 'passive' calorimetric detector even if the mass increases substantially. In the example used this mass increase could be anything up to a factor of 10^6, as the detector heat capacity C, which is proportional to the mass, appears under the square root. Large mass naturally implies higher efficiency for radiation detection.

The price we have to pay, however, is that normal considerations of ionization statistics now limit the overall resolution of a peak in an energy spectrum. This means that we now have a detector which, while not comparable in resolution with a small passive calorimeter, is superior to it in size and efficiency, and is in addition noticeably better in resolution than the conventional semiconductor detector, where both ionization statistics and preamplifier noise must be taken into account. A pulse-height spectrum will now show an FWHM spread corresponding to the ionization statistics, i.e. $2.355 (FEw)^{1/2}$, as in equation (7.23). However, the other noise source could hopefully be small enough and the basic resolution good enough so that two successive pulses making up the spectrum with numbers of charge pairs differing only by one (i.e. 3 eV difference in energy) could be seen as distinct when stored, say, on an oscilloscope. Of course, this is in the future; work on ionization calorimetry is at an early stage, and observed overall detector resolutions are still worse than with conventional semiconductor detectors even down at milligram levels of detector material.

A more recent experiment uses a combination of ionization and phonon production in a rather different way. The ultimate aim here is to produce a large detector for dark matter research (mentioned previously) which will also have the ability to distinguish between genuine events (producing nuclear recoils) and background events (producing electron recoils). The ideal would be to determine, using a single detector, the ionization and phonon production independently. In that way we could distinguish between events producing phonons but little or no ionization (slow nuclear recoils) and others producing both (electron recoils).

Separate ionization and phonon outputs from a single detector would be thus employed. In practice, however, the phonon output, detected as usual by a thermistor, would have energy contributions both from phonon and ionization processes, that is, as in the previous case, $Ve + w$ per particle. Here, however, if we use high collecting voltages the value of w will be completely swamped by the much larger quantity Ve, and with it the phonon signal. So we must use extremely low values of $V(< 1\,V)$ to prevent this.

Even so, for a 60 keV recoil electron yielding, say, 20 000 electron–hole pairs and about 40 keV energy in phonons (with the usual 2:1 division between the lattice and ionization) a collecting voltage of 1 V for the electron–hole pairs would add 20 keV to the phonon signal. The device to be described below therefore uses 0.4 V bias typically.

The detector itself consists of a 60 g germanium absorber, to which is attached a germanium thermistor. A phonon signal is taken off the thermistor via a voltage-sensitive preamplifier, and an ionization signal from the germanium absorber (used as a low-leakage conduction counter at 20 mK and the low operating bias mentioned above) via a charge-sensitive preamplifier. The time-scales of the ionization and phonon signals turn out to be very different, the former on a microsecond scale and the latter on a tens of milliseconds scale. An account of this work together with preliminary results will be found in reference 16.

In conclusion we can say that calorimetry in general seems set to make some remarkable advances in the field of radiation spectrometry — as indeed may some other exotic methods such as superconducting detectors, for which the interested reader may consult references 17 and 18.

References

1. Frank I.M. (1986). On some peculiarities of Vavilov–Cherenkov radiation. *Nuclear Instruments and Methods*, **A248**, 7–12. This and the following three references appear as part of the published proceedings of a seminar on Cherenkov detectors. Some 40 contributions are included in the proceedings, one from Cherenkov himself.
2. Althoff, K.H. (1986). Twenty-five years experience with Cherenkov detectors in Bonn. Ibid., 39–52.
3. Apsimon, R.J., Flower, P.S., Freeston, K., Hallewell, G.D., Morris, J.A.G., Morris, J.V., Paterson, C.N., Sharp, P.H., Uden, C.N., Davenport, M., Eades, J., Newton, D., Coyle, P., Mercer, D., Danaher, S., McClatchey, R.H., and Thompson, L. (1986). A ring imaging Cherenkov detector. Ibid., 76–85. Includes photographs of the actual detector.
4. Moshnikov, O.S. and Kolesnichenko, V.N. (1986). Cherenkov radiation as a means of radioisotope diagnosis of eyeball tumours. Ibid., 252–3.
5. Policarpo, A.J.P.L., Alves, M.A.F., dos Santos, M.C.M., and Carvalho, M.J.T. (1972). Improved resolution for low energies with gas proportional scintillation counters. *Nuclear Instruments and Methods*, **102**, 337–48.
6. Powell, C.F. and Occhialini, G.P.S. (1947). *Nuclear physics in photographs*. Clarendon Press, Oxford. Shows tracks of charged particles in photographic emulsions.

7. Close, F., Marten, M., and Sutton, C. (1987). *The particle explosion*. Oxford University Press, New York. A beautifully illustrated popular-style account. Shows particle tracks in emulsions and other detectors.
8. Durrani, S.A. and Bull, R.K. (1987). *Solid state nuclear track detection*. Pergamon, Oxford.
9. Roy, S.C., Apfel, R.E., and Lo, Y.-C. (1987). Superheated drop detector: a potential tool in neutron research. *Nuclear Instruments and Methods*, **A255**, 199–206.
10. Riel, G., Rao, N., Kerschner, H., and Nelson, M. (1991). Superheated drop, "bubble", neutron dosimeter performance in a work environment. *IEEE Transactions on Nuclear Science*, **38**, 494–6.
11. Moseley, S.H., Mather, J.C., and McCammon, D. (1984). Thermal detectors as X-ray spectrometers. *Journal of Applied Physics*, **56**, 1257–62. Discusses the principles of a single-photon calorimeter.
12. Moseley, S.H., Kelley, R.L., Schoelkopf, R.J., Szymkowiak, A.E., McCammon, D., and Zhang, J. (1988). Advances toward high spectral resolution quantum X-ray calorimetry. *IEEE Transactions on Nuclear Science*, **35**, 59–64.
13. Wang, N., Sadoulet, B., Shutt, T., Beeman, J., Haller, E.E., Lange, A., Park, I., Ross, R., Stanton, C., and Steiner, H. (1988). A 20 mK temperature sensor. Ibid., 55–8.
14. Alessandrello, A., Brofferio, C., Camin, D.V., Cremonesi, O., Fiorini, E., Giuliani, A., Pessina, G., and Previtali, E. (1990). Bolometers. *Nuclear Instruments and Methods*, **A289**, 504–11. Includes discussions on massive calorimeters and the detection of non-ionizing radiation.
15. Luke, P.N., Beeman, J., Goulding, F.S., Labov, S.E., and Silver, E.H. (1990). Calorimetric ionization detector. Ibid., 406–9.
16. Cummings, A., Wang, N., Shutt, T., Barnes, P., Emes, J., Giraud-Heraud, Y., Haller, E.E., Lange, A., Rich, J., Ross, R., Sadoulet, B., Smith, G., and Stubbs, C. (1991). Performance of a 60 gram cryogenic germanium detector. *IEEE Transactions on Nuclear Science*, **38**, 226–30.
17. Sadoulet, B. (1988). Cryogenic detectors of particles: hopes and challenges. *IEEE Transactions on Nuclear Science*, **35**, 47–54. A survey article which also indicates possible new developments in a number of fields.
18. Stodolsky, L. (1991). Neutrino and dark-matter detection at low temperature. *Physics Today*, **44**, No. 8, Part 1, 24–32. A semi-popular account of this topic, which also contains information on superconducting and calorimetric detectors.

Further reading

Fernow, R.C. (1986). *Introduction to experimental particle physics*. Cambridge University Press. Chapter 8 has a good coverage of Cherenkov detectors. It also provides a good introduction to high-energy detectors, like bubble chambers, which lie outside the scope of our book.

General bibliography

Most of the texts listed below have nuclear radiation detectors as a principal topic. Each provides further reading for most or all of the chapters in this book. The references already given at the end of each chapter are normally to books or articles specializing in topics covered in the chapter concerned.

Bromley, D. A. (ed.) (1979). Detectors in nuclear science. *Nuclear Instruments and Methods*, **162**. A two-part special issue covering many aspects of radiation detection at research level up to that time.

Cooper, P. N. (1986). *Introduction to nuclear radiation detectors*. Cambridge University Press. A concise account.

Eichholz, G. G. and Poston, J. S. (1979). *Principles of nuclear radiation detection*. Ann Arbor Science. A general textbook, for which there is a companion laboratory manual for radiation detection courses.

England, J. B. A. (1974). *Techniques in nuclear structure physics*, (2 vols). Macmillan, London. Includes not only detectors but also accelerators, magnetic spectrometers and general instrumentation and techniques not covered in the present book.

Kleinknecht, K. (1986). *Detectors for particle radiation*. Cambridge University Press. Includes high-energy particle detector systems (these are not covered by the present book).

Knoll, G. F. (1989). *Radiation detection and measurement*, (2nd edn). Wiley, New York. Provides a very comprehensive coverage of the whole topic, with particularly extensive references to the research literature.

Leo, W. R. (1987). *Techniques for nuclear and particle physics experiments*. Springer, Berlin. Subtitled 'A how-to approach', and based on a laboratory course given for nuclear physics students.

Mann, W. B., Rytz, A., and Spernol, A. (1988). *Radioactivity measurements: principles and practice*. Pergamon, Oxford. Refers particularly to radiation detection in radioisotope assaying (radioisotope identification and activity measurement) and radiation protection.

Price, W. J. (1964). *Nuclear radiation detection*, (2nd edn). McGraw-Hill, New York. An older book which has been widely used.

Tait, W. H. (1980). *Radiation detection*. Butterworth, London. Provides an introductory approach to the subject.

Tsoulfanidis, N. (1983). *Measurement and detection of radiation*, Hemisphere, Washington and McGraw-Hill, New York. An extensive introduction to the subject.

Index